Sustainability and the New Economics

Stephen J. Williams • Rod Taylor
Editors

Sustainability and the New Economics

Synthesising Ecological Economics and Modern Monetary Theory

Springer

Editors
Stephen J. Williams
Silverwater, NSW, Australia

Rod Taylor
Giralang, ACT, Australia

ISBN 978-3-030-78794-3 ISBN 978-3-030-78795-0 (eBook)
https://doi.org/10.1007/978-3-030-78795-0

This Springer imprint is published by the registered company Springer Nature Switzerland AG
The registered company address is: Gewerbestrasse 11, 6330 Cham, Switzerland

Preface

In October 2021, the United Nations Humans Rights Council voted to recognise a new human right: the right to a “safe, clean, healthy and sustainable environment”.

In a three-page document it acknowledged “the importance of a safe, clean, healthy and sustainable environment as critical to the enjoyment of all human rights”.

The UN High Commissioner for Human Rights, Michelle Bachelet, has called on member states to “take bold actions to give prompt and real effect to the right to a healthy environment”.

While the resolution is yet to go before the UN General Assembly, it is likely to have significant ramifications in legislatures, courts and tribunals around the world. Together with the UN Sustainable Development Goals, there is increasingly nowhere for governments to hide as we confront our long-standing unsustainable practices and the resultant existential risks. Moreover, excessively polluting corporations and acquiescent shareholders will be seen to be violating basic human rights.

This book explores a sustainable pathway so all human rights are maximised.

One of the themes of this book is that national governments have strong powers to control their economies and their environments – if only they would use them – and are not beholden to corporations or financial markets. Government is in the driver’s seat. Even so, we ask you to keep this new human right to a **safe, clean, healthy and sustainable environment** at the front of your mind while reading the book – and reflect on whether you are mostly upholding or violating it.

Silverwater, NSW, Australia — Stephen J. Williams

Giralang, ACT, Australia — Rod Taylor
October 2021

Contents

Chapter 1
Introduction

Stephen J. Williams and Philip Lawn

Abstract This chapter introduces the main themes of the book, centred on sustainability. It gives a brief summary of the modern environmental movement from the publication of Rachel Carson's *Silent Spring* in 1962. It also briefly touches on the history of neoliberalism as both a reaction to the Keynesian Revolution in economics and the environmental movement. In essence, we argue that world leaders have largely ignored the consistent warnings from scientists and ecological economists on the existential perils of unrelenting growth, especially since the early 1970s. However, we acknowledge a growing chorus of new authoritative voices that are calling for significant and urgent change along the lines of the United Nations Sustainable Development Goals, a global Green New Deal or Just Transition. While welcome, we argue all such prescriptions for 'sustainability' will ultimately fail without a recognition of the limits to growth and other key features of sustainability, usually derived from ecological economics. We further argue that the recent emergence of modern monetary theory (a much-improved macroeconomics) provides important insights that complement ecological economics, producing a powerful new economics to guide humanity away from imminent danger towards vastly expanded wellbeing.

This book explores the subject of ecological, economic and social sustainability – or simply sustainability. Sustainability tends to be a motherhood term, in that no sensible person could be against it, but it remains imprecise for most people and therefore easy to use and abuse as the occasion demands. Like words such as 'freedom' and 'democracy', in the hands of the skilled rhetorician, it is particularly useful as it has a positive cachet but no real accountability attached to it.

We define what we mean by sustainability in this book's glossary: we essentially mean limiting humanity's use or demands of the planet's physical endowments to the

S. J. Williams
Silverwater, NSW, Australia

P. Lawn (✉)
Measuring Genuine Progress, Adelaide, Australia

Torrens University, Adelaide, Australia

S. J. Williams, R. Taylor (eds.), *Sustainability and the New Economics*,
https://doi.org/10.1007/978-3-030-78795-0_1

planet's inherent biocapacity (Earth's sustainable resource supplies). Within such limits, humankind can comfortably focus on organising itself and its activities to maximise wellbeing. Should both be successfully accomplished, sustainable development is achieved. Overshooting the planet's biocapacity, if not corrected in time, leads to collapse. As a species, we are presently not living within the planet's biocapacity. We are not even close, needing about 1.6 Earths to sustain production-consumption, when we have only one Earth (Global Footprint Network 2021; Dasgupta Review 2021, p1).[1] Indeed, after some 60 years of warnings, we are still rapidly heading in the wrong direction, seemingly willing to run the most dangerous experiment ever devised, with no plan B. In the words of the United Nations Environment Programme (UNEP Report 2021), we have been "waging a war on nature", such that we are now in a "planetary emergency" with a "need for a clear break with current trends".

Another way to understand sustainability is to acknowledge that humanity must live off the planet's 'income' and not its capital. We destroy the planet's capital when we consume resources faster than they can be replaced by nature, meaning that there is even less income (and capital) to live off in the future; while there is decreasing capacity for the planet to absorb wastes. This leads to three principles of ecological sustainability:

1. Renewable resources: extract them no faster than they can regenerate.
2. Non-renewable resources: extract them no faster than we can cultivate (substitute) renewable resources to replace depleted non-renewables (therefore keeping the total amount of natural capital intact).
3. Waste: generate waste no faster than the ecosphere can safely assimilate it (Lawn 2018).

The publication of Rachel Carson's book, *Silent Spring* (1962), is often viewed as the beginning of the modern environmental movement. In subsequent years came the passage of nascent environmental legislation in the United States (such as the Clean Air Act 1963); Ehrlich's *The Population Bomb* (1968); the establishment of the US Environmental Protection Agency in 1970 (the year of the first Earth Day); The United Nations Conference on the Human Environment, held in Stockholm in 1972; and the publication of the Club of Rome report, *The Limits to Growth* (1972).

Other highlights, among many, were the establishment of the Worldwatch Institute in 1975; the release of the United Nations World Charter for Nature in 1982; the Brundtland Report of 1987 (*Our Common Future*); the Montreal Protocol to protect the ozone layer in 1987; the establishment of the Intergovernmental Panel on Climate Change in 1988; the Rio Earth Summit of 1992 (where the United Nations Framework Convention on Climate Change was formed); the Kyoto Protocol on climate change of 1997; the World Summit on Sustainable Development in 2002; the Paris climate agreement of 2015–16; the adoption of the UN Sustainable Development Goals of

[1] The global ecological footprint was 1.7 times global biocapacity immediately prior to the 2020 coronavirus pandemic. It has since fallen to a factor of 1.6. Since this decline was not the result of explicit measures to reduce humankind's resource demands, amelioration of the pandemic is likely to restore the ratio to 1.7.

2015; and various biodiversity reports that simply go from bad to worse, indicating that we now seem to be in the midst of the sixth mass extinction.[2]

The growing awareness of environmental problems from the early 1960s, and the subsequent enactment of environmental regulations to curb industrial excess, did not go unnoticed by the forces of free-market capital. In fact, those forces had coalesced decades earlier in their opposition to the Keynesian Revolution in the years following the Great Depression, when national income in some countries began to be redistributed more evenly. One manifestation of that opposition was economist Friedrich Hayek's formation of the Mont Pelerin Society in 1947 (Higgs 2014). Another landmark was the manifesto written in 1971 by US business lawyer and soon-to-be Supreme Court judge Lewis Powell, and the free-enterprise propaganda machine that flowed from that (Mitchell and Fazi 2017). These were the antecedents of the free-market ideology known as neoliberalism, which overtly emerged in the 1970s. While helping to rapidly grow the Gross Domestic Product (GDP) of most countries, neoliberalism has played a major role in the depletion of global resource stocks, biodiversity loss, rising greenhouse gas emissions, the widening gap between rich and poor, both within and between countries, and the destruction of the fabric and self-sufficiency of many communities. Neoliberalism still grinds along, perhaps on life support, as we write this introductory chapter in early 2021. Contrary to what many believe, neoliberalism is not about governments taking a backseat in economic affairs. Whether intentional or not, it is a about governments reconfiguring economies in ways that narrowly privatise the gains generated by the many (rent seeking) while broadly socialising the losses inflicted by the powerful and wealthy few (cost shedding).

Countervailing the explosion in environmental destruction of this post-World War II period (the Great Acceleration) and the eventual domination of rudderless hydra-headed capital forever seeking greener pastures (globalisation), was the development of notions of steady-state (non-growing) economies (Daly 1977), the formal emergence of ecological economics in the late-1980s,[3] and ideas promoting planned degrowth in countries that had 'overshot' their sustainable size (e.g. Hickel 2020). This group of ideas are based on 'limits to growth' thinking and focus on resource throughput analysis juxtaposed to the planet's likely capacity to provide ongoing resources and still function in a stable and healthy state. Related ideas include the concept of the Anthropocene, planetary boundaries, doughnut economics, and ecological footprinting (see Glossary, this volume).

When Carson published *Silent Spring* in 1962, Earth's human population was just over 3 billion and Gross World Product (GWP) was about $US1.5 trillion. Material throughput was about 20 billion tonnes a year. As we write this, the world's population is approaching 8 billion, and GWP is close to $90 trillion. Total material use is about 100 billion tonnes a year (Carrington 2020). Conservative estimates

---

[2] A sustainable development 'timeline' (history) from 1962 to 2012, published by the International Institute for Sustainable Development, can be found at https://www.iisd.org/publications/sustainable-development-timeline-2012

[3] Its antecedents go back much earlier than this.

suggest this level of material use is likely to be about twice as much as the planet can sustain, even with vast numbers of people still living in poverty, and that the global economy probably overshot its maximum sustainable scale in the 1970s.

How do we explain humanity's lack of corrective action after almost 60 years of warnings? One answer is that humanity as a whole has not ignored the warnings. Rather, those temporarily in power, particularly in high-income countries, have either suppressed or been wilfully blind to the warnings due to their ideological commitment to growth-based capitalism. This mindset believes that about 3% compound growth every year is normal and healthy for an economy, which amounts to a doubling of the physical size of the economy every 23 years. Moreover, it believes there is no viable, or better, alternative and that it can be continued forever, even as rates of natural resource use rise and pollution builds to such insane levels that the natural world is on the verge of collapsing (see Chap. 2 of this volume).

The absurdity of the assumption that a healthy biosphere is largely irrelevant to human progress seems to escape those with greatest power and influence. Even the Covid-19 pandemic – like other zoonotic diseases associated with habitat destruction – does not dampen the calls for yet more growth. One of the 'solutions' offered by the mainstream to fix this dilemma is the belief that humanity will miraculously decouple economic expansion from environmental harm (fast enough and deeply enough), even though no such adequate decoupling has been demonstrated (see Chap. 5 of this volume). This faith in science-technology-efficiency is deeply ironic given that the optimists continue to ignore loud and continual warnings from scientists about the inevitable ecological impact of continued growth (Ripple et al. 2017; Bradshaw et al. 2021). The better attitude to technology is that societies should operate sustainably at all times and utilise technological advances to increase human wellbeing, not grow the economy further. In other words, a society should not go beyond its sustainable threshold with the hope that a dramatic technological breakthrough will fix the overshoot.

Another explanation of the failure to take corrective action is the difficulty of making any paradigm shift in things that matter, especially economics (Cahill 2011). Institutions and ways of thinking become entrenched, and systems are developed to defend the status quo and the people who benefit most from it. It is how neoliberal ideology and its first cousin, neoclassical macroeconomics, survived the Global Financial Crisis of 2008. Both were seriously weakened and have been further battered by the coronavirus pandemic. In both cases, an economic lifeline was and has been thrown by governments. With impending economic, social, and ecological shocks in sight, both await a coup de grace. Neoliberalism will hopefully be replaced by a more science-based 'ideology' that respects planetary boundaries, such as some form of ecological economics. Neoclassical macroeconomics may be replaced by a more evidenced-based approach, which is likely to be modern monetary theory (MMT).[4] We currently see an increasing convergence between leading

[4] MMT is not a set of policies, but an explanation of how the macroeconomic system works today, for good or ill. Only by understanding it correctly can we hope to improve it for the better.

promoters of limits-based economic theory, such as Kate Raworth (2017), MMT macroeconomic perspectives (Kelton 2020), and progressive voices like economist Mariana Mazzucato (2021). This convergence, if it ever occurs, still has some distance to travel, with a belief in 'green growth' in high-income countries still stalling the urgent policies needed to correct overshoot.

MMT's importance is that it disabuses us of the notion that governments of most major economies cannot afford to urgently build renewable-energy infrastructure, provide adequate schools, hospitals, pensions, and other public goods associated with sustainability. Instead of hoping that profit-making entities will save us by investing their financial capital in appropriate technology, we learn that certain governments are only constrained by the availability of real resources that are either idle or can be redirected into urgently-needed projects. That is to say, there is no inherent budget constraint for these 'monetary-sovereign' governments, but only a real-resource constraint (see Chap. 14 of this volume). These governments have immense power, unlike the supplicant corporations (foreign or domestic) they usually cow-tow to. Of course, given the resource-throughput overshoot that many countries now suffer from, a realisation that there is no inherent financial constraint does not presage a cornucopian future. However, it does allow us to focus on real constraints centred on natural capital rather than mythical constraints that favour private corporations and the financial sector. In short, MMT allows us to see how governments have the ultimate power to advance wellbeing, if only they would use that power.

Just as some leading economists like Paul Samuelson (1915–2009) preferred to propagate myths about budget constraints for fear of politicians spending like drunken sailors and fuelling inflation (see Chap. 14), so too do some in the sustainability community prefer to maintain this myth for fear of accelerating resource throughput. However, maintaining this myth would be a mistake. MMT, with its focus on real resource constraints, provides us with a great opportunity to press for better data and analysis of biocapacity, despite its complexity and uncertainty. This in turn should lead to better policies, such as appropriate caps on resource throughput and the sufficient allocation of the ensuing sustainable resource use rate to the establishment of low resource-intensive and waste-intensive economies.

Emerging policy solutions to the existential threats humanity is now facing include the United Nations Sustainable Development Goals and notions of a global Green New Deal (sometimes called a 'Just Transition' or 'Build Back Better'). These ideas all have merit, but will ultimately fail if the key problem is not addressed: namely, confining physical resource throughput to the planet's biocapacity in an equitable and efficient way – both internationally and intra-nationally. Given that low-income countries need to expand their economies, this will mean that high-income countries need to significantly contract their economies in a well-managed way, while reducing national inequality and increasing resource use efficiency to maintain or improve overall wellbeing. A good way to start would be for high-income countries to stop exploiting poorer nations: that would help achieve both goals. In the final analysis, all countries need to ascertain their individual biocapacities and confine their resource throughput accordingly. International trade within a 'globalised' economy complicates and confuses the issue, but does not absolve humanity from living sustainably while maximising wellbeing. Population growth needs to be addressed almost everywhere in an ethical and evidenced-based way, starting with the provision of family planning services to the 270 million women deprived of modern contraception (Kantorova et al. 2020).

Other signs of a growing awareness of resource limits and overshoot in the mainstream include:

- A major synthesis report from the United Nations Environment Programme titled 'Making Peace With Nature' (UNEP Report 2021). The report clearly warns of global disaster without urgent root-and-branch reform, including assessing limits ('natural capital accounting') and abandoning GDP as a proxy measure of progress and wellbeing.[5]
- A recent vote by the European Parliament's Committee on the Environment, Public Health and Food Safety to accept a report that advocates for materials-use targets and the pursuit of a circular economy[6] (European Parliament 2021).
- A commitment made by more than 50 countries – the High Ambition Coalition for Nature and People – to protect almost a third of nature by 2030 (Greenfield and Harvey 2021), which may become a step towards protecting half of nature via linked corridors in a 'Global Deal for Nature' (Dinerstein et al. 2019).
- A strong statement made by the United Nations Secretary-General in support of a new economic paradigm along the lines of ecological economics, with a focus on 'Earth jurisprudence' (UN Secretary-General 2020).
- A powerful speech made by the President of Ireland in 2020 where he calls for a "radical paradigm shift" in economics away from neoliberalism and "growth *ad infinitum*" towards a steady-state economy – he emphasises that simply putting a "new 'green' lens on the existing orthodox growth model will not suffice" (Higgins 2020).
- The increasing trend of national courts to hold their governments to account on meaningful emissions targets (Meheut 2021; see also Chap. 13 of this volume). Moreover, there are increasing calls for 'ecocide' to be recognised as a crime by the International Criminal Court (ABC RN 2021).
- The call by people's movements to recognise and respect the 'Rights of Mother Earth' (2010). Even school students, led by the extraordinary Greta Thunberg, can see what most global leaders cannot (School strikes for climate 2020). Thunberg's critique of the "fairy tales of eternal economic growth" in the face of the sixth mass extinction is particularly poignant (Good 2019).
- An acknowledgement within the UK-government-commissioned Dasgupta Review (2021) that the current global economic and financial system is unsus-

[5] Note that this report still advocates for "sustainable economic growth and decent work for all" in line with the UN Sustainable Development Goals (SDG8) although this is primarily directed at low-income countries. It also seems to use a conventional macroeconomic framing of limited financial capital, when the actual limit is real resources, sustainably managed. Monetary-sovereign governments have unlimited fiscal capacities (see Chap. 14, this volume).

[6] The feasibility of a truly circular economy is highly questionable due to the mostly linear nature of matter-energy conversion (see Glossary).

tainable and needs urgent root-and-branch reform. The disease it diagnoses, like the UNEP Report (2021), is almost identical to the disease this book highlights.
- Overwhelming support for strong action on the climate emergency according to what is probably the world's largest survey of public opinion on climate change – the 'People's Climate Vote' conducted by the United Nations Development Programme (UNDP 2021).

What this book advocates is a similar paradigm shift in political, economic, and ethical thinking to what has been advocated above. It is doubtful whether sustainability will be achieved without it. Placing band-aids on, or making cosmetic changes to, the current system will not do. Moreover, the pace of change must be swift to avoid catastrophic outcomes, knowing that some adverse outcomes are already 'locked in' for the immediate future (see Chap. 2 of this volume). These unavoidable outcomes are the price of inaction over the past 60 years where, firstly, the costs of GDP growth have been overlooked, and secondly, humankind's resource demands have been allowed to overtake the Earth's sustainable resource supplies. Because of the former, we now have *uneconomic* malignant growth in many places (additional costs exceeding the additional benefits) (see Chap. 11 of this volume). If we are lucky, we have ten short years to turn the ship around.

The focus must now turn away from the purblind pursuit of expanding the physical size of economies to a focus on maximising wellbeing within planetary boundaries. This will mean improving what we produce, producing new goods more efficiently, ensuring national income – and therefore the goods purchased with it – are equitably distributed, and achieving full employment without the need for constant expansion of the economy (for example, by redefining full-time hours, sharing jobs around, and implementing a government Job Guarantee scheme).

A range of wellbeing measurements will need to be improved and focused on. This process is under way, but needs significant support from all world leaders, including the United Nations.[7] Should we move away from obsessing about the size of national economies and seeking prestige at international forums based on quantity instead of quality, international tensions would decrease substantially. Continuing GDP growth requires a globalised economy supported by a range of international institutions and arrangements that severely affects many countries – mostly low-income nations – as a result of the appropriation of their raw materials and the exploitation of their labour. As we write this, the Doomsday Clock is set at 100 seconds to midnight (Bulletin of the Atomic Scientists 2021).

Recent events in the United States, culminating in a violent mob storming the Capitol building to prevent the transfer of power after a legitimate election, highlights the precarious nature of current circumstances. In what is one of the richest countries on Earth, the dangers of inequality, poor education, poor employment

[7] The United Nations developed a Human Development Index in 1990, with an upgrade in 2020 to include more relevant data. There are other worthy indexes, including happiness indexes, the Sustainable Development Index, and the Genuine Progress Indicator (GPI, see Chap. 11 of this volume). The UNEP (2021) has explicitly endorsed the GPI.

prospects, free markets that decimate economic sectors, a degraded democracy where money talks and people lose respect for institutions, truth as an option depending on the circumstances, and an unregulated internet that operates like the Wild West, are all on display.

While the following chapters explore humanity's global predicament, the reader will find many references to Australia where the authors have particular expertise. Australia provides an excellent example of 'greed over human need' and how not to create a sustainable society. It was the home of the oldest continuous culture on Earth – about 60,000 years old – until colonised in 1788. It has taken less than 250 years for a relatively small human population (although growing rapidly in recent decades) to significantly harm the nation's ecology by pursuing a high-growth philosophy (see Chap. 3 of this volume). Australia is a large landmass, yet suffers from a severe biodiversity crisis and other ecological ailments with what may seem to be a relatively modest 26 million human inhabitants.[8]

Successive Australian governments adopted the neoliberal mindset enthusiastically from the 1980s onwards, with tariff reductions, privatisation of state assets, deregulation of financial and labour markets, tax cuts, the valorisation of 'free markets' and obsessions with government budget surpluses. The resultant unemployed were branded as lazy while inequality steadily rose. Governments exploited the egalitarian myth of Australia as the land of the 'fair go', whose inhabitants were the descendants of exceptional military heroes, outstanding yeomen farmers and superhuman athletes. If you didn't succeed in Australia, it was your own fault, not the economic model. This has tended to breed a somewhat self-satisfied population with large pockets of entitlement and ugly nationalism. There is not sufficient recognition that the country's wealth is mostly a product of (once) abundant natural resources that the mainly Anglo-European human population did not create, but either took by force from the indigenous population or later inherited. This natural wealth is rapidly being destroyed through greed and stupidity. All the while, enough Australian voters have tolerated inept politicians who predominantly serve wealthy political donors, including fossil fuel companies, rather than advancing human wellbeing. The wheels are greased by highly concentrated media ownership, with the anti-science Murdoch news empire at the centre.

Australia is also a country whose head of state is actually the monarch of a foreign power (the United Kingdom). It is also unusual in being the only Western democracy without a Bill of Rights (or Charter of Rights, see Chap. 13 of this volume). Such complacency may well be a product of its hedonistic 'she'll be right' lifestyle – a lifestyle that one could admire if it wasn't debased by overconsumption, the sometimes-appalling treatment of refugees and an inability to fully respect the rights and

[8] Australia's population is not really modest when one understands the nature of the mostly arid and semi-arid continent, that was prone to natural disasters like drought, fire and flood even before anthropogenic climate change.

wishes of the indigenous population.[9] Australians are increasingly experiencing the direct terror of climate change (abnormal fire, flood, drought, coral bleaching, coastal erosion) yet, as we write this, our national conservative government is still a cheerleader for fossil fuels and has patronised the rest of the world at US President Biden's Climate summit (coinciding with Earth Day 2021) with transparent inanities about its emissions reductions (Hughes and Steffen 2021). Only strong sanctions from the rest of the world are likely to heal Australia's ailing polity in the short- to medium-term, with improved education a corrective in the longer term.

What are the chances of twenty-first-century humanity avoiding a full-scale disaster, given that civilisational collapse has mostly been the norm rather than the exception over recorded history (Tainter 1988)? We return to this point in this book's concluding chapter. But our collective actions so far on addressing the clear and present danger of climate change does not bode well for even larger reforms (see Chap. 2 of this volume). The nationally determined contributions (emissions targets) of the 2015 Paris Agreement are woefully inadequate, yet even these are unlikely to be met:

> The 2015 Paris Agreement aims to keep global warming by 2100 to below 2 °C, with 1.5 °C as a target. To that end, countries agreed to reduce their emissions by nationally determined contributions (NDCs). Using a fully statistically based probabilistic framework, we find that the probabilities of meeting their nationally determined contributions for the largest emitters are low, e.g. 2% for the USA and 16% for China. On current trends, the probability of staying below 2 °C of warming is only 5%, but if all countries meet their nationally determined contributions and continue to reduce emissions at the same rate after 2030, it rises to 26%. If the USA alone does not meet its nationally determined contribution, it declines to 18%. *To have an even chance* of staying below 2 °C, the average rate of decline in emissions would need to increase from the 1% per year needed to meet the nationally determined contributions, to 1.8% per year (Liu and Raftery 2021, emphasis added).

Note that no amount of warming can be considered safe, even as 1.5 °C of average warming slips away from us and something less than 2 °C of warming seems like an increasingly optimistic, yet dystopian goal (IPCC Special Report 2018, Chap. 3).

The remainder of the book is structured as follows:

Part I of this book outlines the current mess, with a focus on:

- The Earth System, the Great Acceleration and the Anthropocene
- The parlous state of Australia's natural environment (warning to the world)
- Health issues relating to climate change
- Shortcomings with the UN Sustainable Development Goals.

Part II examines issues relating to how the mess evolved, with a focus on:

- The evolution of neoliberalism
- Population growth

---

[9] Lest this sound too harsh, we acknowledge that Australia was an early adopter of voting rights for women; is the birthplace of the original Green political party; has enacted Native Title land rights in some cases; and is home of the world-famous Sydney Gay and Lesbian Mardi Gras festival.

- The Club of Rome's *Limits to Growth* report of 1972
- The role of the fossil fuel industry
- Economic failures of the Intergovernmental Panel on Climate Change.

Part III focuses on solutions:

- An introduction to ecological economics
- Energy systems for sustainable prosperity
- Climate change litigation and human rights
- Paying for a Green New Deal via an understanding of modern monetary theory
- Concluding remarks and policy formulation.

The main takeaway message from the book is that nations should aim for a steady-state economy of an optimum size and urgently abandon the discredited cult of perpetual economic growth. This is best achieved by setting strict limits on natural resource throughput via a cap-auction-trade system (see Chap. 15 and Appendix 3 of this volume) softened by equity-guaranteeing distributional institutions.

References

ABC RN (2021) Ecocide: making environmental damage an international crime. Future Tense, 7 February. https://www.abc.net.au/radionational/programs/futuretense/ecocide:-making-environmental-damage-an-international-crime/13047838. Accessed 12 Feb 2021

Bradshaw CJA, Ehrlich PR, Beattie A et al (2021) Underestimating the challenges of avoiding a ghastly future. Front Conserv Sci. https://doi.org/10.3389/fcosc.2020.615419

Bulletin of the Atomic Scientists (2021) Doomsday clock statement. https://thebulletin.org/doomsday-clock/current-time/. Accessed 1 Feb 2021

Cahill D (2011) Why does neoclassical thinking still dominate economics? The Conversation, 17 October. https://theconversation.com/why-does-neoclassical-thinking-still-dominate-economics-3861. Accessed 9 Feb 2021

Carrington D (2020) World's consumption of materials hits record 100bn tonnes a year. The Guardian, January 22. https://www.theguardian.com/environment/2020/jan/22/worlds-consumption-of-materials-hits-record-100bn-tonnes-a-year. Accessed 2 Feb 2021

Carson R (1962) Silent spring. Houghton Mifflin, Boston

Daly HE (1977) Steady-state economics. W.H. Freeman, San Francisco

Dasgupta Review (2021) The economics of biodiversity: the Dasgupta Review (headline messages). https://assets.publishing.service.gov.uk/government/uploads/system/uploads/attachment_data/file/957629/Dasgupta_Review_-_Headline_Messages.pdf. Accessed 13 Feb 2021

Dinerstein E, Vynne C, Sala E et al (2019) A global deal for nature: guiding principles, milestones, and targets. Sci Adv. https://doi.org/10.1126/sciadv.aaw2869

European Parliament (2021) MEPs call for binding 2030 targets for materials use and consumption footprint. Press release 27 January. https://www.europarl.europa.eu/news/en/press-room/20210122IPR96214/meps-call-for-binding-2030-targets-for-materials-use-and-consumption-footprint. Accessed 1 Feb 2021

Global Footprint Network (2021). https://www.footprintnetwork.org. Accessed 13 Feb 2021

Good JE (2019) Greta Thunberg's radical climate change fairly tale is exactly the story we need. The Conversation, 28 September. https://theconversation.com/greta-thunbergs-radical-climate-change-fairy-tale-is-exactly-the-story-we-need-124252. Accessed 10 Feb 2021

Greenfield P, Harvey F (2021) More than 50 countries commit to protection of 30% of Earth's land and oceans. The Guardian, January 11. https://www.theguardian.com/environment/2021/jan/11/50-countries-commit-to-protection-of-30-of-earths-land-and-oceans. Accessed 3 Feb 2021

Hickel J (2020) Less is more: how degrowth will save the world. Heinemann, London

Higgins MD (2020) Speech at the Engineers Ireland annual conference. https://president.ie/en/media-library/speeches/climate-action-and-the-role-of-engineers. Accessed 23 Apr 2021

Higgs K (2014) Collision course: endless growth on a finite planet. MIT Press, Cambridge

Hughes L, Steffen W (2021) Spot the difference: as world leaders rose to the occasion at the Biden climate summit, Morrison faltered. The Conversation. https://theconversation.com/spot-the-difference-as-world-leaders-rose-to-the-occasion-at-the-biden-climate-summit-morrison-faltered-159295. Accessed 24 Apr 2021

IPCC Special Report (2018) Global warming of 1.5°C (SR15). https://www.ipcc.ch/sr15/. Accessed 13 Feb 2021

Kantorava V, Wheldon MC, Ueffing P et al (2020) Estimating progress towards meeting women's contraceptive needs in 185 countries: a Bayesian hierarchical modelling study. PLoS Med. https://doi.org/10.1371/journal.pmed.1003026

Kelton S (2020) The deficit myth: modern monetary theory and how to build a better economy. John Murray, London

Lawn P (2018) Lecture given to economic students at Adelaide University, October 2018. https://www.youtube.com/watch?v=K7L63x4TBZI (at 23 mins). Accessed 1 Dec 2020

Liu PR, Raftery AE (2021) Country-based rate of emissions reductions should increase by 80% beyond nationally determined contributions to meet the 2 °C target. Commun Earth Environ 2:29. https://doi.org/10.1038/s43247-021-00097-8

Mazzucato M (2021) Tweet of 2 February at https://twitter.com/MazzucatoM/status/1356361775277613059. Accessed 3 Feb 2021

Meheut C (2021) Court faults France over 'ecological damage' from its emissions levels. New York Times, 3 February. https://www.nytimes.com/2021/02/03/world/europe/france-emissions-court.html. Accessed 4 Feb 2021

Mitchell, Fazi (2017) Reclaiming the state: a progressive vision of sovereignty for a post-neoliberal world. Pluto, London

Raworth K (2017) Doughnut economics: seven ways to think like a 21st century economist. Random House, London

Rights of Mother Earth (2010) World's people's conference on climate change and the Rights of Mother Earth. https://pwccc.wordpress.com/category/working-groups/. Accessed 10 Feb 2021

Ripple WJ, Wolf C, Newsome TM et al (2017) World scientists' warning to humanity: a second notice. Bioscience, v67 12, pp. 1026–1028, https://doi.org/10.1093/biosci/bix125

School strikes for climate (2020). https://www.abc.net.au/news/2020-09-25/global-student-strike-for-climate-action/12702434. Accessed 10 Feb 2021

Tainter JA (1988) The collapse of complex societies. Cambridge University Press, Cambridge

UN Secretary-General (2020) Harmony with nature. Report of the Secretary-General pursuant to General Assembly resolution 74/224, 28 July. https://documents-dds-ny.un.org/doc/UNDOC/GEN/N20/199/25/PDF/N2019925.pdf?OpenElement. Accessed 2 Feb 2021

UNDP (2021) People's climate vote. Internal news report 27 January. https://www.undp.org/content/undp/en/home/news-centre/news/2021/Worlds_largest_survey_of_public_opinion_on_climate_change_a_majority_of_people_call_for_wide_ranging_action.html. Accessed 19 Feb 2021

UNEP Report (2021) Making peace with nature: a scientific blueprint to tackle the climate, biodiversity and pollution emergencies (key messages and executive summary). https://wedocs.unep.org/xmlui/bitstream/handle/20.500.11822/34949/MPN_ESEN.pdf. Accessed 19 Feb 2021

Part I
The Current Mess

Chapter 2
The Earth System, the Great Acceleration and the Anthropocene

Will Steffen

Abstract The Earth System is a single, self-regulating system comprised of interacting physical, chemical, biological and human components. For the last 11,700 years, the system has existed in a relatively stable state called the Holocene. However, around the mid-twentieth century, human activities – population, economy, resource use, technologies – underwent a sharp increase that drove correspondingly rapid and, in many cases, unprecedented changes to the structure and functioning of the Earth System. This 'Great Acceleration' became the basis for a proposed new geologic epoch in Earth history – the *Anthropocene.* Perhaps the best-known feature of the Anthropocene is climate change, which threatens, under a worst-case scenario, to drive the planet into *Hothouse Earth* conditions. Tipping points in the Earth System, such as melting of the Greenland ice sheet and the conversion of large forests to less carbon-rich ecosystems, provide the feedback mechanisms for taking the trajectory of the system out of human control and irreversibly into Hothouse Earth. The planetary boundaries framework, based on limits to human perturbation of nine key processes that together define a stable state of the Earth System, provides the planetary 'safe operating space' for avoiding Hothouse Earth and returning the planet to Holocene-like conditions.

2.1 What Is the Earth System?

We are all familiar with our home planet, the Earth, but the concept of the Earth System often catches people by surprise, even some scientists. This wasn't always so. Indigenous cultures around the world often viewed our home planet as a single system, with life itself being an integral part of the system – an active player in the functioning of the whole and not simply a passenger on a planet that happened to have the right conditions for life.

W. Steffen (✉)
Australian National University, Canberra, Australia
e-mail: will.steffen@anu.edu.au

S. J. Williams, R. Taylor (eds.), *Sustainability and the New Economics*,
https://doi.org/10.1007/978-3-030-78795-0_2

Contemporary cultures have largely lost that understanding, thanks in part to the modern scientific approach to studying our planet. We have become experts at taking systems apart and studying their components in ever greater detail, often using cause-effect logic to understand how a part of a system works. This has been an essential tool in helping contemporary societies become wealthier, healthier and more powerful in terms of our ability to manipulate, control and transform the environment around us for the benefit of humans, but often at the expense of the rest of life on the planet.

Regaining the understanding of the Earth as a single system, an understanding that our indigenous colleagues have had for centuries, has been a recent development for our contemporary societies. The renaissance began with Vladimir Vernadsky's classic work 'The Biosphere', which described the active role that life itself played in the evolution of the Earth (Vernadsky 1926). Life helped shape the climate, drove the great element cycles of the Earth, and steered the trajectory of the planet through a variety of epochs and ages. More recently, in 1972 James Lovelock introduced the Gaia hypothesis: the Earth System (or Gaia as he named it) is an entity comprised of the total ensemble of living beings and the environment with which they interact. Lovelock hypothesised that living beings regulate the global environment by generating homeostatic feedbacks (Lovelock and Margulis 1974). 'The Biosphere' and 'Gaia' set the stage for the concept of the Earth System as we know it today.

The concept of the Earth System was further developed by the US National Aeronautics and Space Administration (NASA) global change research program in the 1980s. One of the most influential products of that research effort was a visualisation of the Earth System in the form of a systems diagram (Fig. 2.1). The figure

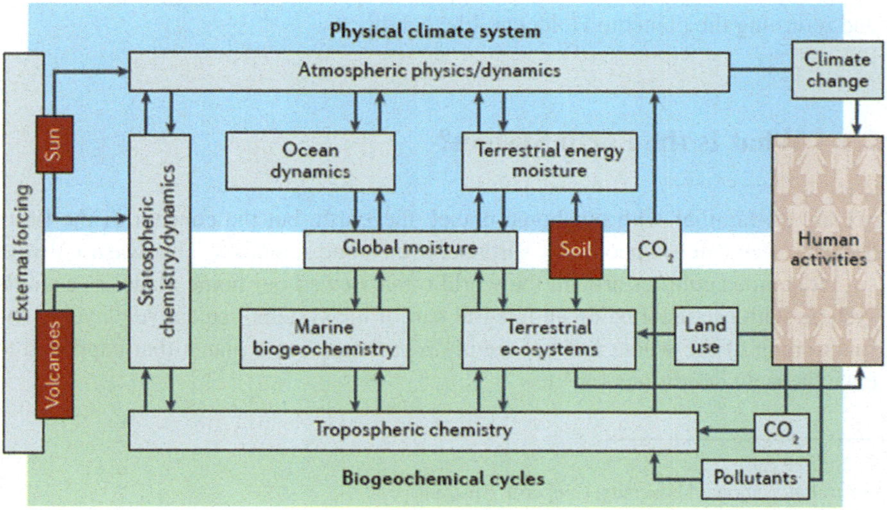

Fig. 2.1 The NASA 'Bretherton' diagram of the Earth System. (NASA 1986, redrawn and published in Steffen et al. 2020)

was revolutionary at the time as it showed the biosphere was as equally important to the Earth System as the physical climate system. In particular, it highlighted the strong role that biogeochemical cycles played in the functioning of the planet as a whole. However, human activities were placed at the side of the figure, with only a few arrows showing our impacts on the Earth System, and only two showing the feedbacks to humans from a changing Earth System. Thus, humans were largely considered to be outside of the Earth System, and not fully integrated into the system. Nevertheless, the Bretherton diagram (as it was called) and the NASA research program were major advances in the study of the Earth System.

Global change research went truly global at the end of the 1980s with the launch of the International Geosphere-Biosphere Programme (IGBP). Spanning the globe in its scientific reach and involving over 50 countries in its research effort, the IGBP partnered with the World Climate Research Programme to build the world's first major study of the Earth System as a whole. A major product of this effort was the IGBP synthesis report (Steffen et al. 2004), accompanied by a large conference in Amsterdam, Netherlands, in 2001. The conference attracted over 1000 participants from more than 100 countries. Its Amsterdam Declaration (published in Steffen et al. 2004) put forward not only a definition of the Earth System, but also summarised the significant impacts of human activities on the structure and functioning of the system:

> The Earth System behaves as a single, self-regulating system comprised of physical, chemical, biological and human components. The interactions and feedbacks between the component parts are complex and exhibit multi-scale temporal and spatial variability … Human activities are significantly influencing Earth's environment in many ways in addition to greenhouse gas emissions and climate change. Anthropogenic changes to Earth's land surface, oceans, coasts and atmosphere and to biological diversity, the water cycle and biogeochemical cycles are clearly identifiable beyond natural variability …

Over the past two decades, Earth System science has blossomed, rapidly building an extensive knowledge base using a wide range of research tools. Observations and experiments study the rapidly changing dynamics of the system, and palaeo studies provide a deep time baseline against which current changes can be compared. Models of various types both interpret the changes being observed and project potential trajectories of the Earth System into the future. Assessments and syntheses evaluate and assemble the mass of information to provide reports and advice to policymakers and the public.

This research effort has led to an updated version of the earlier Bretherton diagram (Fig. 2.2; Steffen et al. 2020). This new conceptualisation of the structure and dynamics of the Earth System now includes humans ('the Anthroposphere') as a prominent, integrated sphere of the system. The core of the Anthroposphere is the production/consumption component – our massive, globalised economy – which is influenced by increasingly homogenised cultures, values and beliefs. Knowledge, science and technology are the proximate drivers of the production/consumption machinery, and institutions and the political economy grease the wheels. The Anthroposphere interacts in many ways with the other two major spheres – the

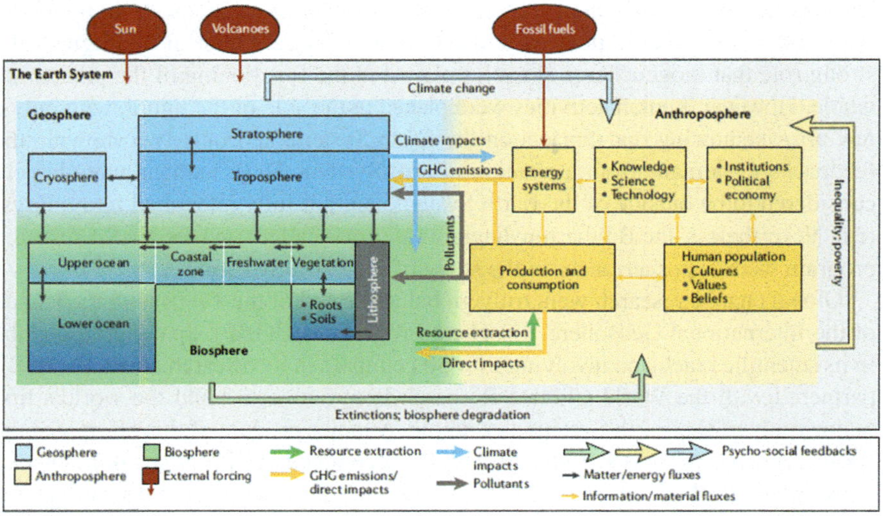

Fig. 2.2 An updated conceptual model of the Earth System (Steffen et al. 2020)

Geosphere and the Biosphere – and has now become the major driver of change to the Earth System as a whole.

2.2 The Great Acceleration

Modern humans have always been part of the Earth System since we evolved about 200,000 years ago in Africa, and for over 90% of that time we existed in small numbers and were hunter-gathers only. Life was challenging, as much of it was spent in the grips of ice ages that were cold and dry. But by the depth of the last ice age, about 20,000–25,000 years ago, we had spread to every continent on Earth except Antarctica (Oppenheimer 2004) and were poised to begin what some would call 'the ascent of man' (Bronowski 2011). Fortunately, the Earth System cooperated by entering into an unusually long interglacial period that stratigraphers have called the Holocene epoch.

The relatively stable, warmer, wetter conditions of the Holocene allowed some humans to develop agriculture from 10,000 to 12,000 years ago, produce excess food, settle in one place, develop villages and then cities, and build more complex technologies and economies. During the early Holocene written language was developed, allowing transfer of larger amounts of knowledge across generations. Domestication of animals and plants increased, along with the use of fire to convert forests into agricultural land. Then, in the late 1700s, the industrial revolution began in England and quickly spread around the world through the 1800s, marking one of the most decisive transitions in human history (McNeill 2001). However, the industrial revolution was soon to be dwarfed by the Great Acceleration of the twentieth and twenty-first centuries (Steffen et al. 2015a; McNeill and Engelke 2016).

Figure 2.3 shows the development of the human enterprise from 1750, before the beginning of the industrial revolution, through 2010, characterised by 12 indicators that describe humans and our activities – population, economy, resource use, transport and communication. The increase in all of these indicators through the industrial revolution is clear, but the post-1950 rise of the Great Acceleration is dramatic. Although population rose sharply, it was easily outpaced by the growth of the global economy (closely linked to resource use) and many of the communication, transport and financial technologies we now take for granted. It could well be described as an

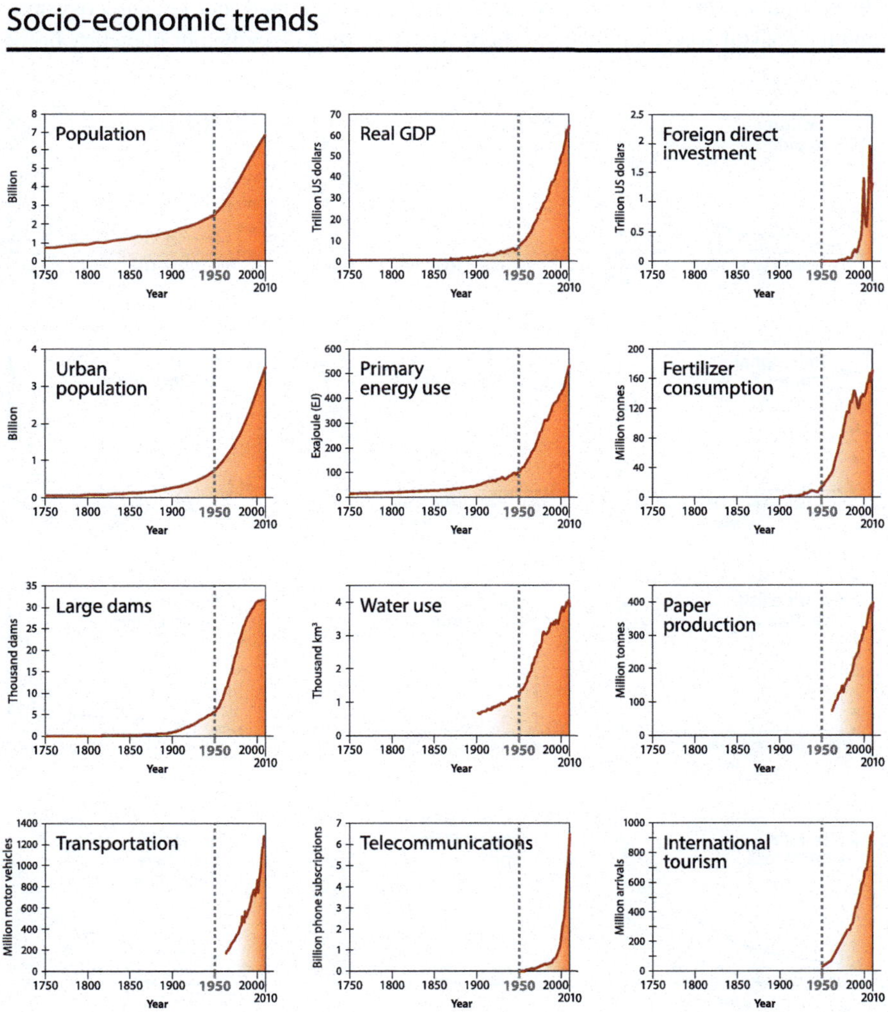

Fig. 2.3 Trends from 1750 to 2010 in globally aggregated indicators for socio-economic development (Steffen et al. 2015a)

explosion of human activity, the likes of which the planet had not experienced previously.

The structures, artefacts, networks, and institutions of the Great Acceleration have allowed many humans to enjoy lives that could only be dreamt about in the past. But the fruits of the Great Acceleration have come at a cost that is becoming increasingly visible and threatening. Figure 2.4 shows a corresponding set of 12 graphs, again from 1750 through 2010, that describe the structure and functioning of the Earth System. The top six graphs describe important components of the geosphere, the non-living part of the system (cf. Fig. 2.2). These include the gaseous composition of the atmosphere, the chemistry of the ocean, stratospheric ozone and the physical climate system. The bottom six figures describe important features of the biosphere -the living part of Earth. These include land and ocean ecosystems and the coastal zone that connects them. Together they describe our planetary home.

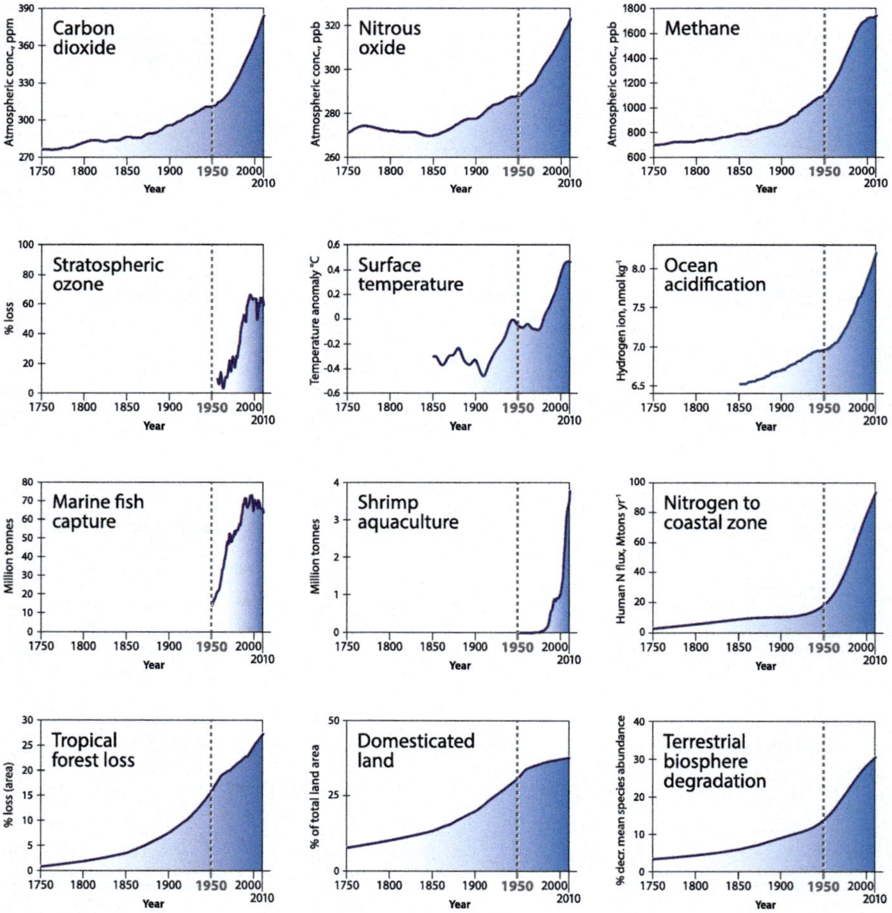

Fig. 2.4 Trends from 1750 to 2010 in indicators for the structure and functioning of the Earth System (Steffen et al. 2015a)

Although not all of these figures show the sharp discontinuity around 1950 or after, they are nevertheless striking for two reasons. First, the indictors for all 12 of these biogeophysical features of the Earth System have moved outside of the Holocene envelope (normal range) of variability. Second, the primary driver of all of these trajectories are human pressures of various types and not the natural variability of the Earth System (Steffen et al. 2015a).

The earlier version of the graphs published in Fig. 2.3 (Steffen et al. 2007), however, was strongly criticised because it lumped together all of 'mankind' or 'humankind'; that is, the graphs masked the fact that the Great Acceleration was not driven equally by all of humanity but rather by the relatively small number of humans living in OECD countries (Malm and Hornborg 2014). The 2015 Great Acceleration paper confirmed the importance of the equity issues raised by Malm and Hornborg by deconvoluting the 12 graphs into the OECD countries, the rapidly developing BRICS countries (Brazil, Russia, India, China and South Africa), and the rest – all poorer countries. The results were striking. Although only 19% of the human population lived in the OECD countries, they accounted for 74% of globally aggregated GDP (Gross World Product), and thus dominated global consumption of goods and services.

2.3 The Anthropocene

Two decades ago, atmospheric chemist Paul Crutzen, who had already been awarded a Nobel Prize for his work on unravelling the chemistry of the ozone hole, was struck by the enormous changes to the Earth wrought by humans in the last few centuries. At a meeting of the IGBP early in 2000, he reacted to the use of the term 'Holocene' as a descriptor of the current state of the Earth System and forcefully proposed that the Earth System had left the Holocene and entered a new epoch he called the Anthropocene (Crutzen and Stoermer 2000; Crutzen 2002). It was an insight that was to change the course of scientific research and to challenge humanity's perception of itself and its role on the planet.

Already the IGBP had promoted the term 'global change' and its use as quite distinct from climate change. Although the latter was attracting much more attention in the public, far more than climate change was happening in terms of human-driven changes to the Earth System. Particularly important were human-driven changes to the biosphere: rapidly increasing extinction rates, transforming large areas of Earth's land surface and filling the oceans with pollutants (IPBES 2019). The changes had become so profound that they prompted the suggestion that we are now in the third stage of evolution of life on Earth (Williams et al. 2015). The rapid rise of the 'Technosphere' (Haff 2014) and its accelerating role in driving change to the Earth System as a whole reinforced the argument that the Earth had been driven out of the Holocene and was now in the Anthropocene.

Crutzen's proposal for the Anthropocene included a direct challenge to the geological research community by specifically proposing that the Anthropocene become a new epoch in the Geologic Time Scale, coming after the Holocene. This challenge was met by the stratigraphic research community with the formation in 2009 of the Anthropocene Working Group (AWG). Led by Jan Zalasiewicz and Colin Waters, the AWG has spent a decade systematically examining the evidence that the Earth has in fact left the Holocene and is now in a new epoch, the Anthropocene. But the AWG has taken a much broader and inclusive approach to the Anthropocene, including a wide range of scholars, in addition to stratigraphers, in its membership – Earth System scientists, governance experts, humanities scholars, legal experts, social scientists and historians. The group has generated a prolific output of papers in its decade of work, and in 2019 came to a formal, binding vote on the ultimate questions around which it was formed:

- Should the Anthropocene be formalised in the Geologic Time Scale?
- Should the base of the Anthropocene (start date) be placed around the mid-twentieth century?

These questions had been debated vigorously in the AWG for a decade, with many strong feelings expressed and argued throughout that period. The outcome was a surprisingly decisive vote, with a 29-4 response of "yes" to each question (AWG 2019). This binding vote constitutes a recommendation from the AWG to the bodies above it that will need to ratify the vote to formalise the Anthropocene and enter it into the Geologic Time Scale, ending the Holocene around 1950.

Regardless of the ultimate outcome of the formalisation process, the Anthropocene has generated an enormous amount of interest and discussion beyond the natural science community. It has prompted a deeper examination of human impacts on the Earth through the Holocene, for example, prompting one proposal that early agriculture from about 7000 years ago may have emitted enough carbon dioxide to delay the onset of the next ice age (Ruddiman et al. 2015). Nevertheless, the evidence for a mid-twentieth century start date for the Anthropocene is overwhelming. Potential markers for the beginning of the Anthropocene abound. They include radioactive nuclides from atomic bomb tests; hydrogen and oxygen isotopes in ice cores; nitrates in coastal ocean sediments; plastics; synthetic fibres; technofossils; extinction rates; city strata including metro systems; concrete along the Spanish coast; and even the change in size of chicken bones from 1950 to the present. The Anthropocene has also triggered a large and creative response from the arts community, becoming the theme of several films, cartoons and songs, as well as lending its name to cafés and pubs, and their products, around the world.

2.4 Climate Change

The most well-known phenomenon associated with the Anthropocene is climate change, which is ultimately caused by the disruption of the energy balance at the Earth's surface. The primary driver of climate change is the increase in concentration of heat-trapping gases in the atmosphere, which raises the planet's surface temperature and leads to a large range of other changes in the structure and functioning of the Earth System via a complex network of impacts and feedbacks.

The most important of the heat-trapping gases, or 'greenhouse gases', is carbon dioxide (CO_2) which is emitted primarily by the burning of fossil fuels – coal, oil and gas. Additional human-induced CO_2 emissions arise from land clearing and land degradation. Anthropogenic emissions of CO_2 have been rising since the beginning of the industrial revolution and in 2019 reached a record level of 11.7 GtC (billion tonnes of carbon), or 42.9 $GtCO_2$. About 87% of this total originated from the burning of fossil fuels, with the remainder arising from land-use change (Friedlingstein et al. 2019).

While CO_2 is the most important of the greenhouse gases, methane (CH_4) and nitrous oxide (N_2O) are also important, along with a few others. As these two gases absorb different amounts and wavelengths of infrared radiation (heat emitted from the Earth's surface) and have different lifetimes in the atmosphere, estimating their impact on the climate requires a conversion of their global warming potential to a standard based on the warming potential of CO_2 over a 100-year period. The resulting overall human impact on the Earth's greenhouse effect is thus standardised to a 'CO_2-equivalent' concentration, or CO_2-e for short. The current (November 2020) value for the CO_2 concentration is 413 ppm (parts per million) (NOAA 2020) and the CO_2-e concentration, which captures human forcing of the climate system more completely, reached 491 ppm in 2017. When the cooling effect of human emissions of aerosols (fine particles that scatter incoming sunlight) is included, the CO_2-e concentration in 2017 becomes 454 ppm (European Environment Agency 2020).

The links between human emissions of greenhouse gases, the changing concentrations of these gases in the atmosphere, and the resulting changes in the climate system lie at the heart of climate policy. In particular, the link between emissions and the resulting rise in global average temperature is crucial. The 'carbon budget' approach is a scientifically robust approach to estimating the level of greenhouse gas emission reductions required to meet a desired temperature target, for example, the Paris Accord 1.5 °C or 'well below' 2 °C targets (Collins et al. 2013). The approach is based on the approximately linear relationship between the cumulative amount of CO_2 emitted from all human sources since the beginning of industrialisation (often taken as 1870) and the increase in global average surface temperature (IPCC 2013). Once the carbon budget has been consumed, emissions need to be net zero[1] to meet the desired temperature target.

There are several key areas of uncertainty that influence the carbon budget required to meet a temperature target:

[1] 'Net zero emissions' means the magnitude of carbon dioxide emissions to the atmosphere is matched by the magnitude of carbon dioxide removal from the atmosphere by, for example, 'carbon capture and storage' technologies (sometimes called 'Negative Emission Technologies') or storing more carbon in vegetation and soils. At present these technologies are speculative or are in the early development stage, and none are technologically or commercially viable.

- *Probability of meeting the target.* Higher probabilities of meeting a given temperature target (e.g. 1.5 °C) imply a smaller carbon budget. Normally a 66% or greater probability is adopted.
- *Accounting for other greenhouse gases.* The other gases that contribute to CO_2-e (e.g. CH_4 and N_2O) are assumed to be reduced at the same rate as CO_2. If these gases are not reduced, or reduced more slowly than CO_2, then the CO_2 budget must be reduced accordingly. That is, CO_2 emissions need to be reduced even further to compensate for the continued emission of non-CO_2 gases.
- *Accounting for feedbacks in the climate system.* Carbon cycle feedbacks, such as permafrost melting or the shift of the Amazon rainforest to a savanna, which emit carbon to the atmosphere, are not accounted for in the carbon budget approach. Including estimates for these reduces the budget further (Ciais et al. 2013).

Here we apply the carbon budget to the lower Paris target of 1.5 °C, based on the budget remaining from the beginning of 2020 (Table 2.1). We adopt a 66% probability of meeting the temperature target and use the IPCC SR1.5 (IPCC 2018) carbon budget analysis as our framework, noting that, in Table 2.1, the IPCC SR1.5 budget begins from 1 January 2018, and not from the beginning of the industrial revolution, as for the IPCC AR5 budget.

We estimate that accounting for non-CO_2 greenhouse gases, e.g. CH_4 and N_2O, would reduce the remaining budget by 25 GtC. The IPCC (2018) gives a range for this adjustment, which also includes cooling aerosols, of −110 to +55 GtC. Here we assume that it will be more difficult to achieve deep reductions in non-CO_2 gases compared to CO_2 itself; however, we assume rather optimistically that some reductions in these gases will be achieved, so that the net adjustment to our budget is only 25 GtC.

The current rate of human emissions of CO_2 is about 11.5 GtC per year (Friedlingstein et al. 2019), so the remaining 1.5 °C budget must be reduced by 23 GtC to bring it to the beginning of 2020. Carbon cycle feedbacks were estimated to be 110 GtC by 2100 for a 2 °C temperature rise (Steffen et al. 2018), and this estimate has been linearly downscaled for a 1.5 °C forcing to give 70 GtC as a reasonable estimate of the budget reduction due to carbon emissions from Earth System feedbacks. A recent analysis of the current states of these and other feedback processes shows that they have already been activated at a ~ 1.1 °C temperature rise (Lenton et al. 2019), and so our estimate for a 1.5 °C forcing seems reasonable compared to both observations and modelling studies.

Table 2.1 Carbon budget for a 66% probability of restricting temperature rise to no more than 1.5 °C, based on the IPCC SR1.5 approach (IPCC 2018)

Budget item/Process	GtC
Base budget from 1 Jan 2018	155
Accounting for non-CO_2 greenhouse gases (Estimated from Table 2.2 of IPCC SR1.5 (2018))	−25
Historical emissions for 2018 and 2019 (Friedlingstein et al. 2019)	−23
Carbon cycle feedbacks (Steffen et al. 2018)	−70
Remaining budget to net zero emissions	37

These adjustments reduce the final budget to 37 GtC, or less than 4 years of emissions at current rates. Thus, even if global emissions reached net zero in 2030, the remaining budget would be overspent by 20.5 GtC, assuming a linear rate of emission reduction from now to 2030. Assuming a linear rate of emission reduction would require emissions to reach net zero in 6.5 years (mid-2026 from the time of writing), equivalent to an emission reduction rate of about 15% per year. By comparison, estimates suggest that the COVID-19 crisis could reduce 2020 emissions by 5–7% (Le Quéré et al. 2020).

The only reasonable conclusion from this analysis is that the lower Paris target of 1.5 °C is now out of reach. Past inaction, and in particular, failure to begin significant emission reductions before 2020, have cost us dearly.

However, we emphasise that when carbon budgets become as small as the remaining budget for the 1.5 °C lower Paris target, they are strongly influenced by assumptions made in constructing the budget, and thus become less reliable indicators of the feasibility of meeting a temperature target. For example, if we assume that non-CO_2 greenhouse gases will be reduced at the same rate as CO_2 and if we ignore feedbacks, the remaining budget becomes 132 GtC, or about 11.5 years at current rates. Assuming a linear rate of emission reduction would mean that we would need to reach net-zero by 2043, a very big challenge but perhaps achievable.

Is it still possible to meet the upper Paris target of somewhere above 1.5 but below 2.0 °C? Let's assume a temperature target of 1.8 °C and use the same analysis as that in Table 2.1 (including feedbacks and slower reduction of non-CO_2 gases). The base budget now becomes about 280 GtC, and the estimated carbon cycle feedbacks become about 98 GtC. This leaves us with a remaining budget of 134 GtC, or about 11.5 years at current emission rates. Assuming a linear rate of emission reduction implies that we must reach net zero emissions in 23 years, or about 2043.

These carbon budget analyses, although with large uncertainties, suggest that the Green New Deal targets, adopted by several countries, of reaching net zero emissions by 2050 are too late, unless a very high rate of emission reductions is achieved over the next decade with a long 'tail' of very low emissions stretching out to 2050. In any case, the cumulative emissions must remain under about 130 GtC in total. A linear reduction of emissions from 2020 to 2050 would emit a cumulative total of 172 GtC, well above the remaining 1.5 °C target of 37 GtC (Table 2.1) and also above the 'well below' 2 °C target of 134 GtC.

It is clear from observations of climate change-related impacts in Australia alone – the massive bushfires of the 2019–2020 Black Summer; the third mass bleaching of the Great Barrier Reef in only 5 years; and long-term cool-season drying of the country's southeast agricultural zone – that even a 1.1 °C temperature rise has put us into a dangerous level of climate change. Furthermore, as noted above, the 1.5 °C Paris target is now out of reach, meaning that we now do not have a 'safe' climate future. The best outcome we can now achieve is to stay somewhat below 2 °C: an outcome that would present significant challenges to human societies as well as severe impacts on the natural world.

In the next section we describe a science-based approach to defining a planetary 'safe operating space' for humanity, arguing that we have already departed a 'safe climatic space' and are moving further into a high-risk zone.

2.5 Planetary Boundaries

With the growing body of global change research showing that human activities were increasingly destabilising the Earth System as a whole, culminating in the Anthropocene concept in 2000, there was increasing concern about these human impacts and what they meant for the future of our planetary home. The 1972 United Nations Conference on the Environment, with its 'One Earth' discourse, was the first major meeting to address this concern; ~~but~~ others followed, such as the Rio Conference in 1992. Much of this early effort was focussed on documenting the problem and calling for action, but in the first decade of the new millennium, some solutions began to emerge. The two most prominent examples are the Millennium Development Goals (MDGs 2000) and the more recent Sustainable Development Goals (SDGs 2015).

In terms of the Earth System approach, one of the most innovative proposals for dealing with global change is the Planetary Boundaries (PB) framework, introduced by Swedish researcher Johan Rockström in 2009 (Rockström et al. 2009). The framework is based directly on Earth System science, rather than on human needs, and its reference point is the stable, 11,700-year Holocene state during which technologically-advanced human civilisations developed. The approach is to identify the minimum number of processes or sub-systems that define the state of the Earth System, select control variables for how these processes are influenced or driven, and identify the limit to perturbations of the processes, beyond which the Earth System was likely to have left a Holocene-like state.

The structure of the PB framework is shown in Fig. 2.5. The figure is drawn as a series of concentric circles that represent various levels of human perturbation of the PB processes: the inner green circle represents the 'safe operating space' where none of the PBs has been transgressed, while yellow is a zone that accounts for uncertainty about the precise position of the boundary. Because precise knowledge about threshold positions and acceptable levels of degradation does not yet exist for all of the PB processes, there is uncertainty, in some cases considerable, around the placement of the boundary. The approach has been to minimise risk by placing the boundary at the lower level of the uncertainty range. The red zone is an area of high risk, beyond the uncertainty range, where the risk of a highly degraded, destabilised or transformed process or state is high. The control variables radiate out from the centre of the figure forming a set of nine grey wedges, one for each of the planetary boundaries. The figure is normalised so that the outer limit of the green circle is at the point where each of the nine control variables is precisely at its boundary position, and the outer limit of the yellow zone is at the upper end of the zone of

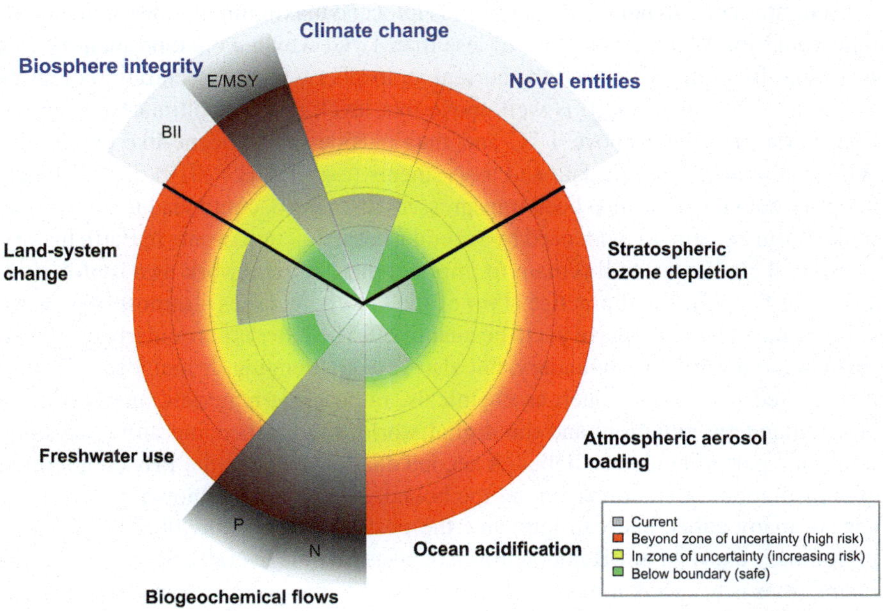

Fig. 2.5 The status ca. 2015 of the nine planetary boundaries (grey wedges) with the proposed three core boundaries labelled in blue. BII means 'biodiversity intactness index' and E/MSY means 'extinctions per million species-years (Steffen et al. 2015b)

uncertainty. Thus, the inner green circle represents the safe operating space for the Earth System as a whole.

Some, but not all, of the processes have thresholds beyond which the process shifts to a substantially different mode of operation. In these cases, the boundary itself is placed well below the point of the threshold to account for uncertainties in precisely where the threshold might lie. Examples of processes that are likely to have thresholds are climate change (see next section), ocean acidification, some components of the land system change PB (e.g., the Amazon forest) and the nitrogen biogeochemical flow (e.g., freshwater eutrophication). In other cases where no threshold has been shown to exist, the boundary is placed at a level of degradation below which research shows that the process is likely to operate in a Holocene-like mode (Rockström et al. 2009; Steffen et al. 2015b). The climate change boundary is a good example of how the system works. The control variable is atmospheric CO_2 concentration, and the boundary is set at 350 ppm, with an uncertainty range (or range of increasing risk) from 350 to 450 ppm. Concentrations above 450 ppm lie in the zone of high risk – the red circle. The CO_2 concentration in 2015 was just under 400 ppm, as indicated by the position of the grey wedge in Fig. 2.5.

The three PB processes labelled in blue in Fig. 2.5 are proposed to be 'core' planetary boundaries, implying that the transgression of any one of these three boundaries, on its own, would push the Earth System out of a Holocene-like state. For example, the climate change boundary is set at an atmospheric CO_2

concentration of 350 ppm, well above the Holocene maximum of ~280 ppm but one that would limit temperature rise to less than 1 °C, which we would identify as a Holocene-like state. However, the current atmospheric CO_2 concentration of about 413 ppm (November 2020) is well above that level, and will ultimately lead to a temperature rise well above 1 °C and thus outside of Holocene-like conditions. Based on a large body of data and assessments (e.g., IPBES 2019), the biosphere integrity boundary has also been transgressed, and it, too, could lead to a new state of the Earth System, as evidenced by several mass extinction events in Earth history.

Several challenges still remain to improve the PB framework as a useful guidance tool for policy and practice. Two of the nine boundaries – atmospheric aerosols and novel entities – have yet to be quantified. The latter is particularly important as a potential third core boundary, but also very challenging. It includes new entities created by humans, such as chemicals, radioactive materials, new forms of electromagnetic radiation, and the digital world that has become an increasingly important part of human societies. A second challenge is that the PBs do not operate in isolation; interactions among the boundary process are known to be significant in many cases, and can influence the position of the individual PB itself. A preliminary analysis of PB interactions suggests that the overall effect of including interactions is to shrink the 'safe operating space' – the green area in Fig. 2.5 (Lade et al. 2019).

The PB framework has already proven to be a useful tool for managing human impacts on the Earth System at a range of scales, from local enterprises to nation states. One of the most innovative applications, though, is the 'Common Home of Humanity' initiative, which aims to establish legal status for a well-functioning Earth System state – basically a Holocene-like state – as the intangible natural heritage of all humanity (Magalhães et al. 2016). The PB framework serves as the scientific definition of a 'well-functioning Earth System' in this approach.

2.6 Future Trajectories of the Earth System

Understanding our home planet – the Earth System – as a single system with a 4.5-billion-year history that displays a variety of states and trajectories raises an obvious question: given the rapidly increasing human pressures on the system, what is its future trajectory? Where on Earth are we going?

The answer to that question is complex and plagued with considerable uncertainties. We now know with a high degree of certainty that the Earth System is on a rapid trajectory away from the 11,700-year stability domain of the Holocene and is already well into the trajectory of the Anthropocene (Steffen et al. 2004, 2016; Section 3; and Fig. 2.4). We will not go back to the Holocene, even with the most vigorous efforts to stabilise the climate – that is, achieving the lowest emission scenario of the IPCC – and our best efforts to slow and then halt the degradation of the biosphere. So just what sort of future do we face in terms of our ability to persist and thrive as a species on our home planet?

Multiple answers to that question have been proposed, but here we adopt a complex systems approach (Scheffer 2009) to suggest where the Earth System might be headed (Steffen et al. 2018). This approach is based on the theory that complex systems like the Earth System exist in well-defined states with transitions between them; intermediate conditions/states are not possible. The mechanisms behind this behaviour are feedback processes internal to the system that can destabilise it and push it into another stable state, or 'basin of attraction'.

In the context of the Earth System, this type of behaviour is illustrated in Fig. 2.6, a cartoon that is sometimes called a stability landscape (Steffen et al. 2018). Human pressures – greenhouse gas emissions and biosphere degradation – have already pushed the Earth System out of the Holocene and into the Anthropocene, a trajectory into hotter conditions. We postulate that ahead lies a 'fork in the road'. The current trajectory is accelerating the system towards that bifurcation, with the increasing risk that our pressures will push the system onto the 'Hothouse Earth' trajectory. The critical point here is that there is a point beyond which we lose control of the system and its own internal feedbacks drive it past a global threshold and irreversibly into a much hotter state.

The 'planetary threshold' in Fig. 2.6 is formed by a group of tipping points and feedbacks within the Earth System. These are sub-systems that appear stable under minor stresses, but when a critical point is reached, internal feedback processes within the system dominate and drive it into a new state. Think of a kayak – tip it a little and it is stable, but tip it too much and you rapidly find yourself underwater.

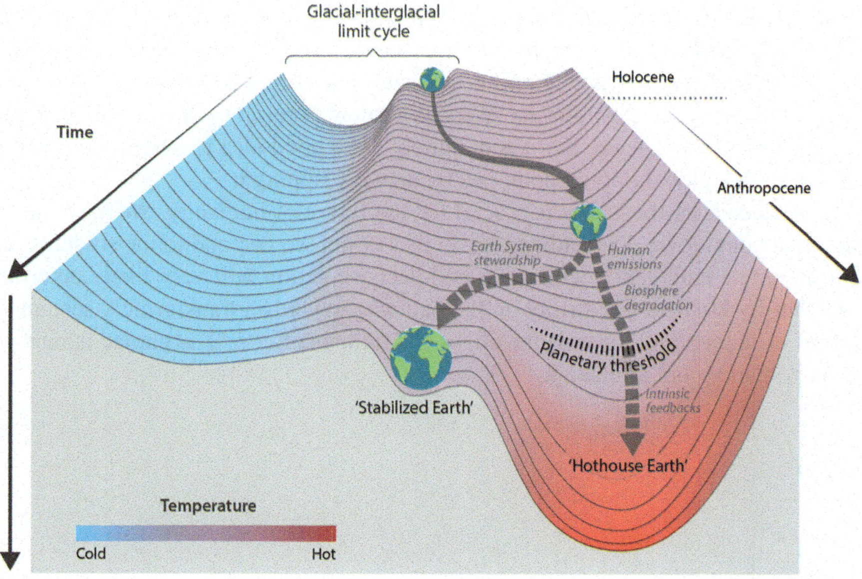

Fig. 2.6 Stability landscape showing the pathway of the Earth System out of the Holocene and into the Anthropocene. Two potential pathways into the future are shown as 'Stabilized Earth' and 'Hothouse Earth' (Steffen et al. 2018)

An Earth System example is the Greenland ice sheet, which is beginning to melt primarily from its surface, which at the centre of Greenland is several kilometres above sea level. As the ice surface melts, it is lowered into a warmer climatic zone, which increases the melt rate further, which lowers the surface even more, and so on. At some point, the process becomes irreversible – the system has been ‘tipped’ – and the ice sheet eventually disappears, even if the original warming from climate change is eliminated.

Tipping points in the Earth System were first systematically identified and studied over a decade ago (Lenton et al. 2008). They exist in three major forms – (i) large bodies of ice, such as the Antarctic ice sheets and the floating ice over the Arctic Ocean; (ii) major biomes such as the Amazon rainforest and the vast boreal forests across the northern high latitudes; and (iii) ocean and atmospheric circulation systems, such as the North Atlantic thermohaline circulation and the El Niño-Southern Oscillation. But these tipping points don’t act in isolation; they can be connected via changes in temperature or through coupling by atmospheric and oceanic circulation. For example, melting Arctic sea ice uncovers darker ocean water that in turn absorbs more sunlight leading to increased warming in the region. This raises the temperature over Greenland, increasing the rate of ice melt. Another example is a slowdown in North Atlantic Ocean circulation, which reduces rainfall over the Amazon basin and pushes the rainforest closer to a tipping point. These and other such links can generate a widespread tipping cascade, a critical phenomenon in the transition of the Earth System from one state to another.

A recent assessment of observations showed that many of these tipping elements have already been activated and are moving towards their tipping points (Lenton et al. 2019): ice loss is accelerating in Greenland and West Antarctica; drought is increasing in the Amazon rainforest; the Atlantic circulation has been slowing since the 1950s; permafrost is beginning to thaw; and boreal forests are experiencing increasing fires and pest attacks. We don’t know precisely when a global tipping cascade will be triggered, but it is definitely not a good idea to determine precisely where such a global tipping point lies by actually tipping it.

The decade of the 2020s is crucial for the future of humanity. We have entered the final countdown. Depending on whether or not we respond to the students’ call for a climate emergency with rapid and decisive action will very likely determine whether by 2030 we are firmly on the pathway towards a transformed, sustainable world or descending towards the chaos of Hothouse Earth. The choice is ours and time is running out to make it.

References

AWG (Anthropocene Working Group) (2019) Results of binding vote by AWG. Released 21 May 2019. http://quaternary.stratigraphy.org/working-groups/anthropocene/

Bronowski J (2011) The ascent of man. Penguin Books, London

Ciais P et al (2013) Carbon and other biogeochemical cycles. In: Stocker TF et al (eds) Climate change 2013: the physical science basis, contribution of working group I to the fifth assessment report of the Intergovernmental Panel on Climate Change. Cambridge University Press, Cambridge/New York, pp 465–570

Collins M et al (2013) Long-term climate change: projections, commitments and irreversibility. In: Stocker TF et al (eds) Climate change 2013: the physical science basis, contribution of working group I to the fifth assessment report of the Intergovernmental Panel on Climate Change. Cambridge University Press, Cambridge, pp 1029–1136

Crutzen PJ (2002) Geology of mankind: the anthropocene. Nature 415:23

Crutzen PJ, Stoermer EF (2000) The "Anthropocene". IGBP Newslett 41:17–18

European Environment Agency (2020). https://www.eea.europa.eu/data-and-maps/indicators/atmospheric-greenhouse-gas-concentrations-6/assessment-1. Accessed 16 June 2020

Friedlingstein P et al (2019) Global carbon budget 2019. Earth Syst Sci Data 11:1783–1838. https://doi.org/10.5194/essd-11-1783-2019

Haff PK (2014) Humans and technology in the Anthropocene. Six rules. Anthropocene Rev 1:126–136

IPBES (Intergovernmental Science-Policy Platform on Biodiversity and Ecosystem Services) (2019) Global assessment report on biodiversity and ecosystem services. https://www.ipbes.net/news/ipbes-global-assessment-summary-policymakers-pdf

IPCC (Intergovernmental Panel on Climate Change) (2013) Summary for policymakers. In: Stocker TF, Qin D, Plattner G-K, Tignor M, Allen SK, Boschung J, Nauels A, Xia Y, Bex V, Midgley PM (eds) Climate change 2013: the physical science basis, contribution of working group I to the fifth assessment report of the Intergovernmental Panel on Climate Change. Cambridge University Press, Cambridge/New York, pp 3–29

IPCC (Intergovernmental Panel on Climate Change) (2018) Special report on global warming of 1.5°C. http://ipcc.ch/report/sr15/

Lade SJ, Steffen W, de Vries W, Carpenter SR, Donges JF, Gerten D, Hoff H, Newbold T, Richardson K, Rockström J (2019) Earth system interactions amplify human impacts on planetary boundaries. Nat Sustain 3:119–128

Le Quéré C et al (2020) Temporary reduction in daily global CO_2 emissions during the COVID-19 forced confinement. Nat Clim Chang. https://doi.org/10.1038/s41558-020-0797-x

Lenton TM et al (2008) Tipping elements in Earth's climate system. Proc Natl Acad Sci U S A 105:1786–1793

Lenton TM, Rockström J, Gaffney O, Rahmstorf S, Richardson K, Steffen W, Schellnhuber HJ (2019) Climate tipping points – too risky to bet against. Nature 575:593–596

Lovelock J, Margulis L (1974) Atmospheric homeostasis by and for the biosphere: the Gaia hypothesis. Tellus 26:2–10

Magalhães P, Steffen W, Bosselmann K, Aragão A, Soromenho-Marques V (2016) The safe operating space treaty: a new approach to managing our use of the earth system. Cambridge Scholars Publishing, Newcastle upon Tyne

Malm A, Hornborg A (2014) The geology of mankind? A critique of the Anthropocene narrative. Anthropocene Rev 1:62–69

McNeill JR (2001) Something new under the sun. W.W. Norton, New York/London

McNeill JR, Engelke P (2016) The great acceleration: an environmental history of the Anthropocene since 1945. Harvard University Press, Cambridge

MDGs (UN Millennium Development Goals) (2000). https://www.un.org/millenniumgoals/ Accessed 12 June 2020

NASA (National Aeronautics and Space Administration) (1986) Earth system science overview. A program for global change. Prepared by the Earth System Sciences Committee, NASA Advisory Council

NOAA (National Oceanographic and Atmospheric Administration) (2020). https://www.esrl.noaa.gov/gmd/ccgg/trends/. Accessed 16 June 2020

Oppenheimer S (2004) Out of Eden: the peopling of the world. Constable and Robinson, London

Rockström J, Steffen W, Noone K, Persson Å, Chapin FS III, Lambin EF, Lenton TM, Scheffer M, Folke C, Schellnhuber HJ, Nykvist B, de Wit CA, Hughes T, van der Leeuw S, Rodhe H, Sörlin S, Snyder PK, Costanza R, Svedin U, Falkenmark M, Karlberg L, Corell RW, Fabry VJ, Hansen J, Walker B, Liverman D, Richardson K, Crutzen P, Foley JA (2009) A safe operating space for humanity. Nature 461:472–475

Ruddiman WF et al (2015) Late Holocene climate: natural or anthropogenic? Rev Geophys. https://doi.org/10.1002/2015RG000503

Scheffer M (2009) Critical transitions in nature and society. Princeton University Press, Princeton

SDGs (Sustainable Development Goals) (2015). https://www.un.org/sustainabledevelopment/. Accessed 12 June 2020

Steffen W, Sanderson A, Tyson PD, Jäger J, Matson P, Moore B III, Oldfield F, Richardson K, Schellnhuber HJ, Turner BL II, Wasson RJ (2004) Global change and the earth system: a planet under pressure, The IGBP book series. Springer, Berlin/Heidelberg/New York

Steffen W, Crutzen PJ, McNeill JR (2007) The Anthropocene: are humans now overwhelming the great forces of Nature? Ambio 36:614–621

Steffen W, Broadgate W, Deutsch L, Gaffney O, Ludwig C (2015a) The trajectory of the Anthropocene: the great acceleration. Anthropocene Rev. https://doi.org/10.1177/2053019614564785

Steffen W, Richardson K, Rockström J, Cornell SE, Fetzer I, Bennett EM, Biggs R, Carpenter SR, de Vries W, de Wit CA, Folke C, Gerten D, Heinke J, Mace GM, Persson LM, Ramanathan V, Reyers B, Sörlin S (2015b) Planetary boundaries: guiding human development on a changing planet. Science 347. https://doi.org/10.1126/science.1259855

Steffen W, Leinfelder R, Zalasiewicz J, Waters CN, Williams M, Summerhayes C, Barnosky AD, Cearreta A, Crutzen P, Edgeworth M, Ellis EC, Fairchild IJ, Gałuszka A, Grinevald J, Haywood A, Ivar do Sul J, Jeandel C, McNeill JR, Odada E, Oreskes N, Revkin A, Richter D deB, Syvitski J, Vidas D, Wagreich M, Wing SL, Wolfe AP, Schellnhuber HJ (2016) Stratigraphic and earth system approaches to defining the Anthropocene. Earth's Future 4:324–345. https://doi.org/10.1002/2016EF000379

Steffen W, Rockström J, Richardson K, Lenton TM, Folke C, Liverman D, Summerhayes CP, Barnosky AD, Cornell SE, Crucifix M, Donges JF, Fetzer I, Lade SJ, Scheffer M, Winkelmann R, Schellnhuber HJ (2018) Trajectories of the earth system in the Anthropocene. Proc Natl Acad Sci U S A. https://doi.org/10.1073/pnas.1810141115

Steffen W, Richardson K, Rockström J, Schellnhuber HJ, Dube OP, Dutreuil S, Lenton TM, Lubchenco J (2020) The emergence and evolution of earth system science. Nat Rev Earth Environ 1:54–63

Vernadsky VI (1998 [1926]) The biosphere (complete annotated edition: foreword by Lynn Margulis and colleagues, introduction by Jacques Grinevald, translated by David B Langmuir, revised and annotated by Mark AS McMenamin). Copernicus (Springer-Verlag), New York

Williams M, Zalasiewicz J, Haff PK, Schwägerl C, Barnosky AD, Ellis EC (2015) The Anthropocene biosphere. Anthropocene Rev 2(3):196–219. https://doi.org/10.1177/2053019615591020

Chapter 3
Australia's Natural Environment: A Warning for the World

Chris R. Dickman and David B. Lindenmayer

Abstract The immense size, varied land systems, different climatic regimes, and great antiquity of the Australian landmass have fostered the development of an extraordinarily rich and endemic biota. Yet, despite the importance of this unique natural heritage, evidence is mounting that Australia's living systems are under increasing assault. On land, the major threats faced by the biota include land clearing and fragmentation, pressures from livestock production, pest species and pathogens, consumption and extraction of natural resources, altered fire regimes, climate change and climate variability, and interactions between these drivers. Equivalent pressures face the biota of aquatic environments, with climate change interacting with other threats to produce poor prognoses for the future of freshwater resources and major marine ecosystems such as the Great Barrier Reef. Despite the success of some conservation programs, the varied threats to Australia's fragile environment are exacerbated by the demands of the nation's burgeoning human population and the generally apathetic response of government to the manifest problems. We suggest that the Australian environment can be seen as a 'canary in the climate coal mine'—an early herald of global environmental decline—and that more recognition of the problems, more courage, more resources and stronger laws will be crucial if any semblance of ecological sustainability is to be achieved.

C. R. Dickman (✉)
University of Sydney, Sydney, Australia
e-mail: chris.dickman@sydney.edu.au

D. B. Lindenmayer
Australian National University, Canberra, Australia
e-mail: David.Lindenmayer@anu.edu.au

S. J. Williams, R. Taylor (eds.), *Sustainability and the New Economics*,
https://doi.org/10.1007/978-3-030-78795-0_3

3.1 Introduction

The continental land area of Australia covers some 7,595,342 km^2, with the island of Tasmania adding 64,520 km^2 to this total and the country's 5442 smaller islands a further 33,000 km^2. Spanning latitudes between 10° 40′ and 43° 39 S and longitudes between 113° 09′ and 153° 39′ E, Australia encompasses an extraordinarily wide range of land systems and climate regimes. The tropical north has a pronounced wet season with associated cyclones between November and April. In stark contrast, southerly latitudes mostly experience temperate seasonal conditions that can include winter snow at higher elevations, although parts of the south and south west are characterised by a Mediterranean-type climate of hot, dry summers and cool, moist winters. Approximately 70% of continental Australia is arid (annual rainfall <375 mm), with summer rainfall dominating in northern arid regions and winter rainfall in the south. Much of the continent is characterised by gently undulating landscapes, but more rugged topographic relief—and the continent's highest peak (Mt Kosciuszko, 2228 m above sea level)—occur along the Great Dividing Range that runs for more than 3500 km from north Queensland to western Victoria. Renowned as the world's driest inhabited continent, Australia has relatively few major rivers or large freshwater lakes. Indeed, the largest water body, Lake Eyre, is saline and ephemeral, reaching its maximum area of up to 10,000 km^2 only after heavy and prolonged rains throughout its vast catchment area (Dickman 2018). Australia's jurisdiction also encompasses seven external territories, the most expansive of which covers 42% of Antarctica, almost 5.9 million km^2.

Beyond the margins of the land, Australia's environment includes vast areas of near shore, mid and deep ocean systems. National jurisdiction—the Exclusive Economic Zone (EEZ)—extends 200 nautical miles (370 km) seaward from all land areas except in the north due to the closeness of New Guinea, and encompasses some 14 million km^2. Australia's sovereign rights in the EEZ include the conservation and management of fisheries and of natural resources in waters superjacent to the seabed as well as of the seabed and its subsoil (http://www.ga.gov.au/scientific-topics/marine/jurisdiction/maritime-boundary-definitions#eez).

As might be expected for an immense land mass with many satellite islands and claims over vast expanses of ocean, Australia is home to extraordinarily rich assemblages of animals, plants, fungi and other organisms. Chapman (2009) estimated that about 570,000 species occur in Australia, with 147,579 of these being formally described. Although impressive, the estimated figure is already known to be too low: increasing use of molecular taxonomic techniques has revealed thousands of new cryptic species over the last decade, and recent studies indicate further that bacterial diversity has been grossly under-estimated, both in Australia, and globally (Dickman 2018). Australia meets the criteria for being 'megadiverse' in that it is home to more than 5000 endemic plant species and has marine ecosystems within its borders (Mittermeier et al. 1997). Only 16 other countries so qualify. In addition to simple counts of numbers of species, Australia contains eight of the world's 14 major terrestrial biomes, 10 freshwater and 23 marine biomes as well as myriad ecosystems and ecological communities that constitute these biomes (Dickman 2018).

Despite Australia's astonishing biological riches, evidence is mounting that this rich living heritage is fading rapidly, that complex ecological systems are failing, and even such iconic and well-loved national symbols as the Great Barrier Reef are in terminal decline (Van Dijk 2020). Alarm bells were rung by astute naturalists such as John Gould and Gerard Krefft about faunal declines as early as the nineteenth century, and by the turn of the twentieth century, Royal Commissions were under way to inquire into the factors leading to losses of vegetation and the degradation of vast tracts of land.

In 1996, the first independent national assessment of Australia's overall environment was made, and its report card was mixed. On the one hand the report noted that "Some aspects of the Australian environment are in relatively good condition by international standards", but on the other that "… Australia has some very serious environmental problems" (State of the Environment Advisory Council 1996, p ES-4). State of the Environment reports have been produced at five-yearly intervals since 1996. The most recent, published in 2016, was notably less sanguine (Cresswell and Murphy 2017), presenting no evidence that the major pressures on biodiversity had decreased over the entire period of reporting.

Australia and its biota are unique in many ways, but also harbingers of extreme climatic events, catastrophic fires, environmental degradation, species extinctions and losses of ecological function that will increasingly afflict all life on all parts of planet Earth (e.g., Bradshaw et al. 2021). The continent can therefore be seen in many ways as a canary in the climate coal mine – an early exemplar of the kinds of environmental problems to be faced globally if we continue on our present trajectory. In this chapter we begin by briefly sketching some of the historical factors that have shaped Australian ecosystems to appreciate the importance and uniqueness of the biota, and why endemic organisms are particularly vulnerable to the rapidly changing conditions. We then describe the main drivers of biodiversity loss in the country's terrestrial, freshwater and marine systems before offering some concluding thoughts on solutions to some of Australia's many environmental problems.

3.2 Australia: An Historical Perspective

Since its separation from eastern Antarctica 38–45 million years ago, the continental plate that supports the Australian land mass has drifted slowly northward, at about the pace of growth of one's fingernails, propelled by convection currents that that lie deep below the Earth's crust. Around 30 million years ago the Australian plate collided with the Pacific plate to its north, leading to buckling and upthrusts of land along the active edges of the plates that now form part of the cordillera of New Guinea. These titanic deep-time events had profound implications for the establishment of Australia's biota. On land, some elements of the biota hark back 250 million years to the early Triassic period when the world's continents were a single land mass, Pangaea. Following the sundering of Pangaea, these archaic taxa, including many flowering plants and invertebrates, have evolved in splendid isolation on the Australian landmass but share deep affinities with relatives elsewhere in the world. The ancestors of other taxa, such as the marsupials, migrated to Australia from South America when these great southern continents were still connected by the Antarctic land bridge, perhaps 60–65 million years ago. With the establishment of land to Australia's north, other taxa such as the first Australian rodents arrived in the continent by island-hopping through Melanesia. Still other species-groups, including the ancestors of nasute termites and some spiders, migrated to Australia by rafting across expanses of ocean (Dickman 2018). The sundering of Antarctica from Australia and, soon after, from South America, had further implications for marine biota. The isolation of Antarctica allowed formation of the circumpolar current, with the increased productivity of the southern oceans then shaping the evolution of large marine mammals, including filter-feeding whales.

Because of the early arrival, varied origins and subsequent isolation of many taxa, Australia's biota is phylogenetically diverse. For example, the continent boasts such divergent species as the Wollemi pine (*Wollemia nobilis*), platypus (*Ornithorhynchus anatinus*), koala (*Phascolarctos cinereus*) and 'dinosaur ant' (*Nothomyrmecia macrops*) that have no close living parallels. The prolonged isolation of these and many other taxa has led to very high levels of endemism in the biota. In well-studied groups such as flowering plants and terrestrial vertebrates, over 90% of species that occur in Australia are found nowhere else (Chapman 2009). Endemic species sometimes act as symbols of national identity and pride, with charismatic endemic animals such as koalas and emus (*Dromaius novaehollandiae*) and plants like waratahs (*Telopea* spp.) achieving economic, cultural, aesthetic and other values. However, endemism also comes at a price: if large components of a nation's biota occur nowhere else, there is an absolute imperative to ensure the persistence of that biota by using the environment wisely and sustainably. As we show below, this is a price that Australians have too often been unwilling to pay.

Once in Australia, ancestral organisms had to adapt to their new physical environment and to the prevailing climatic conditions. Physically, soils over much of the continent are shallow or nutrient deficient as two of the major processes that renew soil—glaciation and orographic activity (mountain-building)—have occurred to a lesser extent in Australia than in other parts of the world. Prolonged weathering also has ensured that these soils are poor, with nutrients such as phosphorus and nitrogen being leached from the soil profile over vast areas. The nutrient deficiency of the soils has had profound evolutionary consequences (Orians and Milewski 2007). For example, it has fostered nitrogen-fixing adaptations in many members of the pea family, such as *Acacia* spp., and has probably driven the widespread association between green plants and mycorrhizal fungi that surround plant roots and assist with nutrient uptake. Amongst animals, there is considerable evidence that nutrient deficiency in food selects for slower growth and reproduction. For example, lineages of frogs, birds and mammals with a long evolutionary history in Australia generally have slower life histories than ecologically similar vertebrates in other parts of the world, and slower life histories than related species that have arrived in Australia more recently (Dickman 2007). Slow growth and breeding likely place old endemic taxa at greater risk than newly introduced species in Australia's rapidly changing—and deteriorating—environment.

3.3 Australia: Current Perspective

The five State of the Environment reports published since 1996 paint a picture of inexorable decline in the diversity and condition of Australia's natural systems. The reports also lament the paucity of programs that have been established to either track biodiversity or quantify the effects of management interventions, although all reports have highlighted bright spots in the gloom where they could. As of December 2020, the national *Environment Protection and Biodiversity Conservation Act* (EPBC Act) *1999* listed 1900 species and subspecies as threatened with extinction (including 93 taxa that are extinct), as well as 87 threatened ecological communities. Despite these large numbers of entities on the EPBC Act, the lists are incomplete owing to limited information on poorly studied taxa, taxonomic revisions that have identified many new cryptic species, and recent events such as the drought and bushfires in 2019–2020 that will have put many further species and communities at risk. In addition, being listed as threatened on the EPBC Act guarantees little. An independent review of the Act concluded recently that Australia's "… current environmental trajectory is unsustainable" and that "The EPBC Act is ineffective … and is not fit to address current or future environmental challenges" (Samuel 2020: 1). Despite this review being commissioned by the Australian Government, there is currently little sign that its recommendations will be effectively implemented (Burnett 2021).

Other independent assessments of Australia's biota also indicate marked declines. From 2018 to 2020, for example, the Threatened Species Recovery Hub of the government-funded National Environmental Science Program collated information on threatened plants, birds and mammals, and showed declines of varied magnitude since 1985 or 1995 in each group. In view of their high historical rate of extinction in Australia, mammals (34 species extinct since 1800) fared surprisingly well. Since 1995 threatened mammals collectively have declined by ~35% in relative population size (Fig. 3.1a). By contrast, threatened birds have declined by more than 50% over the same period (Fig. 3.1b), and threatened plants by an extraordinary 72% (Fig. 3.1c). Freshwater systems such as the vast Murray-Darling Basin frequently run dry due to water diversion for agriculture, threatening both aquatic organisms and riparian communities such as those dominated by river red gums. In the marine environment, a recent assessment by the International Union for Conservation of Nature (IUCN) noted a decline in the condition of the Ningaloo Coast and Shark Bay World Heritage sites from 'Good' in 2017 to 'Good with some concerns' in 2020, with the condition of the Great Barrier Reef declining from 'Significant concern' to 'Critical' over the same period (Osipova et al. 2020). The same report also

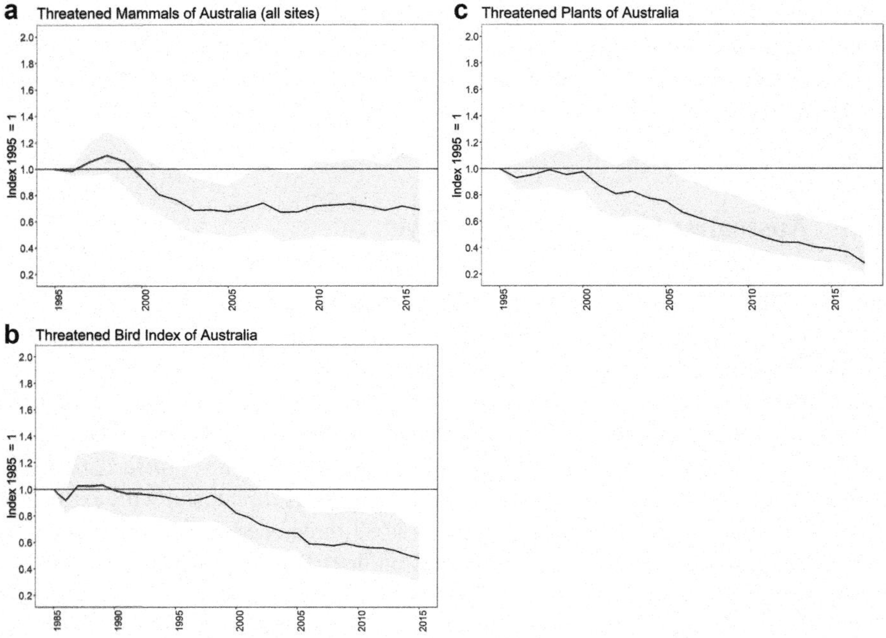

Fig. 3.1 Threatened Species Index results for Australian (**a**) mammals, (**b**) birds and (**c**) plants. The index for each group is based on multiple datasets derived from research projects, government surveys, citizen science reports, targeted monitoring surveys and other sources averaged across those threatened species for which data exist. The lines represent the mean index score and the shaded area the 95% confidence limits, with the trends representing the change over time in relative population size from a base year (1985 for birds, 1995 for mammals and plants). The index results are compiled and maintained under the auspice of the Threatened Species Recovery Hub of the government-funded National Environmental Science Program. Figures are taken from the Threatened Species Index (https://tsx.org.au/, accessed 1 Feb 2021)

downgraded the condition of the Gondwana Rainforests of Australia and Greater Blue Mountains Area to being of 'Significant concern'; no sites were assessed as having improved in conservation outlook in the 3 years to 2020 (Osipova et al. 2020).

Of course, not all components of Australia's environment have fared poorly in recent years. State government programs such as Bounceback in South Australia and Western Shield in Western Australia have made great strides in removing major biodiversity threats, restoring ecological processes and reintroducing species that had become locally extinct. Many Australian islands have also been subject to effective conservation management; for example, invasive species such as feral cats and black rats have been removed recently from several offshore and oceanic islands, thus facilitating partial reconstruction of the original fauna (Algar et al. 2020). The case of sub-Antarctic Macquarie Island provides a further inspiring example of successful intervention with substantial conservation gains (Garnett et al. 2018). Effective conservation management in other contexts has saved many threatened species from likely extinction (e.g. Bolam et al. 2021). Nationally, some 151 million hectares (~20% of Australia's land area) lie within the nation's protected area system, with land being added incrementally when it becomes available as Indigenous Protected Areas or to meet the criteria of reserves being comprehensive, adequate and representative (https://www.environment.gov.au/land/nrs). Over 300 million hectares of ocean are gazetted as marine protected areas (https://mpatlas.org/countries/AUS). Although the term 'protected' seems to be variously interpreted under different jurisdictions (Cockerell et al. 2020), six Australian World Heritage sites are rated independently to be still in 'Good' condition by the International Union for Conservation of Nature, with another five 'Good with some concerns' (Osipova et al. 2020).

The above examples show that environmental wins can be made (Garnett et al. 2018), even though the overall trajectory of Australia's environmental health continues to deteriorate. It should be possible to achieve more wins and retain functioning natural systems with their current levels of biodiversity, but this will require greater community engagement and more resourcing, commitment and bravery from different levels of government than has been shown hitherto (Wintle et al. 2019). The national government, for example, has been apathetic in its response to the overarching problems posed by climate change over the last decade but enthusiastic in its support for the continued use of fossil fuels. Despite welcome support for initiatives such as the National Environmental Science Program, overall funding for initiatives that promote environmental sustainability has dwindled, and all support has been terminated for organisations such as the Environmental Defenders Office[1] that seek to assist communities and their local environments to resist destructive developments. Most recently, the government's Job-Ready Graduates Bill slashed support for environmental courses by 29%, a reduction of almost Aust$10,000 that will be received by universities per student in environmental disciplines each year (Turner 2020). This is one of the largest funding cuts to any university discipline, and will surely reduce national expertise in environmental management at the same time that pressures on the environment are escalating. In the next section we outline the main such pressures that need to be addressed.

3.4 Drivers of Environmental Change

Natural systems, their constituent species and ecological communities face similar pressures in many parts of the world (e.g. Commission for the Human Future 2020; Osipova et al. 2020; Secretariat of the Convention on Biological Diversity 2020). In Australia, the major pressures have been listed in the most recent State of the Environment Report (Cresswell and Murphy 2017). We describe these pressures here, focussing firstly on the overarching menace imposed by human

[1] The Environmental Defenders Office, now a national registered charity with state branches, receives some funding from state governments while relying on donations.

overpopulation. We go on to discuss threats to terrestrial environments in some detail, and then review briefly the additional threats that face freshwater and marine environments.

3.5 Human Overpopulation

Much debate attended the likely limits to growth of the human population during the 1960s and 1970s, but the problem of overpopulation seems to have faded from sight even as the world population has doubled since the early concerns were raised. However, there is no doubt that rapid and inexorable increases in the human population—and the excessive demand for resources that people make—are rapidly depleting the natural world and its ability to regenerate (Goldie and Betts 2014). On the face of it, overpopulation in Australia should be less problematic than in other parts of the world as the nation's population density is relatively low: 3.3 people/km^2 compared with the global average of >53 people/km^2 (Dickman 2014). However, such a comparison is misleading. First, the thin and nutrient-deficient soils and erratic low-rainfall climate that characterise much of Australia cannot sustain high levels of productivity. Intensive farming attempts have resulted in immense damage to landscapes over vast areas. Second, Australia's per capita usage of resources is high compared with most other countries, with resource use elevated by production of food, mineral and fuel products for large markets overseas. In a recent analysis, continued growth along our current trajectory was shown to be unsustainable either for the nation's natural environment or for human society (Turner 2019).

With respect to wildlife, for every addition of a million people in Australia over the last 50 years, almost one (0.95) native vertebrate species has become extinct (Fig. 3.2). Depending on how rapidly the human population continues to grow, Dickman (2014) predicted that Australia's tally of extinct vertebrates could range from 77 to 103 vertebrate species by 2100. Of course, relationships such as that shown in Fig. 3.2 do not imply that people are directly extirpating other species, although this may be the ultimate fate of taxa such as flying-foxes (*Pteropus* spp.) that are now actively culled in many areas to protect human interests. Rather, increasing human population size has many and varied indirect effects on other species and their habitats, as we note here. In view of the difficulty of achieving meaningful environmental sustainability in the short to medium term with burgeoning human population growth, it is imperative that we initiate honest and rational debate about how we should move forward on issues associated with population size, both in Australia and globally (Goldie and Betts 2014; Chap. 7 of this volume).

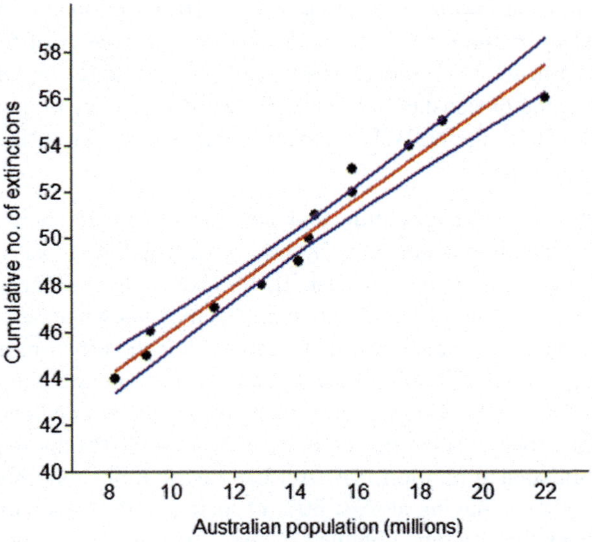

Fig. 3.2 Human population size in Australia from 1950 to 2009 plotted against the cumulative number of extinctions of terrestrial vertebrate species over the same period. Between 1800 and 1950 at least 43 species of terrestrial vertebrates became extinct at different times. Between 1950, when the Australian population was 8.3 million, and 2009 when the population had reached 22 million, four species of native frogs and nine species of mammals became extinct. The graph shows the 13 species that disappeared over this period (dots), the linear regression (straight line) and 95% confidence intervals (curved lines). The regression equation is $y = 36.5 + 0.95x$ ($R^2 = 0.96$), indicating that for every million people added to the Australian population since 1950, almost one (0.95) native vertebrate species has become extinct. (Redrawn from Dickman (2014))

3.6 Terrestrial Environments

Land Clearing and Fragmentation These processes are perhaps the major drivers of decline of Australia's terrestrial ecosystems, with half of all species listed as threatened on the EPBC Act at risk due to habitat loss. Land clearing is conducted to allow new housing developments, mining, construction of roads, easements and other built infrastructure, but most significantly it facilitates agricultural expansion. Most clearing (>70%) now occurs in areas that have regrown after previous clearing, and rates of clearing have largely stabilised or even decreased in recent years (Cresswell and Murphy 2017). However, the states of Queensland and New South Wales (NSW) stand as exceptions. In the late 1990s, land clearing in Queensland exceeded 400,000 hectares a year, and was reined in by legislative amendments in 2004 only after the government had been embarrassed by a World Wide Fund for Nature report showing that land clearing in Queensland killed about 100 million vertebrates each year. Subsequent re-weakening of environmental legislation and relaxation of compliance in 2012–2013 saw more than a tripling of wooded areas being cleared, from 78,000 hectares a year in 2009–2010, to 300,000 hectares a year in 2014–2015 (Cogger et al. 2017). Similarly, in NSW, laws that had previously kept land clearing in check were replaced in 2016 by the inappropriately named *Biodiversity Conservation Act 2016*, leading to a near-tripling of the area of land cleared in the north-west of the state where data were available (Taylor and Dickman 2018). Whereas the Queensland Parliament passed new laws to once again check land clearing in 2018, the NSW Parliament to date has shown little interest in further legislative reform.

Broadscale clearing for agriculture and major developments such as new housing estates provide conditions that are highly detrimental to most native species, although some generalists may persist. If, however, the original habitat is fragmented and not completely cleared, the remnants may allow the persistence of species that have small area requirements or can exploit the urban-remnant bushland matrix. Garden skinks, rainbow lorikeets and brushtail possums are some of the most well-known of these latter species. Intriguingly, many threatened species also occur in city environments, with a study by Ives et al. (2016) finding that 503 species listed as threatened on the EPBC Act had distributions that intersected urban areas, and 51 of these having at least 30% of their distributions within city limits. However, species that become confined to small patches of habitat do not always persist long term, and eventually disappear due to threats from the surrounding environment, limited resources and opportunities to disperse; this postponement of loss is the 'extinction debt'. As fragmentation increases, so too does the magnitude of the debt that must be paid.

Pressures from Livestock Production Some of the land clearing noted above is to prepare for the sowing of cereal, vegetable, fruit or textile crops such as cotton, but much current clearing is done to increase pasture acreage for livestock, notably sheep. There is abundant evidence of the historically damaging effects of overgrazing in low rainfall rangeland systems (Lunney 2001), and several studies have documented increases in vertebrate diversity and abundance following livestock removal. Livestock deplete fragile soil crusts and can remove palatable plant species, but in many ways the most pervasive effects of livestock on biodiversity arise from inappropriate management. To run sheep in the rangelands, for example, dingoes must be culled or fenced out and species such as kangaroos and rabbits that would have been kept in check by dingoes then have to be controlled. There is seldom any opportunity for regeneration of shrubs or trees due to the constancy of grazing, and landscapes suffer increasingly from salinization, loss of organic material or soil erosion. There has been much debate about the potential benefits of reintegrating dingoes into rangeland systems, with increased use of guardian dogs or other appropriate means of reducing the impacts of dingoes, but resistance from the agricultural sector has prevented any meaningful progress.

Pest Species and Pathogens These species represent another major pressure on Australia's natural environment, and are the most frequently nominated class of threat to species listed on the EPBC Act. Cresswell and Murphy (2017) pointed out further that 12 of 21 key threatening processes (KTPs) listed on the Act concern invasive species such as rabbits, feral cats, red foxes, cane toads, escaped garden plants and yellow crazy ants, whereas three KTPs identify threats arising from pathogens. These threats vary in the numbers of native species and ecological communities that they affect, their geographical scope, and the extent to which they can be effectively managed. For example, feral cats occupy >99.8% of continental Australia and Tasmania as well as many offshore islands and island territories. They prey upon species in all vertebrate classes, and have been implicated in 27 of the 47 extinctions that have occurred among native vertebrates since European settlement (Woinarski et al. 2019). Despite much inquiry into how best to reduce cat-impacts, success has been limited and research into this question continues. Yellow crazy ants, by contrast, are more limited in their current range in Australia. Although these ants are not yet known to have caused any extinctions and can be eradicated in the early stages of a new infestation, they have wreaked havoc on insular ecosystems such as Christmas Island and can severely deplete populations of native invertebrates and small, less mobile vertebrates when in large numbers. The current costs of selected vertebrate pests to production industries in Australia have been estimated at almost Aust$600 million per year (McLeod 2016), with existing and projected costs to biodiversity increasing this total many fold.

The effects of three EPBC Act-listed pathogens—all fungi—are pervasive but varied in the severity of impact and the species they infect. Chytrid fungus disease occurs only in amphibians, killing its hosts by disrupting the all-important physiological functioning of the skin. The fungus has been implicated in the extinction of four Australian species of frog, dramatic declines in range of 10 more species and in population reductions of many more species. By contrast, myrtle rust fungus causes defoliation, stunted growth and death of shrubs and trees in the family Myrtaceae. Although this fungal pathogen was identified in Australia in only 2010, it has spread rapidly from its point of origin in coastal NSW to occupy areas in all east coast states, as well as sites in the Northern Territory. As Myrtaceae includes eucalypts, bottle brushes, paperbarks and many other trees that are floristically and structurally dominant in Australia, myrtle rust has the potential to alter the integrity and ecological function of many Australian landscapes. Root rot fungus—*Phytophthora cinnamomi*—likewise can deplete large areas of native vegetation, choking the vascular system of susceptible plant species. This pathogen currently covers hundreds of thousands of hectares across Australia's south-eastern states and, most worrying, over a million hectares in Western Australia where it threatens the integrity of the extraordinarily rich Southwest Botanical Province (Cresswell and Murphy 2017).

Consumption and Extraction of Natural Resources Logging, forestry and mining are among the most pervasive threats in this category, although illegal harvesting of plants such as cycads and orchids for horticulture and of animal species for the exotic pet trade can be detrimental for particular species. With respect to forestry, there are comparatively limited areas of wood production forest in Australia and almost all are relatively close to the wetter, coastal margins of the continent. Nevertheless, there have been significant impacts of forestry on the nation's wooded environments. For example, highly valuable species such as Red Cedar (*Toona* *ciliata*) were heavily logged and lost from large parts of their initial distribution at the time of European settlement. Elsewhere, extensive logging of native forests has resulted in the loss of key structures such as large old trees and tree ferns, altered the composition of understorey and ground cover plants, and led to major modification of soil environments. Species such as cavity-dependent arboreal marsupials have declined significantly in wood production native forests (Lindenmayer et al. 2020). Logging has other impacts on forested environments such as fragmenting otherwise intact landscapes and altering key ecological processes such as hydrological and fire regimes, as well as generating large amounts of greenhouse gas emissions.

Mining, mineral prospecting and exploration occur in most parts of the continent and vary in the nature and spatial extent of their effects. While many individual mines are small and have localised effects, large open-cut coal, iron ore and bauxite mines result in considerable loss of habitat. Habitat restoration may be attempted after mines have closed, and this can be quite successful, but in some instances, damage caused by the extraction process is irreparable. The blasting of the Juukan Gorge in the Pilbara in 2020—a site of immense cultural significance to Puutu Kunti Kurrama traditional owners—provides one recent example. Destruction of habitat of the endangered Black-throated Finch (*Poephila cincta cincta*) in the Galilee Basin in Queensland for a series of new mega-coal mines is another example. In addition, conventional mining often requires the use of large quantities of fresh water to separate minerals from the uneconomic fraction of ore deposits, and creates tailing piles, tailing ponds and networks of service roads that can be together inimical to the persistence of local biota. More recent means of resource extraction, such as hydraulic fracturing ('fracking') for oil and gas, do not result in large open-cut areas, but may damage swamp, spring and wetland habitats due to pollution of the aquifers on which they depend.

Large-scale resource extraction is subject to legislative oversight, but individual or small-scale prospecting activities are less regulated. This laissez-faire approach can have surprisingly destructive effects on biodiversity. For example, in just one small opal prospecting region near Coober Pedy, South Australia, approximately 1–2 million prospecting shafts have been dug and remain permanently open. Sometimes over 30 metres deep, these shafts capture and doom 10–28 million reptiles to a slow death each year, including individuals of the threatened bronzeback legless lizard (*Ophidiocephalus taeniatus*) (Pedler 2010). If this appalling situation is extrapolated to those parts of other States where unregulated mining exploration occurs, it is possible that hundreds of millions of small vertebrates are killed each year, every year, with potentially major effects on local biodiversity.

Altered Fire Regimes Fire is a key ecological process in almost all Australian terrestrial ecosystems. However, it is clear that fire regimes are changing in many parts of the continent. Fire regimes encompass key attributes of a sequence of fires over time – the frequency, timing, severity, seasonality and patchiness of fire (among other characteristics). In States such as Victoria, there is compelling evidence that fires are burning more frequently, more extensively, and more severely than they have done historically. For example, some wet temperate forest ecosystems have burned three-to-four times in the past 25 years when the natural fire regimes should see a fire every 75–150 years. This has, in turn, led to losses in old growth cover of up to 85% in some ecosystems with corresponding negative impacts on species that are strongly associated with long unburnt areas. Indeed, for groups such as birds, altered fire regimes are considered to be the second most important key threatening process in Australia after land clearing.

Climate Change and Variability It has long been understood that Australia's natural environments and associated biodiversity are highly susceptible to the effects of climate change (Steffen et al. 2009). Climate change has already contributed to species extinctions, such as the Bramble Cay Melomys (*Melomys rubicola*), although policy and management incompetence also played their part in its demise. Indeed, the profound effects of climate change have been highlighted by recent analyses that document the extent of ecosystem collapse in many terrestrial and marine ecosystems across (and adjacent to) the Australian continent (Bergstrom et al. 2021). In a terrestrial context, part of the susceptibility is based on the lack of topographic relief throughout much of the Australian continent, thereby limiting the potential for climate refugia in the face of rapidly increasing temperatures. Other factors that have major impacts on ecosystems and biodiversity are indirectly associated with a changing climate. Fire regimes are an important example, with extreme fire days (as reflected by the Forest Fire Danger Index) now ten times more prevalent than they were in the 1960s. Climate and weather are the primary drivers of fire behaviour, and as indicated by the estimated three billion animals killed by fires in the 2019–2020 fire season in south-eastern Australia, fire has direct and highly significant impacts on biodiversity and ecosystem integrity.

Interactions Between Drivers Many Australian environments are susceptible to multiple stressors. However, there also can be significant interactions between these stressors that in turn have marked effects on ecosystem condition and on biodiversity. For example, forests can be subject to both logging and fire as well as interactions between them. Forests that have been burnt can be subject to post-fire logging (sometimes also called salvage logging). Salvage logging has pronounced impacts on soil nutrients and structure, and on native plants and animals. The reverse kind of interaction also can occur, whereby logged and regenerated areas are subsequently susceptible to elevated fire severity in the event of reburning.

There are many other instances of key interactions in ecosystems that influence environmental conditions and biodiversity. For instance, feral predators such as cats actively seek out recently burnt areas to hunt, most likely because foraging efficiency is higher there and prey are easier to capture because of the lack of vegetation cover. Similarly, burnt areas that are regenerating after fire can be highly susceptible to high levels of grazing and browsing by both native and introduced herbivores, thereby impairing post-disturbance recovery of these ecosystems.

3.7 Freshwater and Marine Environments

The same drivers of environmental change described for land environments, or their analogues, also affect the biodiversity and function of freshwater and marine ecosystems. For example, the draining of wetlands for agriculture and new housing estates, the damming and diversion of rivers for human settlements, agriculture and industry have become increasingly necessary to support burgeoning human populations, while the rise of super-trawlers for mid-ocean, benthic and demersal fisheries has led to the depletion of some fish stocks, reduced areas of intact seagrass and coral, and damage to sea mounts. Run-off of fertilisers, pesticides, herbicides and eroded topsoil is a major source of pollution in many Australian river systems, and carpets of sediment are now posing large problems for near-shore communities, including large parts of the Great Barrier Reef. Introduced pests and pathogens are also problematic. European carp (*Cyprinus carpio*) infest freshwaters in all States and Territories except the Northern Territory, dominating and outcompeting native fish and detrimentally affecting water quality by their habit of stirring up silt and mud. On the Great Barrier Reef, the crown of thorns starfish (*Acanthaster planci*) remains the best-known pest. Although native to the Indo-Pacific, outbreaks of this prolific coral-feeder appear to be facilitated by influxes of fertilisers and other pollutants into the reef system and may result in massive loss of coral. Climate change, similarly, is having dramatic effects on aquatic environments. The boom and bust cycles that characterise much of inland Australia—long droughts that are punctuated by episodic deluges of rain—are becoming more exaggerated and will likely see river systems running dry for extended periods between floods (Larkin et al. 2020). In marine systems, warming and acidifying ocean waters are predicted to have increasingly negative effects on most organisms, including kelps, seagrasses, corals and hard-shelled molluscs (Cresswell and Murphy 2017).

Two recent reports exemplify the parlous state of Australia's aquatic environments. In the first, the Productivity Commission reviewed progress that had been made since 2004 when the Federal, State and Territory governments agreed to a joint National Water Initiative (NWI) to secure freshwater resources (Productivity Commission 2021). Although acknowledging that progress had been made in national water reform and management, the Commission's report emphasized that the current NWI was insufficient to meet the demands of Australia's increasing population, industry and agriculture, the escalating challenges imposed by climate change, and also failed to meet the aspirations of First Nations peoples for greater control of water resources. The report argued for improvements to be made in new water infrastructure and water delivery, and also for reforms in management of water for environmental sustainability. However, the report was not clear on whether positive biodiversity outcomes could be achieved with its recommended reforms, and was silent on the key question of whether the world's driest inhabited continent will have sufficient fresh water in future to sustain its development goals and its natural systems. Strong government action is needed to ensure the wise use of national water resources. Unfortunately, in 2014 the Federal Government scrapped the National Water Commission—the agency charged with ensuring that agreed reforms are implemented—signaling a lack of interest, if not a policy failure, in providing the leadership needed for effective water management (Productivity Commission 2021).

In the second report, Bestley et al. (2021) reviewed current understanding of penguins, other seabirds and marine mammals in the Antarctic and Southern Ocean. As these taxa are predators that depend largely on the productivity of small prey such as krill and fish, Bestley et al. (2021) noted that the predators are useful indicators of the general state of marine resources and ocean health. The effects of climate change were identified as an overarching menace to the Southern Ocean system, acting in concert with other threats such as expanding commercial fisheries, global pollutants, invasive species and diseases, and human activities including tourism and research. Many seabirds and marine mammals are currently threatened; recovering these species and assuring the future health of the Antarctic and Southern Ocean ecosystems will require effective management that is coordinated by both the Australian and other national governments.

3.8 Ecological Sustainability – Is It Possible?

The conservation of biodiversity is a critical part of ecological sustainability both in Australia and globally. In Australia, it is clear that we spend a tiny fraction (less than 10%) of what is needed to conserve the nation's biodiversity (Wintle et al. 2019). It is also clear that our legislation at both State and Federal levels is highly deficient and promotes rather than prevents the loss of ecosystems, habitats, threatened species and communities. Significant legal reform and appropriate levels of funding will be critical to tackling Australia's environmental problems and setting out on a new pathway that is clearly different from the highly unsustainable trajectory that the nation is currently on. In many cases, the management interventions required to conserve parts of Australia's environment and/or its associated species are well understood, but levels of resources are woefully inadequate to ensure that effective management can occur. In other cases, it has become clear that there are success stories in environmental and conservation management, and we need to more extensively publicize these to catalyze much greater efforts by governments and others (Garnett et al. 2018).

3.9 Conclusions

The Australian environment and the nation's biodiversity face a wide range of extraordinary pressures, many of which appear to be increasing in the magnitude of their effects (e.g. climate change, expanding human population size, and altered fire regimes). Australia is, sadly, a world leader in its appalling record on biodiversity loss. In many cases, we know what needs to be done to arrest these declines and better conserve the rich and often unique natural heritage that characterizes Australia. Levels of resources must be substantially increased and allied legal reforms to strengthen environmental regulations are urgently needed to boost future prospects for the conservation of Australia's natural environments.

References

Algar D, Johnston M, Tiller C et al (2020) Feral cat eradication on Dirk Hartog Island, Western Australia. Biol Inv 22:1037–1054. https://doi.org/10.1007/s10530-019-02154-y

Bergstrom DM, Wienecke BC, van den Hoff J et al (2021) Combating ecosystem collapse from the tropics to the Antarctic. Glob Chang Biol. 27:1692–1703. https://doi.org/10.1111/gcb.15539

Bestley S, Ropert-Coudert Y, Nash SB et al (2021) Marine ecosystem assessment for the Southern Ocean: birds and marine mammals in a changing climate. Front Ecol Evol 8:566936. https://doi.org/10.3389/fevo.2020.566936

Bolam FC, Mair L, Angelico M et al (2021) How many bird and mammal extinctions has recent conservation action prevented? Cons Lett 14:e12762. https://doi.org/10.1111/conl.12762

Bradshaw CJA, Ehrlich PR, Beattie A et al (2021) Understanding the challenges of avoiding a ghastly future. Front Cons Sci 1:615419. https://doi.org/10.3389/fcosc.2020.615419

Burnett P (2021) A major report excoriated Australia's environment laws. Sussan Ley's response is confused and risky. The Conversation 29 Jan 2021. https://theconversation.com/a-major-report-excoriated-australias-environment-laws-sussan-leys-response-is-confused-and-risky-154254. Accessed 1 Feb 2021

Chapman AD (2009) Numbers of living species in Australia and the world, 2nd edn. Australian Government Department of the Environment, Water, Heritage and the Arts, Canberra

Cockerell B, Pressey RL, Grech A et al (2020) Representation does not necessarily reduce threats to biodiversity: Australia's Commonwealth marine protected area system, 2012–2018. Biol Conserv 252:108813. https://doi.org/10.1016/j.biocon.2020.108813

Cogger H, Dickman C, Ford H et al (2017) Australian animals lost to bulldozers in Queensland 2013–15. WWF-Australia, Sydney

Commission for the Human Future (2020) Surviving and thriving in the 21st century. Commission for the Human Future, Canberra

Cresswell ID, Murphy HT (2017) Australia State of the Environment 2016: biodiversity. Australian Government Department of the Environment and Energy, Canberra

Dickman CR (2007) A fragile balance: the extraordinary story of Australian marsupials. Mallon Publishing, Melbourne

Dickman CR (2014) Whither wildlife in an overpopulated world? In: Goldie J, Betts K (eds) Sustainable futures: linking population, resources and the environment. CSIRO Publishing, Melbourne, pp 13–23

Dickman CR (2018) Biodiversity in Australia: an overview. In: Pullaiah T (ed) Global biodiversity, vol 4 selected countries in the Americas and Australia. Apple Academic, Oakville, pp 515–558

Garnett S, Latch P, Lindenmayer DB, Woinarski J (eds) (2018) Recovering Australian threatened species: a book of hope. CSIRO Publishing, Melbourne

Goldie J, Betts K (eds) (2014) Sustainable futures: linking population, resources and the environment. CSIRO Publishing, Melbourne

Ives CD, Lentini PE, Threlfall CG et al (2016) Cities are hotspots for threatened species. Glob Ecol Biogeog 25:117–126. https://doi.org/10.1111/geb.12404

Larkin ZT, Ralph TJ, Tooth S et al (2020) Identifying threshold responses of Australian dryland rivers to future hydroclimatic change. Sci Rep 10:6653. https://doi.org/10.1038/s41598-020-63622-3

Lindenmayer DB, Blanchard W, Blair D et al (2020) The response of arboreal marsupials to long-term changes in forest disturbance. Anim Conserv. https://doi.org/10.1111/acv.12634
Lunney D (2001) Causes of the extinction of native mammals of the Western Division of New South Wales: an ecological interpretation of the nineteenth century historical record. Rangel J 23:44–70. https://doi.org/10.1071/RJ01014
McLeod R (2016) Cost of pest animals in NSW and Australia, 2013–14. eSYS Development-NSW Natural Resources Commission, Sydney
Mittermeier RA, Gil PR, Mittermeier CG (1997) Megadiversity: Earth's biologically wealthiest nations. Cemex, Mexico City
Orians GH, Milewski AV (2007) Ecology of Australia: the effects of nutrient-poor soils and intense fires. Biol Rev 82:393–423. https://doi.org/10.1111/j.1469-185X.2007.00017.x
Osipova E, Emslie-Smith M, Osti M et al (2020) IUCN World Heritage outlook 3: a conservation assessment of all natural World Heritage sites, November 2020. IUCN, Gland
Pedler RD (2010) The impacts of abandoned mining shafts: fauna entrapment in opal prospecting shafts at Coober Pedy, South Australia. Ecol Manage Restor 11:36–42. https://doi.org/10.1111/j.1442-8903.2010.00511.x
Productivity Commission (2021) National water reform 2020 draft report. Productivity Commission, Canberra
Samuel G (2020) Independent review of the EPBC Act – final report. Australian Government Department of Agriculture, Water and the Environment, Canberra
Secretariat of the Convention on Biological Diversity (2020) Global biodiversity outlook 5. Secretariat of the Convention on Biological Diversity, Montreal
State of the Environment Advisory Council (1996) Australia state of the environment 1996. CSIRO Publishing, Melbourne
Steffen W, Burbidge AA, Hughes L et al (2009) Australia's biodiversity and climate change. CSIRO Publishing, Melbourne
Taylor MFJ, Dickman CR (2018) Native animals lost to tree-clearing in NSW 1998–2015. WWF-Australia, Sydney
Turner GM (2019) Is a sustainable future possible? J Proc Roy Soc NSW 152:47–65
Turner P (2020) Clear-felling environmental expertise. Advocate 27:16–17
Van Dijk A (2020) Australia's environment summary report 2019. Australian National University, Canberra
Wintle BA, Cadenhead NCR, Morgain RA et al (2019) Spending to save: what will it cost to halt Australia's extinction crisis? Cons Lett 12:e12682. https://doi.org/10.1111/conl.12682
Woinarski JCZ, Legge SM, Dickman CR (2019) Cats in Australia: companion and killer. CSIRO Publishing, Melbourne

Chapter 4
Climate Change and Human Health

Colin D. Butler, Ben Ewald, Forbes McGain, Karen Kiang, and Ann Sanson

Abstract This chapter focuses on the consequences to human health from climate change. We argue that the prospects for global human health are more ominous than most informed activists for global health realise. Unsurprisingly, risks to human health increase with the severity of climate change. After reviewing the history of the concerns about climate change and health, the chapter reviews some of the "primary" (heatwaves and other extreme weather events), "secondary" (kidney disease) and "tertiary" (large-scale violence) health risks. The mental health effects of some of these issues, which can arise in association with any of these categories, are also discussed. We also mention some possible effects, specific to children. Healthcare's own considerable carbon footprint and resource consumption are described, with suggestions for improvement. While the focus of the chapter is mostly global, we briefly mention air pollution from coal-fired electricity generation in Australia, where a comparatively high proportion of electricity is generated by this method.

C. D. Butler (✉)
National Centre for Epidemiology & Population Health, Australian National University, Canberra, Australia
e-mail: colin.butler@anu.edu

B. Ewald
University of Newcastle, Newcastle, Australia
e-mail: bene2@spin.net.au

F. McGain
Western Health, Melbourne, Australia

University of Melbourne, Melbourne, Australia

University of Sydney, Sydney, Australia
e-mail: Forbes.McGain@wh.org.au

K. Kiang
University of Melbourne, Melbourne, Australia

University of Sydney, Sydney, Australia

Royal Children's Hospital, Melbourne, Australia

A. Sanson
University of Melbourne, Melbourne, Australia
e-mail: annvs@unimelb.edu.au

S. J. Williams, R. Taylor (eds.), *Sustainability and the New Economics*,
https://doi.org/10.1007/978-3-030-78795-0_4

4.1 Introduction

Many indicators of global population health have improved in the post-World War II era,[1] including global life expectancy, infant mortality risk, and the risk of dying in childbirth. From a human-centred position, this is a triumph. Commentators with a "glass half-full" perspective celebrate this progress and see no compelling reason for it to cease (Rosling et al. 2018). The authors of this chapter are less confident. In fact, without a comprehensive policy shift along the lines of a Green New Deal that effectively deals with unsustainable production-consumption, population health at the global scale may falter. The COVID-19 pandemic may be the curtain-raiser for a much darker period of history; we may already have passed "peak health" (Brijnath et al. 2012).

It is under-recognised, including by most commentators on health, that the health improvements that have occurred in the last two centuries have been substantially underwritten by the global exploitation of natural resources, including the burning of cheap and non-renewable fossil fuels and by the clearing of forests (Butler 2017; Frumkin 2020). If this continues, humanity will exceed additional planetary boundaries (see Chap. 2). Beyond these boundaries lies peril.

The adverse effects of global environmental change manifest through pathways such as climate change, biodiversity loss, resource scarcity and contamination of the biosphere with persistent pollutants, including plastic. These manifestations, many of which interact, are already having noticeably harmful effects on human health. In this chapter, we briefly describe some of the existing and likely health consequences, particularly from climate change and air pollution.

4.2 Climate Change, the Anthropocene, and Human Development

Burning coal to create useful work[2] kickstarted the industrial revolution, and, according to some definitions, the "Anthropocene", the human-dominated era (Waters et al. 2016). The output of the "dark satanic mills" powered by steam was impressive; the smoke and dust seemed to blow away. Although London had, even by 1800, been known as the "big smoke" for centuries (Brimblecombe 2012 (1987)), it is still hard for some politicians to accept that particulate air pollution from the burning of coal harms people nationally. Many people still do not grasp that the invisible, odourless gas carbon dioxide (CO_2), also a major combustion product of coal, would prove an even worse pollutant, harming people and ecosystems globally.

[1] Many scientists now call this dangerous era the Anthropocene (see Chap. 2).

[2] In the thermo-dynamic sense.

4.3 Coal, Air Pollution and Health in Australia

In Australia in 2020, coal-fired power stations – many reaching the end of their useful life – still generate over 50% of the electricity consumed, far higher than the OECD average of 22% (OECD 2020). The combustion products from this burning are typically dispersed via 100-metre-high chimneys, lowering local concentrations of noxious pollutants but allowing wide atmospheric spread of particulate matter, including to large cities once thought safely distant, with serious health impacts. The pollutants of concern for health are particles such as sulphur dioxide (SO_2), nitrogen oxides, and mercury, as well as the greenhouse gas carbon dioxide (CO_2). Analysis of fine particles (PM 2.5) in Sydney air shows that 14% to 18% are sulphate particles originating in one of the five NSW coal-fired power stations (Cohen et al. 2012), with the largest fraction from the two located at Lake Macquarie, 100 km north (Crawford et al. 2018). This long-range transport was only understood in about 2010, but should be triggering a reconsideration of where and how coal-fired plants are allowed to operate. In many cities, especially in the global South, coal-fired stations still operate within city boundaries. This not only causes pollution through the air, but also by the transport of large volumes of coal into densely-populated urban areas, a problem that also occurs in cities that export coal, including Newcastle, Australia.

Particulate air pollution has diverse and serious health effects that extend far beyond lung conditions such as bronchitis, asthma and increased risk of lung cancer. Non-respiratory consequences include heart disease, stroke, diabetes, dementia and reduced foetal growth, effects thought to be mediated by systemic inflammation and increased oxidative stress. Hundreds of studies have now documented that more acute health events occur on "bad air days", including heart attacks, cardiac arrests and hospital admissions for respiratory disease. It has also been shown that people enduring poor air quality have worse health, even after adjusting for socio-economic status and health-damaging behaviours such as smoking. These chronic effects outweigh acute effects by a factor of five. While the risk of dying increases by about 1% on a day with an extra 10µg/m3 of fine particle air pollution (Di et al. 2017), the risk of death from chronic exposure to fine particles increases by 6% if air pollution is worsened by an extra 10µg/m3 of fine particles averaged over a year (Pope et al. 2002).

Applying these known risks to observed air quality in the state of NSW means that coal-fired electricity generation has an annual health burden of 279 deaths, 233 babies born under 2.5 kg, and 361 people who will develop type II diabetes (Ewald 2018). These health effects are due to secondary sulphate particles formed in the air from the primary pollutant SO_2. Similar results have been found elsewhere, such as an analysis finding that installing pollution controls to reduce particulate matter from nine coal-fired power stations in the US state of Texas would lead to 678 to 1760 fewer deaths per year (Thurston 2017). Elsewhere in the world, most power stations are required to capture the SO_2 from flu gases in "scrubbers", a well-established technology for post-combustion capture that can remove 90% of the

SO_2. These are neither fitted nor required for any power station in Australia under current legislation. While coal-fired power stations are regulated to operate under environmental licences, the allowed levels of SO_2 are so high that scrubbers have never been installed. The licence conditions for NSW power plants, for example, allow 1760 mg/m3 of SO_2, while in Europe new plants are allowed only 5% of this amount (up to 85 mg/m3).[3] These Australian licences were written before the long-range transport of air pollutants was understood, but that time has well and truly passed.

Coal-fired power stations in NSW do, however, have fabric filters that, if maintained properly, are very effective at capturing primary particle pollution. Brown-coal power stations in the state of Victoria do not use filters, because brown coal is considered so dirty that it would overwhelm them. Instead, they rely on cheaper and much less effective electrostatic precipitators, with adverse health consequences that are not quantified.

Mercury is another pollutant of concern. Coal contains small amounts of mercury which ends up either in the post-combustion ash or the flu gases. Globally, coal burning is the largest source of mercury contamination of the marine food web (Giang et al. 2015). Mercury is so toxic that even tiny amounts cause health effects, primarily to the neurodevelopment of children. The exposure pathway is complex, with long-range transport followed by biologic transformation to methyl mercury in aquatic environments, then accumulation through food chains. This happens over thousands of kilometres and multiple years so the regulatory effort should be to reduce mercury release world-wide. The Minamata Convention is an international agreement to do just this – Australia has signed the agreement but it has yet to ratify it. Australia's National Pollutant Inventory has been collecting data on industrial pollution since 1999. In this period, 17.1% of mercury release has been from the coal-fired electricity sector, exceeded by non-ferrous metal production (33%) and alumina production (19.5%). In 2018, total Australian mercury release was 7100 kg of which 28% was from the electricity sector. Mercury can be removed from flu gases but ends up in ash dams, from where it can still enter the aquatic environment if the dams are poorly managed. However, no Australian power station is equipped to capture mercury from flu gases.

Coal-fired electricity generation is a highly polluting industry with direct health effects as well as significant climate effects. While there seems to be no economically feasible way to eliminate CO_2 emissions while continuing to burn coal, other coal pollutants can be significantly reduced using proven technology, although at some cost.

[3] Until 2014, coal-fired power stations in NSW were only permitted to emit up to 175 mg/m^3 of SO_2.

4.4 Climate Change and Health

Climate change has, since its earliest appearance in the health literature, been recognised, by some, as an existential threat[4] to humanity (Anonymous 1989). Yet, for almost two decades, recognition of the immense threat to global health from global heating became almost a footnote in the climate and health literature, even as evidence of climate change accumulated (Butler 2018). This drought in the health literature was eventually broken by a well-publicised breakthrough paper in 2009 which argued that climate change could be the "biggest global health threat of the twenty-first century" (Costello et al. 2009). Even so, in a generally neoliberal political period where markets were said to need minimum regulation (see Chap. 6), policymakers in high-income countries, supported by fossil-fuel interests, continued to suppress and ignore the risk to our civilisation of climate change and other aspects of "planetary overload"[5] (McMichael 1993).

4.4.1 Classifying the Health Effects of Climate Change, Including Mental Health

There are many classifications of the adverse effects on human health from climatic change. One of the oldest and simplest recognises three broad categories, labelled "primary, secondary and tertiary" (Haines and Parry 1993). Analogous terms are "direct", "indirect" and "systemic".

There is now strong evidence of the significant, wide-ranging and increasing impacts of climate change on mental health (Clayton et al. 2017; Fritze et al. 2008; Cianconi et al. 2020). These diverse mental health effects of climate change can be considered compounding because primary, secondary and tertiary health effects each have a mental health dimension (Butler et al. 2014).

Heatwaves and Health

Perhaps the best-known example of a primary (direct) health effect is from heatwaves and more humid conditions, which lower the effectiveness of sweating as a way to regulate body temperature. The combination of heat and humidity in some settings, already, is making manual work in a non-airconditioned environment impossible for more than a few minutes (Raymond et al. 2020).

[4] "At the worst, global ecosystems will be unable to adapt to a rapidly changing climate, world food production will collapse, and coastal areas will be inundated by rapidly rising sea levels. Such end-of-the-world sketches are familiar, usually in relation to AIDS or nuclear war, but there is an inevitability about global warming, which stems not from human behaviour or human error but from the radiative properties of atmospheric releases and the fundamental laws of physical science" (Anonymous 1989).

[5] A term coined by McMichael who is widely recognised as securing the place of health in the reports of the Intergovernmental Panel on Climate Change (IPCC) from 1996 onwards.

Other health consequences of excessive heat exposure include heat stress, heat stroke, accidental injuries, violence and the worsening of some chronic diseases, such as multiple sclerosis. Kidney diseases may be an important consequence of occupational exposure to excessive heat and humidity, especially if worsened by insufficient access to drinking water.

Chronic Kidney Disease of Unknown Cause

Epidemics of chronic kidney disease (CKD) of unknown cause have been reported in Central America,[6] Sri Lanka, and India and suspected in particular "hot spots" in the Middle East, Africa, and North America. Retrospective analysis suggests that this phenomenon started to emerge in Central America in the 1970s (Johnson et al. 2019). Those affected are primarily young male labourers from rural communities required to perform strenuous work under very hot conditions, such as sugarcane harvesters and workers in Sri Lankan rice paddies. Workers at higher altitudes, such as in coffee plantations, are far more protected (Barraclough et al. 2017; Johnson et al. 2019). Sufferers typically present with few or no symptoms, but progress to end-stage kidney disease, requiring dialysis if they can afford it. Construction workers in the Middle East are also at risk (Ahmadalipour and Moradkhani 2018).

While the underlying cause of CKD remains controversial, there is general agreement that heat stress is likely to be a key contributor, possibly exacerbating damage from an environmental toxin found in pesticides, soil or drinking water. Beyond these "proximal" causes, however, lie poor regulation and exploitation, conditions where many manual labourers have few rights and are effectively "disposable" (Bales 1999).

Extreme Weather Events

Another long-recognised primary danger, to which climate change contributes, is extreme weather events (EWEs) including floods, wildfires, droughts, and storms. Coastal populations[7] are especially vulnerable to these, including from sea level rise (Kulp and Strauss 2019).

While adaptation and development have, in some places,[8] greatly reduced the mortality of these EWEs, thresholds of damage still exist beyond which great human suffering is inevitable. Adaptation strategies have their limits.

---

[6] Especially in Nicaragua and El Salvador, where the rate has been found to be over 20 times that of climatically comparable, geographically proximal countries such as Costa Rica and Cuba (Ordunez et al. 2014).

[7] Even under conservative, low-carbon emission scenarios, and where the Antarctic ice sheets remain stable this century, 150 to 250 million people (90% confidence interval) currently occupy global land below projected high tide lines for 2100. Under high emissions, the upper bound of this confidence interval increases to 630 million, in calculations that do not factor either population growth or coastal in-migration (Kulp and Strauss 2019).

[8] In 1970, cyclone Bhola, a category 4 storm, killed at least 300,000 Bangladeshis and led to civil war (Hossain 2018). Adaptation to storms in coastal Bangladesh has been a spectacular success, in relative terms. Nonetheless, ongoing sea level rise, with other factors, threatens this.

Extreme Weather Events and Mental Health

It has long been recognised that EWEs can have profound mental health effects arising from the psychological trauma which they cause, with up to 40% being affected (Galea et al. 2007). Variation in risk and severity is primarily due to differential disaster exposure and previous vulnerability, including indigeneity, age, poverty and lack of social connectedness. Recognised mental health effects of EWEs include post-traumatic stress disorder (PTSD), depression, anxiety, aggression, sleep disruption, interpersonal difficulties, worsened psychotic symptoms, substance abuse, and suicide (Clayton et al. 2017; Cianconi et al. 2020; Berry et al. 2010; Hanigan et al. 2012). While many people show resilience, positive coping, and effective adaptation, or experience only minor stress and distress symptoms, or acute reactions which remit over time (Norris et al. 2002; Neria et al. 2008; Cianconi et al. 2020), others experience serious effects which can persist for years. For some people, mental health difficulties emerge only months or years after a disaster (Newnham et al. 2020).

Taking flooding as an example, a systematic review documented significant increases in cases of PTSD, anxiety, depression and overall lower mental-health-related quality of life, with impacts being similar in developed and developing countries (Fernandez et al. 2015). Other psychological effects of exposure to EWEs and their after-effects include feelings of distress, grief, anger, shock, despair, numbness, loss of identity, and feelings of helplessness and hopelessness (Clayton et al. 2017; Sanson et al. 2019).

By increasing the frequency, severity and duration of EWEs, and creating longer-term environmental damage, climate change can have both acute and chronic effects on mental health, through both direct and indirect mechanisms (Berry et al. 2010). The cumulative stress brought on by slower-onset but longer-lasting climate-related changes like severe drought or sea level rise has also been recognised as contributing to serious mental health problems, including depression and suicidality (American Psychiatric Association 2017).

Secondary Health Effects of Climate Change

An example of a secondary (indirect) health effect of climate change is from alterations in the distribution and "dynamics"[9] of some important infectious diseases, particularly malaria (Akpan et al. 2019), by far the most significant vector-borne[10] disease, and one for which recent control measures have faltered (Alonso and Noor 2017). The epidemiology of many other infectious diseases, especially vector-borne, is also altered by climate change, principally via changes in rainfall and temperature (Caminade et al. 2019). These shifts are not always adverse; for example, rising temperatures in some places may reduce vectorial capacity (Murdock et al. 2016). However, temperature rise in such places may also reduce the capacity of such areas to support human life.

[9] For example, in warmer temperatures, the life cycle of some pathogens accelerates, potentially meaning a more intense epidemic in the same location, due to one or more additional cycles of transmission.

[10] In this context the term "vector" generally refers to insects such as mosquitoes and ticks.

Many infectious diseases have mental health effects, including depression, impaired cognitive function, pain, and fatigue, among other mental health disturbances. The patterns and locations of some of these conditions are influenced by climate change, including Lyme disease (Cochat Costa Rodrigues et al. 2017) and West Nile virus (Murray et al. 2007).

Another secondary health effect is from air pollution due to smoke from forest and grass fires, due to heat, drought and deliberate land-use change (Johnston et al. 2016).

Tertiary Health Effects of Climate Change

Tertiary (systemic) health effects of climate change include phenomena such as famine, civil war and forced migration. These arise through the interaction of social and environmental factors and have profound health effects, including mental health. Another common way to think of the relationship of these events to climate change is to conceptualise global warming as a "risk multiplier",[11] or 'threat accelerant'. Formulated over a decade ago,[12] these concepts are now accepted by the UN Security Council, due to the likelihood of struggles over scarce resources (including water) with displacement and forced migration increasing the potential for intra- and inter-state violence (Abel et al. 2019). The serious and diverse impacts of violence on mental health are well-documented (Sanson and Burke 2019).

Although recognised from the onset of the climate change and health literature, the risk of these events in relation to climate change were, for over two decades, largely marginalised (Butler 2018). Today, such views are far more acceptable, although still disputed by some conservatives (Butler and Kefford 2018). Tertiary effects will arguably have the greatest adverse impact upon health.

The number of people forcibly displaced by a range of factors, many of which are exacerbated by climate change, continues to rise (Berchin et al. 2017). Forced migration leads to a range of trauma symptoms and adjustment problems associated with loss of home, place and social networks (Pfefferbaum et al. 2016; Sanson et al. 2018) in addition to more complex psychopathological consequences, such as identity disorders, long-term personality changes and dissociative syndromes (Woodbury 2019).

War, which is a major cause of forced migration, contributes to PTSD and major depression. An estimated 1.45 billion individuals worldwide have experienced war between 1989 and 2015 and were still alive in 2015, including one billion adults. Of these, over 350 million adult war survivors are estimated to suffer from PTSD and/or major depression (Hoppen and Morina 2019). Of these wars, only a small fraction[13] can be attributed in part to climate change, but this proportion is bound to increase (Bowles et al. 2015).

[11] Climate change is neither necessary nor sufficient for such tertiary effects to arise, but can increase the likelihood.

[12] By the German Advisory Council on Global Change, in 2007.

[13] As well as the Syrian war, the conflict in Darfur, Sudan, was partly fuelled by climatic change (UNEP 2007).

Human Carrying Capacity and Climate Change

The links between forced migration and climate change have, for decades, been fiercely contested. A concept that is even more contested, also relevant to climate change, is that of human "carrying capacity", or the number of people who can be supported in good health on an ongoing basis by a given set of resources, including, crucially, the human resource of co-operation (Butler 2004). Leading climatologist Hans Joachim Schnellhuber[14] has long concluded that threats of climate instability and other aspects of adverse global environmental change mean that Earth's human carrying capacity may soon be exceeded – if it isn't already – leading to the premature deaths of hundreds of millions or even billions (Clark et al. 2005). More recently, Schnellhuber has warned "Climate change is now reaching the endgame, where very soon humanity must choose between taking unprecedented action, or accepting that it has been left too late and bear the consequences" (Spratt 2019). Climate change is likely to have profound effects on food production, potentially leading to severe food shortages.

Impacts of Climate Change on the Health of Children

Children, because of their more sensitive physiological, behavioural and developmental requirements, are especially vulnerable to the effects of climate change (Sanson et al. 2019). Vulnerability is increased among indigenous children and those in low-income countries. Mechanisms include undernutrition, diarrhoeal diseases and physical and mental trauma following exposure to disasters.

After experiencing EWEs, children and youth typically show more severe disturbances than adults (Alisic et al. 2014; Peng et al. 2011). Problems include PTSD, depression, phobias and panic, sleep disorders, social withdrawal, attachment disorders and substance abuse, as well as cognitive deficits and learning problems (Garcia and Sheehan 2016). Such symptoms are more likely to persist and impair children than adults (Norris et al. 2002).

Many of the mental health consequences of exposure to EWEs in children are exacerbated by the resulting stress on parents and compromised family functioning (Clayton et al. 2017). For example, following Hurricane Floyd in North Carolina, families were found to be vulnerable to higher child-abuse risk (Fritze et al. 2008).

Solastalgia and the Mental Health Impacts of Vicarious and Envisaged Climate Change

Climate change harms the economic and social fabric of communities, lessening community wellbeing and hence indirectly affecting mental health (Berry et al. 2010; Torres and Casey 2017), increasing levels of anxiety, depression, interpersonal difficulties, and aggressive behaviours (Chaudry and Wimer 2016). The destruction of landscapes diminishes people's sense of belonging and connectedness to the land (Higginbotham et al. 2007), creating distress and loss that has been termed 'solastalgia' (Albrecht et al. 2007). For indigenous people in particular, connection to country is essential for good mental health (Bowles 2015). In Inuit

[14] Recently retired director of the Potsdam Institute for Climate Impact Research.

communities, climate change is compromising traditional lifestyles and resulting in displacement, loss of identity and isolation, hence harming mental health. Surveys around the world, including in high-income countries that have not yet suffered many direct effects of climate change, indicate that most people are worried about such a serious threat. For example, a 26-country survey on eight potential threats found that climate change was ranked highest in 13 countries (Fagan and Huang 2019). Overall, 68% of those surveyed regarded it as a major threat (an increase of 11% since 2013).

Further, knowledge about the impending climate crisis has its own impacts on mental health, even without first-hand experience (Sanson et al. 2019). Gifford and Gifford (2016) characterise anxiety about the looming crisis as "pre-traumatic stress disorder". Burgeoning rates of "eco-anxiety" (described as severe and debilitating worry which can elicit loss of appetite, sleeplessness, and panic attacks), and "ecological grief" are commonly described (Albrecht et al. 2007).

Many children and young people, especially in high-income countries, express worry, fear, anxiety, sadness, anger, and a sense of powerlessness about the impact of climate change on their lives (Sanson et al. 2019). In a survey of Australians aged between seven and 24, 96% considered climate change a serious problem, and 89% were worried about its effects (Chiw and Ling 2019). In-depth interviews in the USA (Strife 2012) found that 82% of ten-to-12-year-olds expressed strong feelings of fear, sadness, and anger. Many young people believe the world may end during their lifetime due to climate change and other global threats (Tucci et al. 2007).

Other Health Threats in the Anthropocene

This chapter has focused on the numerous risks to health from climate change, but there are other important health threats associated with environmental degradation that are outside the scope of this chapter. Many of these threats are associated with pollution, including traditional pollutants (say from cars and trucks) and novel pollutants from industry and agriculture. The effect of microscopic amounts of plastic in human bodies – and the food chain in general – is yet to be determined. The other obvious threat is the increasing incidence of zoonotic epidemics and pandemics caused by the continual encroachment of humans into animal habitats. Covid-19 is but the latest pandemic to emerge. For example, there are increasing chances of interaction between humans and bats. Deforestation has likely contributed, including by altering bat behaviour, and perhaps by lowering their immunity (Afelt et al. 2018). Drought and other climate anomalies may also have contributed to viral "spillover" from bats (Banerjee et al. 2020). If so, heatwaves, worsened by climate change, may have, too (Butler 2020). There is also widespread contamination of the food chain, for both humans and other species, not only by plastics but by novel molecules that humans have created, some of which act as endocrine disruptors, and some of which are carcinogenic.

4.5 Conclusion

Urgent climate stabilisation is vital to reduce risks to human health. Even then, some degree of adaptation will be necessary to reduce the worst health effects of climate change. The impact of climate change on the Earth system is now so significant that the next ice age will likely be delayed by at least 50,000 years (Ganopolski et al. 2016). If humans do not change course, collectively, this may be their most enduring legacy. On its current trajectory, civilisation may not survive as long, perhaps not even for another century (Ripple et al. 2020). This prospect is realistic not only because of climate change, but also the erosion of biodiversity, the contamination of the food chain with persistent pollutants, the growing scarcity of both renewable and non-renewable resources, and the invention and accumulation of thousands of nuclear weapons. These elements interact. For example, aquifer scarcity in Syria, compounded by drought, contributed to that country's vicious civil war. If scarcity exceeds thresholds in South Asia, then either Pakistan or India could be tempted to use nuclear weapons (Robock et al. 2019). As artificial "intelligence" spreads, an increasing number of weapons systems will be automated, at risk of being hacked, or simply mis-coded, perhaps triggering accidental annihilation (Turchin and Denkenberger 2018). Former US Defence Secretary William Perry has recently argued that it is legitimate, even vital, to describe catastrophic scenarios as a form of prevention (Mecklin 2019). That is, while too much bad news is depressing and counter-productive, Perry argues that some bad news is necessary in order to motivate the changes in public opinion that are needed to shift the course of civilisation to a safer plane.

Climate change threatens the underlying social, economic and environmental determinants of health, including mental health (Burke et al. 2018). Climate change is already, and will increasingly, pose a significant challenge to health services, including those specific to mental health. There is an urgent global need to develop sustainable physical and mental health care as climate change disasters and longer-term impacts become more intense, complex and frequent (Newnham et al. 2020). Given the huge scale, a community focus in approaches to health care is needed. Restoring social cohesion of communities and families immediately after calamities is critical for reducing suffering and promoting effective recovery (Stanke et al. 2012).

The mental health impacts of climate change are unlikely to be felt equally by all, but will fall disproportionately on those who are already vulnerable, especially children, older people, indigenous people and those living in developing countries who will bear the brunt of adverse climate change. Women, economically disadvantaged people and people with pre-existing psychiatric illnesses have also been shown to be more vulnerable to mental health problems after disasters (Clayton et al. 2017; WHO 2014).

Although ecological grief and anxiety are rational responses to the threat of climate change (Cunsolo and Ellis 2018), many people need support to manage these. An effective antidote to anxiety is often action. Young people who become actively engaged in fighting for action on climate change become better able to manage their anxiety about the future and maintain their optimism (Sanson et al. 2019). They are, simultaneously, working to limit the severity of future climate impacts they will face. Hart et al. (2014) concluded that "taking action through playing a meaningful role in the face of adversity can offer psychological protection by helping children to feel more in control, more hopeful and more resilient" (p. 93). Similarly, taking action has proved helpful to many adults. There is growing awareness by health care workers that the imperative "first do no harm" is breached by the high ecological footprint of the healthcare industry, which pollutes the atmosphere, water, and soil (see Box 4.1).

Box 4.1 Some Issues Relating to the Environmental Footprint of Healthcare

- Recent work estimates that approximately 4.4% of global anthropogenic CO_2 equivalent (CO_2e) emissions are due to healthcare activities (Pichler et al. 2019).
- The proportion of national CO_2e emissions due to national healthcare can be much higher; up to 10% in the USA (Eckelman and Sherman 2016).
- The UK Sustainable Development Unit (SDU) has led the world in calculating and reducing healthcare emissions, including by motivating staff. Since 2008, despite increasing healthcare-related activity, England's CO_2e emissions have declined by 18% (Sustainable Development Unit 2018).
- The UK SDU has documented 'carbon hotspots', particularly within hospitals, finding that pharmaceutical procurement alone has a carbon footprint equivalent to England's entire direct hospital energy use. Other carbon-intense foci include anaesthetic gases (nitrous oxide and desflurane use in particular), and medical devices (Sustainable Development Unit 2018).
- Preventive medicine (general practice and other forms of primary health care) have a much lower ecological cost than hospital care (Malik et al. 2018; Eckelman and Sherman 2016). Preventative medical strategies, such as those relating to diabetes, smoking, nutrition and renal function have many benefits beyond the individual.
- The pollution from healthcare, primarily from fossil fuel-based electricity leading to air pollution and climate change, has a significant burden of disability-adjusted-life-years lost (Eckelman and Sherman 2016).
- The evidence base underpinning more environmentally sustainable hospital care has grown rapidly from a low baseline only a decade ago (McGain et al. 2020). Ecological footprint studies, using life-cycle assessment, are beginning to take hold in a range of medical specialties.

(continued)

Box 4.1 (continued)

- The avoid, reduce, reuse, recycle, and research framework is a simple and useful basis to achieve more sustainable healthcare (McGain and Naylor 2014).
- The source of energy (e.g., coal vs. fossil gas vs. solar) has profound effects on the direct CO_2e emissions not only from hospital heating, ventilation and air conditioning (Sherman and Lagasse 2018) but also from healthcare equipment, which can paradoxically have higher CO_2e emissions if reusable rather than single use (McGain et al. 2020). That is, the use of renewable energy to produce medical equipment will lower its carbon footprint, a benefit multiplied if its use is reusable. In other words, "go renewable" not "go disposable".[15]
- Doctors and other healthcare workers have multiple roles, at multiple levels, to rapidly decarbonise healthcare, including as role models.
- In countries with highly polluting electricity sources (e.g. Australia) advocacy to transition hospital energy sources away from fossil fuels (including gas for heating) is a good starting point.
- In countries where electricity is primarily renewable or nuclear sourced (e.g. New Zealand, much of Europe) the focus of healthcare pollution reduction advocates should begin with increased procurement of reusable items.
- Several national and international groups can assist healthcare workers in the transition to environmentally safer healthcare. The Global Green and Healthy Hospitals network has been active in this sphere for more than a decade (see https://www.greenhospitals.net/). National groups including Doctors for the Environment Australia (http://dea.org.au/) also have an integral role to play in this transition.

Herman (1992) concluded that participation in social action offers "a source of power that draws upon … initiative, energy, and resourcefulness" (p. 207), and yields "protection against terror and despair", serving as "the strongest antidote to traumatic experience" (p. 214).

Widespread mental health consequences of knowledge of the reality of the climate crisis are potentially deleterious. While emotional responses to the climate crisis are rational, and can empower people to take action, they can also be debilitating. Especially for children and young people, a lack of confidence in a future habitable planet can result in lack of investment in long-term goals, nihilism and despair.

[15] The worst-case scenario for reusable equipment is to sterilise it using brown coal fired-electricity, such as is still generally the case in the Australian state of Victoria. In the adjacent state of Tasmania (where electricity is now over 100% renewable, with the surplus exported) there is no environmental case for using disposable equipment. The problem with the manufacture of equipment designed for single use is the feedstock. For example, all plastics are based on polymerising ethane or propane, which require (with present technology) fossil fuels (oil or gas). However, this feedstock could also be sourced from renewable sources (e.g. algae), powered by renewables.

For this and other reasons, policymakers should do everything they can to reduce the risk of crises and expand positive action.

Acknowledgements We wish to thank Drs Katherine Barraclough, George Crisp and Kingsley Faulkner.

References

Abel GJ, Brottrager M, Crespo Cuaresma J et al (2019) Climate, conflict and forced migration. Glob Environ Chang 54:239–249. https://doi.org/10.1016/j.gloenvcha.2018.12.003

Afelt A, Frutos R, Devaux C (2018) Bats, coronaviruses, and deforestation: toward the emergence of novel infectious diseases? Front Microbiol 9. https://doi.org/10.3389/fmicb.2018.00702

Ahmadalipour A, Moradkhani H (2018) Escalating heat-stress mortality risk due to global warming in the Middle East and North Africa (MENA). Environ Int 117:215–225. https://doi.org/10.1016/j.envint.2018.05.014

Akpan GE, Adepoju KA, Oladosu OR (2019) Potential distribution of dominant malaria vector species in tropical region under climate change scenarios. PLoS One 14:e0218523. https://doi.org/10.1371/journal.pone.0218523

Albrecht G, Sartore G-M, Connor L et al (2007) Solastalgia: the distress caused by environmental change. Australas Psychiatry 15:S95–S98. https://doi.org/10.1080/10398560701701288

Alisic E, Zalta AK, van Wesel F et al (2014) Rates of post-traumatic stress disorder in trauma-exposed children and adolescents: meta-analysis. Br J Psychiatry 204:335–340. https://doi.org/10.1192/bjp.bp.113.131227

Alonso P, Noor AM (2017) The global fight against malaria is at crossroads. Lancet 390:10112. https://doi.org/10.1016/S0140-6736(17)33080-5

American Psychiatric Association (APA) (2017) Mental health and our changing climate: impacts, implications, and guidance. Washington, DC. https://www.apa.org/news/press/releases/2017/03/mental-health-climate.pdf. Accessed 25 Dec 2020

Anonymous (1989) Health in the greenhouse. Lancet 333:819–820. https://doi.org/10.1016/S0140-6736(89)92275-7

Bales K (1999) Disposable people: new slavery in the global economy. University of California Press, Berkeley, Los Angeles, London

Banerjee A, Baker ML, Kulcsar K et al (2020) Novel insights into immune systems of bats. Front Immunol 11:26. https://doi.org/10.3389/fimmu.2020.00026

Barraclough KA, Blashki GA, Holt SG et al (2017) Climate change and kidney disease—threats and opportunities. Kidney Int 92:526–530. https://doi.org/10.1016/j.kint.2017.03.047

Berchin II, Valduga IB, Garcia J et al (2017) Climate change and forced migrations: an effort towards recognizing climate refugees. Geoforum 84:147–150. https://doi.org/10.1016/j.geoforum.2017.06.022

Berry HL, Bowen K, Kjellstrom T (2010) Climate change and mental health: a causal pathways framework. Int J Public Health 55:123–132. https://doi.org/10.1007/s00038-009-0112-0

Bowles DC (2015) Climate change and health adaptation: consequences for indigenous physical and mental health. Ann Glob Health 81:427–431. https://doi.org/10.1016/j.aogh.2015.06.004

Bowles DC, Butler CD, Morisetti N (2015) Climate change, conflict, and health. J Roy Soc Med 108:390–395. https://doi.org/10.1177/0141076815603234

Brijnath B, McMichael AJ, Butler CD (2012) Rio+20: don't forget health in sustainability talks. Nature 486:191. https://doi.org/10.1038/486191a

Brimblecombe P (2012 (1987)) The big smoke. A history of air pollution in London since medieval times. Routledge, London

Burke S, Sanson A, Van Hoorn J (2018) The psychological effects of climate change on children. Curr Psychiatry Rep 20:35. https://doi.org/10.1007/s11920-018-0896-9

Butler CD (2004) Human carrying capacity and human health. PLoS Med 1:192–194. https://doi.org/10.1371/journal.pmed.0010055

Butler CD (2017) Limits to growth, planetary boundaries, and planetary health. Curr Opin Environ Sustain 25:59–65. https://doi.org/10.1016/j.cosust.2017.08.002

Butler CD (2018) Climate change, health and existential risks to civilization: a comprehensive review (1989–2013). Int J Environ Res Public Health 15:2266. https://doi.org/10.3390/ijerph15102266

Butler CD (2020) Plagues, pandemics, health security, and the war on nature. J Hum Sec 16:53–57. https://doi.org/10.12924/johs2020.16010053

Butler CD, Bowles DC, McIver L et al (2014) Mental health, cognition and the challenge of climate change. In: Butler CD (ed) Climate change and global health. CABI, Wallingford

Butler CD, Kefford BJ (2018) Climate and conflict: magnifying risks. Nature 555:587. https://doi.org/10.1038/d41586-018-03795-0

Caminade C, McIntyre KM, Jones AE (2019) Impact of recent and future climate change on vector-borne diseases. Ann N Y Acad Sci 1476:157–173. https://doi.org/10.1111/nyas.13950

Chaudry A, Wimer C (2016) Poverty is not just an indicator: the relationship between income, poverty, and child well-being. Acad Pediatr 16:S23–S29. https://doi.org/10.1016/j.acap.2015.12.010

Chiw A, Ling HS (2019) Young people of Australia and climate change: perceptions and concerns. A report for Millennium Kids. https://www.millenniumkids.com.au/wp-content/uploads/2019/02/Young-People-and-Climate-Change.pdf. Accessed 25 Dec 2020

Cianconi P, Betro S, Janiri L (2020) The impact of climate change on mental health: a systematic descriptive review. Fr Psychiatr 11 article number 74. https://doi.org/10.3389/fpsyt.2020.00074

Clark WC, Crutzen PJ, Schellnhuber HJ (2005) Science for global sustainability. Toward a new paradigm. KSG working paper no. RWP05-032. https://ssrn.com/abstract=702501 or https://doi.org/10.2139/ssrn.702501. Accessed 30 June 2020

Clayton S, Manning C, Krygsman K et al (2017) Mental health and our changing climate: impacts, implications, and guidance. American Psychological Association and ecoAmerica, Washington, DC. http://ecoamerica.org/wp-content/uploads/2017/03/ea-apa-psych-report-web.pdf. Accessed 26 Dec 2020

Cochat Costa Rodrigues MC, Moreira I, Peixoto MJ et al (2017) The new great imitator – neuropsychiatric symptoms of Lyme disease. Eur Psychiatry 41:S232–S232. https://doi.org/10.1016/j.eurpsy.2017.01.2242

Cohen D, Crawford J, Stelcer E et al (2012) Application of positive matrix factorisation, multilinear engine and back trajectory techniques to the quantification of coal-fired power station pollution in Sydney. Atmos Environ 61:204–211. https://doi.org/10.1016/j.atmosenv.2012.07.037

Costello A, Abbas M, Allen A et al (2009) Managing the health effects of climate change. Lancet 373:1693–1733. https://doi.org/10.1016/S0140-6736(09)60935-1

Crawford J, Cohen D, Atanacio A (2018) The impact of closure of coal-fired power stations on aerosol concentrations in the Sydney Basin. Atmos Pollut Res 9:1167–1176. https://doi.org/10.1016/j.apr.2018.05.002

Cunsolo A, Ellis N (2018) Ecological grief as a mental health response to climate change-related loss. Nature Clim Change 8:275–281. https://doi.org/10.1038/s41558-018-0092

Di Q, Dai L, Wang Y (2017) Association of short-term exposure to air pollution with mortality in older adults. JAMA 318:2446–2456. https://doi.org/10.1001/jama.2017.17923

Eckelman MJ, Sherman J (2016) Environmental impacts of the U.S. health care system and effects on public health. PLoS One 11:e0157014. https://doi.org/10.1371/journal.pone.0157014

Ewald B (2018) The health burden of fine particle air pollution from electricity generation in NSW. Environment Justice Australia, Melbourne. https://www.envirojustice.org.au/wp-content/uploads/2018/11/Ewald_B_2018_The_health_burden_of_fine_particle_pollution_from_electricity_generation_in_NSW.pdf. Accessed 26 Dec 2020

Fagan M, Huang C (2019) A look at how people around the world view climate change. FactTank, Pew Research Center. Available at: https://www.pewresearch.org/fact-tank/2019/04/18/a-look-at-how-people-around-the-world-view-climate-change/. Accessed 12 Dec 2020

Fernandez A, Black J, Jones M et al (2015) Flooding and mental health: a systematic mapping review. PLoS One 10:1–20. https://doi.org/10.1371/journal.pone.0119929

Fritze JG, Blashki GA, Burke S et al (2008) Hope, despair and transformation: climate change and the promotion of mental health and wellbeing. Int J Mental Health Systems 2:13. https://doi.org/10.1186/1752-4458-2-13

Frumkin H (2020) Sustaining life: human health-planetary health linkages. In: Al-Delaimy WK, Ramanathan V, Sorondo MS (eds) Health of people, health of planet and our responsibility. Springer Open, New York

Galea S, Brewin CR, Gruber M et al (2007) Exposure to hurricane-related stressors and mental illness after Hurricane Katrina. Arch Gen Psychiatry 64:1427–1434. https://doi.org/10.1001/archpsyc.64.12.1427

Ganopolski A, Winkelmann R, Schellnhuber HJ (2016) Critical insolation–CO2 relation for diagnosing past and future glacial inception. Nature 529:200–203. https://doi.org/10.1038/nature16494

Garcia DM, Sheehan MC (2016) Extreme weather-driven disasters and children's health. Int J Health Services 46:79–105. https://doi.org/10.1177/0020731415625254

Giang A, Stokes LC, Streets DG et al (2015) Impacts of the Minamata Convention on mercury emissions and global deposition from coal-fired power generation in Asia. Environ Sci Technol 49:5326–5335. https://doi.org/10.1021/acs.est.5b00074

Gifford E, Gifford R (2016) The largely unacknowledged impact of climate change on mental health. Bull At Sci 72:292–297. https://doi.org/10.1080/00963402.2016.1216505

Haines A, Parry ML (1993) Climate change and human health. J R Soc Med 86:707–711

Hanigan IC, Butler CD, Kokic PN et al (2012) Suicide and drought in New South Wales, Australia, 1970–2007. PNAS 109:13950–13955. https://doi.org/10.1073/pnas.1112965109

Hart R, Fisher S, Kimiagar B (2014) Beyond projects: involving children in community governance as a fundamental strategy for facing climate change. In: UNICEF Office of Research (ed) The challenges of climate change: children on the frontline. UNICEF Office of Research, Florence, pp 92–97

Herman J (1992) Trauma and recovery: the aftermath of violence—from domestic abuse to political terror. Basic Books, New York

Higginbotham N, Connor L, Albrecht G et al (2007) Validation of the environmental distress scale. EcoHealth 3:245–254. https://doi.org/10.1007/s10393-006-0069-x

Hoppen TH, Morina N (2019) The prevalence of PTSD and major depression in the global population of adult war survivors: a meta-analytically informed estimate in absolute numbers. Eur J Psychotraumatol 10:1578637. https://doi.org/10.1080/20008198.2019.1578637

Hossain N (2018) The 1970 Bhola cyclone, nationalist politics, and the subsistence crisis contract in Bangladesh. Disasters 42:187–203. https://doi.org/10.1111/disa.12235

Johnson RJ, Wesseling C, Newman LS (2019) Chronic kidney disease of unknown cause in agricultural communities. New Engl J Med 380:1843–1852. https://doi.org/10.1056/NEJMc1907676

Johnston FH, Melody S, Bowman DMJS (2016) The pyrohealth transition: how combustion emissions have shaped health through human history. Phil Trans R Soc B 371:20150173. https://doi.org/10.1098/rstb.2015.0173

Kulp SA, Strauss BH (2019) New elevation data triple estimates of global vulnerability to sea-level rise and coastal flooding. Nature Commun, 10, Article number: 4844. https://doi.org/10.1038/s41467-019-12808-z

Malik A, Lenzen M, McAlister S et al (2018) The carbon footprint of Australian health care. Lancet Planet Health 2:e27–e35. https://doi.org/10.1016/S2542-5196(17)30180-8

McGain F, Muret J, Lawson C et al (2020) Environmental sustainability within anaesthesia and critical care. Br J Anaesth 125:680–692. https://doi.org/10.1016/j.bja.2020.06.055

McGain F, Naylor C (2014) Environmental sustainability in hospitals – a systematic review and research agenda. J Health Serv Res Policy 19:245–252. https://doi.org/10.1177/1355819614534836

McMichael AJ (1993) Planetary overload. Global environmental change and the health of the human species. Cambridge University Press, Cambridge UK

Mecklin J (2019) Former Defense Secretary William Perry: why we must describe doomsday to keep it from happening. Bull At Sci 75:265–268. https://doi.org/10.1080/00963402.2019.1680046

Murdock CC, Sternberg ED, Thomas MB (2016) Malaria transmission potential could be reduced with current and future climate change. Sci Rep, 6, Article number: 27771. https://doi.org/10.1038/srep27771

Murray KO, Resnick M, Miller V (2007) Depression after infection with West Nile virus. Emerg Inf Dis 13:479–481. https://doi.org/10.3201/eid1303.060602

Neria Y, Nandi A, Galea S (2008) Post-traumatic stress disorder following disasters: a systematic review. Psychol Med 38:467–480. https://doi.org/10.1017/S0033291707001353

Newnham EA, Titov N, McEvoy P (2020) Preparing mental health systems for climate crisis. Lancet Plan Health 4:E89–E90. https://doi.org/10.1016/S2542-5196(20)30036-X

Norris FH, Friedman MJ, Watson PJ et al (2002) 60,000 disaster victims speak: part I. An empirical review of the empirical literature, 1981–2001. Psychiatr Interpers Bio Proces 65:207–223. https://doi.org/10.1521/psyc.65.3.207.20173

OECD (2020) Monthly OECD electricity statistics

Ordunez P, Martinez R, Reveiz L et al (2014) Chronic kidney disease epidemic in Central America: urgent public health action is needed amid causal uncertainty. PLoS Negl Trop Dis 8:e3019. https://doi.org/10.1371/journal.pntd.0003019

Peng M, Liu A, Zhou J et al (2011) Association between posttraumatic stress disorder and preflood behavioral characteristics among children aged 7–15 years in Hunan, China. Med Princ Pract 20:336–340. https://doi.org/10.1159/000323757

Pfefferbaum B, Jacobs AK, Van Horn RL et al (2016) Effects of displacement in children exposed to disasters. Curr Psychiatry Rep 18:71. https://doi.org/10.1007/s11920-016-0714-1

Pichler P-P, Jaccard IS, Weisz U et al (2019) International comparison of health care carbon footprints. Environ Res Lett 14:064004. https://doi.org/10.1088/1748-9326/ab19e1

Pope C, Burnett R, Thun M (2002) Lung cancer, cardiopulmonary mortality, and long-term exposure to fine particulate air pollution. JAMA 287:1132–1141. https://doi.org/10.1001/jama.287.9.1132

Raymond C, Matthews T, Horton RM (2020) The emergence of heat and humidity too severe for human tolerance. Sci Adv 6:eaaw1838. https://doi.org/10.1126/sciadv.aaw1838

Ripple WJ, Wolf C, Newsome TM et al (2020) World scientists' warning of a climate emergency. BioSci 70:8–12. https://doi.org/10.1093/biosci/biz088

Robock A, Toon OB, Bardeen CG et al (2019) How an India-Pakistan nuclear war could start—and have global consequences. Bull At Sci 75:273–279. https://doi.org/10.1080/00963402.2019.1680049

Rosling H, Rosling O, Rönnlund AR (2018) Ten reasons we're wrong about the world – and why things are better than you think. Flatiron Books, New York

Sanson AV, Burke SEL (2019) Climate change and children: an issue of intergenerational justice. In: Balvin N, Christie D (eds) Children and peace: from research to action. Springer Open Peace Psychology Book Series, New York, pp 342–363

Sanson AV, Van Hoorn J, Burke SEL (2019) Responding to the impacts of the climate crisis on children and youth. Child Dev Perspect 13:201–207. https://doi.org/10.1111/cdep.12342

Sanson AV, Wachs TD, Koller SH et al (2018) Young people and climate change: the role of developmental science. In: Verma S, Petersen A (eds) Developmental science and sustainable development goals for children and youth. Springer, New York, pp 115–138

Sherman JD, Lagasse R (2018) How healthy is health care? JAMA Netw Open 1:e181000-e. https://doi.org/10.1001/jamanetworkopen.2018.1000

Spratt D (2019) Revisiting the climate collapse: the view from Nuuk in the year 2070. Bull At Sci 75:280–285. https://doi.org/10.1080/00963402.2019.1680050

Stanke C, Murray V, Amlôt R et al (2012) The effects of flooding on mental health: outcomes and recommendations from a review of the literature. PLoS Currents 4:1–25. https://doi.org/10.1371/4f9f1fa9c3cae

Strife SJ (2012) Children's environmental concerns: expressing ecophobia. J Environ Educ 43:37–54. https://doi.org/10.1080/00958964.2011.602131

Sustainable Development Unit (2018) Reducing the use of natural resources in health and social care, 2018 report. Available from https://www.sduhealth.org.uk/documents/Policy. Accessed 11 Dec 2020

Thurston G (2017) Written report of George D. Thurston regarding the public health benefits of EPA's proposed rulemaking regarding best available retrofit technology for Texas sources under the regional haze rule. Environmental Protection Agency. Available via https://www.sierraclub.org/sites/www.sierraclub.org/files/blog/%20Texas%20Regional%20Haze%20Pollution%20Health%20Impacts.pdf. Accessed 11 Dec 2020

Torres JM, Casey JA (2017) The centrality of social ties to climate migration and mental health. BMC Public Health 17:1–10. https://doi.org/10.1186/s12889-017-4508-

Tucci J, Mitchell J, Goddard C (2007) Children's fears, hopes and heroes. Australian Childhood Foundation and the National Research Centre for the Prevention of Child Abuse, Monash University, Melbourne, Australia, Melbourne

Turchin A, Denkenberger D (2018) Classification of global catastrophic risks connected with artificial intelligence. AI Soc 35:147–163. https://doi.org/10.1007/s00146-018-0845-

UNEP (2007) Post-conflict environmental assessment. UNEP, Nairobi

Waters CN, Zalasiewicz J, Summerhayes C et al (2016) The Anthropocene is functionally and stratigraphically distinct from the Holocene. Science 351:aad2622. https://doi.org/10.1126/science.aad2622

WHO (2014) Gender, climate change and health. WHO, Geneva

Woodbury Z (2019) Climate trauma: toward a new taxonomy of trauma. Ecopsychology 11:1–8. https://doi.org/10.1089/eco.2018.0021

Chapter 5
How Sustainable Are the UN Sustainable Development Goals?

Kerryn Higgs

Abstract The UN SDGs, while including indispensable environmental goals for the first time, skate over crucial contradictions between these and genuine solutions to the unaddressed deficits still experienced by billions of people. Of particular concern is the injunction to foster economic growth, defined as growth in per capita gross domestic product (GDP). The 'growth solution' remains at the heart of the 2015 goals, ignoring the extraction and pollution effects inevitably attached to it. The rich world continues to ignore claims for redistributive justice; however, global sustainability is not feasible without at least minimal redistribution. The relentless increase in material flows in the world economy continues to defy all reduction targets and all undertakings. Growth in these flows needs to cease, unless that growth is meeting actual human needs. Given that we cannot abandon significant increases in material flows where the social goals of the SDGs demand them, it will be necessary to reverse growth in the rich world to accommodate this.

5.1 Introduction

The United Nations (UN) Sustainable Development Goals (SDGs) of 2015 (see Box 5.1) are rightly celebrated as a major achievement: an agreement between nations on a comprehensive plan to tackle worldwide social and environmental crises. However, they rely on elements that are likely to undermine their success, and on trade-offs where some SDGs will have to be sacrificed to achieve others. Of particular concern is the injunction to foster economic growth, defined as growth in per capita gross domestic product (GDP). For individual nations, the goals are voluntary, though the UN reports on progress towards goals and targets at the global level.

K. Higgs (✉)
University of Tasmania, Hobart, Australia

Club of Rome, Sydney, Australia

S. J. Williams, R. Taylor (eds.), *Sustainability and the New Economics*,
https://doi.org/10.1007/978-3-030-78795-0_5

Box 5.1: Summary of the Sustainable Development Goals

1. End Poverty;
2. End Hunger;
3. Health/Well-being;
4. Education for all;
5. End discrimination against women;
6. Water/Sanitation;
7. Energy;
8. Growth/Jobs;
9. Industry/Infrastructure;
10. Reduce Inequalities, within and between countries;
11. Cities/settlements;
12. Production & Consumption;
13. Climate Action;
14. Oceans;
15. Land;
16. Peace/Justice;
17. Partnerships.

The SDGs include specific goals for conservation, protection and restoration of land, sea and climate for the first time. A fourth goal (SDG12), 'sustainable consumption and production patterns', also implies environmental limits. These four goals are an advance on the 2001 Millennium Development Goals which, though they talked of 'sustainable development' in general terms, otherwise ignored the Earth system that supports all life, including human life.

5.2 Development Based on Growth and Debt, 1980–2015

In line with decades of development theory and practice, economic growth remains the centrepiece of solutions to world problems, both social and environmental (Higgs 2014: 105–162). The World Conference on Environment and Development (WCED) or Brundtland Commission (1983–87), while endorsing ongoing economic growth, also drew attention to the need for redistribution. It warned that the needs of the poor must be given 'overriding priority'; that developing countries must 'reap large benefits'; that the requirements of future generations must be taken into account; and that the people of the rich world, who were living beyond the 'world's ecological means', must reduce their consumption (WCED 1987). In light of these concerns, and the failure of the twentieth century's astounding economic growth[1] to make a significant difference to the people of the developing world, the Commission made a modest proposal: a *small proportion of future increases* in wealth should be redistributed.

Soon afterwards, a new economic orthodoxy, known as neoliberalism, captured thinking across the world (Higgs 2014: Chap. 6). In the reports of the Rio Earth Summit (UNCED 1992), just 5 years after the WCED concluded, hardly a hint of redistribution of wealth can be found; instead, solutions to both social and environmental ills were confidently anticipated through market liberalisation, economic growth and free trade. Actual funding provisions for the Rio agenda were vague. Various avenues were to be 'explored' or 'encouraged'. The only explicit proposal asked developed countries to meet the UN aid target of 0.7% of GDP, established many decades earlier. There had been a minimal record of success at reaching this target in the past, but all participants agreed to reach it 'as soon as possible' (UNCED 1992).

After Rio, however, aid did not increase; funding did not materialise. Aid was estimated to have averaged 0.51% of GDP in the late 1960s. By 2009, the average volume of aid from North to South had declined to approximately 0.3% of northern GDP (Riddell 2009), while the flow of payments in the opposite direction has been vast. On the basis of figures from the Organisation for Economic Co-operation and Development (OECD), political scientist Susan George (1992: p xv–xvi) calculated that a net amount of $418 billion flowed back to the North in debt service payments alone during the period from 1982 to 1990—an avalanche of payments that did little to defray the debts on which they were levied (Higgs 2014). As George (1992) pointed out, the total flow of funds to the rich world was (and remains) greater still when 'royalties, dividends, repatriated profits, underpaid raw materials and the like' are added. Financial assets in tax havens were estimated at well over $US21 trillion for 2010, much of it siphoned off from corporate profits in the South (Henry 2012; Shaxson et al. 2012). Thus, aid from North to South is dwarfed by the financial flows in the opposite direction, many of them resulting from the structure of the global economy. Jason Hickel's paper (2019b) on the imperative of redistribution identifies the vast illicit financial flows involved in trade. Overall, far from facilitating appropriate development where it was needed, economic growth swelled the coffers of investors.

The International Monetary Fund's Anne Krueger encapsulated the favoured approach in 2004:

> Our job at the Fund is to support governments in their efforts to deliver the sustained and rapid growth needed to raise living standards and reduce poverty.... [T]he solution is more rapid growth—not a switch of emphasis towards more redistribution. Poverty reduction is best achieved through making the cake bigger, not by trying to cut it up in a different way (Krueger 2004).

'Making the cake bigger' has been the key development strategy of UN institutions for many decades.

[1] Industrial production was multiplied 50 times, 80% of that after 1950.

5.3 SDGs Retain the Growth Strategy

Though far more multi-faceted and specific about environmental objectives than previous development goals, the SDGs (UNDESA 2015) rest implicitly on the 'bigger cake' strategy. They first lay out essential goals to meet basic human needs (SDGs 1, 2, 3, 4, 6). To these are added gender equality (5), meeting energy needs (7), provision of infrastructure (9), urban development (11) and reducing inequalities within and between countries (10). The SDGs then go on to address explicitly the conservation of terrestrial and marine environments and the need to halt global warming (SDGs 13, 14 and 15), as well as an aim to make production and consumption sustainable (SDG12).

Although several goals—such as gender equality, peace (16) and partnership (17)—need not demand much material extraction, there is little acknowledgement of the inherent contradictions between the many goals that require an increase in material extraction, and the ecological goals that seek to prevent deterioration of the biosphere. Indeed, the underlying assumption appears to be that GDP growth is indispensable and ecological objectives may be traded off if they threaten it.

The 'planetary boundaries' researchers (Steffen et al. 2015b) have identified at least two areas where humanity has already exceeded its 'safe operating space' (biodiversity loss and biogeochemical disturbances)[2] and two others where we are very close (climate and land system change). Steffen (2019 pers. comm.) concedes that climate data more recent than that used in 2015 may require climate to be moved to the 'already unsafe' category. In addition to these four areas and others not yet considered unsafe, the team has yet to quantify boundaries for atmospheric aerosols, novel entities (chemical pollution in earlier reports) and the functional role of biosphere integrity. Ecological footprint research also confirms an overshoot situation, where the global rates of extraction and pollution already exceed the capacity of natural systems to replace resources and absorb waste (Global Footprint Network 2019). In the SDG framework, the idea that economic growth can be cultivated without encroaching further on these planetary boundaries is assumed to be feasible, though that idea has been challenged for decades by ecological economists such as Daly (1990) and Victor (2019).

---

[2]This phrase refers to the radical disruption of the nitrogen and phosphorous cycles primarily. In nature, nitrogen is largely inert in the atmosphere, though some is mobilised by bacteria and leguminous plants. As a fertiliser, nitrogen has greatly expanded food production, but is now cascading through rivers, groundwater and continental shelves, initiating algal blooms and dead zones (Rockström et al. 2009). In the case of phosphorous, the other ubiquitous fertiliser, there is an added threat—phosphate rock is a declining resource, commanding a rising price; this has grim implications for future agriculture (Cordell et al. 2009), especially where populations lack adequate finance to import it.

5.4 SDG2: 'Safe, Nutritious and Sufficient Food'

Many of the SDGs that address unmet human needs call for increases in the flow of material resources. In a context of rising populations and rising affluence, SDG2 (to supply 'safe, nutritious and sufficient food' to everyone), is likely to threaten SDGs 14 and 15, which aim to restore and preserve terrestrial and marine ecosystems and SDG13 which pledges urgent action on climate change.

SDG2 speaks of sustainable agriculture but does not examine the predominant practices of industrial agriculture: clearing forests, draining wetlands and peatlands, and relying on significant inputs of fertiliser, herbicides, pesticides and energy, still sourced largely from fossil fuels. The injunction at SDG2.3 to double agricultural productivity for small scale farmers means increasing output per farmer without reference to output per hectare or megalitre. Targets 2.2b and 2.2c focus on impediments to global agricultural trade and commodity markets, factors which are secondary for the small farmers who feed the world at the local level. These targets, once again, reflect corporate and growth-oriented strategies and ignore the situation of the world's smallholders, who largely work outside the global marketplace.

Rockström et al. (2017) have argued that land-use emissions must be reduced to zero by 2050, if we are to meet the Paris climate target of 'well below 2°C'. Industrial farming is a crucial pressure on all four planetary boundaries judged to be in trouble; it affects biodiversity loss, nitrogen and phosphorous pollution, land-use change, and climate. If conventional agriculture is expanded to meet SDG2, this will inevitably sacrifice terrestrial ecological integrity (Crist et al. 2017). Adoption of regenerative agriculture is essential if SDG2 is to be made compatible with SDGs 13–15.

Smallholders can assist in this process. Target 2.3 refers to ensuring secure and equal access to land, a critical element. Using systems of agro-ecology that minimise external inputs and mimic ecological processes, the world's small farmers already produce the majority of the world's food on less than a quarter of its farmland (GRAIN 2014). In a situation where land and water are the limiting factors—rather than labour—doubling productivity per person is less important than increasing productivity per unit of land and water (Higgs 2014: p 146–7; UNEP 2019: p 13). In fact, focussing on labour productivity may encourage consolidation of land under the control of larger farmers and agribusiness corporations, adding to the numbers of landless and the diversion of land for biofuel, animal food or luxury crops for export (Daniel and Mittal 2009). Secure access to land is indispensable, but will involve reversing the trend towards consolidation, a trend that is evident in the Green Revolution (Higgs 2014: p 46–7), the quadrupling in acreage of the industrial crops (soy, canola, sugar cane and oil palm) from 1960 to 2011 (GRAIN 2014), and in 'land grabs' amounting to hundreds of millions of hectares this century (Magdoff 2013).

Peasant farmers resist the growth model, which many of them believe has been imposed on them, deepening inequality and accelerating environmental decline. The smallholders who contributed to the Food and Agriculture Organisation's

(FAO) report for Rio + 20 (Wolfenson 2013) described themselves as 'a driving force towards socially fair and ecologically sustainable agriculture systems'. For these farmers, 'the over-arching paradigm of economic growth, considered the highway to secure development, has left the social and environmental dimensions of sustainable development behind'. Wolfenson (2013: p 26) goes on to note that, alongside agro-ecology, smallholders champion

> food sovereignty … a system that returns the land to its social function as the producer of food, puts the people who produce, distribute and consume food at the centre of decisions about food systems and policies, as opposed to the demands of markets and corporations.

Where land is consolidated into larger farms, and smallholders lose access to their communal land and water, they may (or may not) get employment, but wages are poor and productivity gains are not shared with them. Food, fuel and profits go to markets overseas and investors, largely foreign (Daniel and Mittal 2009). The emphasis in SDG2.2c on regulating commodity markets will be of limited relevance where small farmers primarily produce local food for local people, which is the most promising strategy if we hope to provide 'safe, nutritious and sufficient food' for all, without trading off protection of land, sea and climate.

5.5 SDG11, SDG9: Urban Development, Infrastructure and Industry

SDG11, which addresses urban development, will jeopardise SDG13 (climate) unless towns and settlements are designed and built along low-carbon lines. The word 'sustainable' appears in the SDG11 targets, stated alongside 'safe', 'affordable', 'inclusive' and 'accessible'. In line with its ambiguity since the Brundtland Commission, the meaning of 'sustainable' remains unclear. Current town and city development needs radical reform to become sustainable, in particular an explicit goal of low-carbon design and a rejection of urban sprawl, which also impinges on SDG15. Neither of these essentials is mentioned in the SDGs. Meanwhile, the mass of human-made items, such as cities, roads and infrastructure, is now found to exceed the mass of all living things, having doubled every 20 years in recent times (Elhacham et al. 2020).

SDG9 sets a goal of expanding infrastructure and industry. This too will be hard to implement without jeopardising environmental goals. Historically, industrial expansion is linked to escalating energy demand and land-system conversion; associated emissions have risen steadily for two centuries, accelerating since 1950 (Steffen et al. 2015a). If this trend is replicated, future emissions can be expected to keep rising (UNEP 2016c). Unless new infrastructure is designed and built along low-carbon lines, the Paris Agreement's goals will elude us. No SDG or target aims explicitly for low-carbon options, an omission that reflects the failure of the SDGs to address the fundamental challenge of providing essential socio-economic reforms without sacrificing environmental goals.

At the time of writing, China's Belt and Road Initiative (BRI) is an example of infrastructure and industrial development which pays little attention to carbon emissions or ecological damage (Losos et al. 2019), though it does not differ in this respect from the development advocated by the European powers for centuries. While the Chinese government promotes a domestic energy transition, China exports coal-fired technology across the BRI. The pace of this activity has fluctuated, and has slowed since its peak in 2010 (Ren Peng et al. 2017), but power plants have life spans of 50 years or more, and roads will carry traffic for decades, so both will affect greenhouse gas emissions for many years. According to the Global Development Policy Center (2020), 70% of BRI energy investments are fossil fuel based and, despite China's pre-eminence in solar technology, only 1.6% of BRI energy investments are solar based. In addition to the climate implications, the BRI has impacts on land use, wildlife and habitat, and water; it will also expand industrial pollution. According to ecologist William Laurance (2017) the BRI's 'dizzying variety of resource extraction, energy, agricultural, and infrastructure projects … are wreaking unprecedented damage to ecosystems and biodiversity' across the world.

5.6 SDG12: Sustainable Production and Consumption

In recommending sustainable production and consumption, SDG12 emphasises efficiency, management and the reduction of waste. Such strategies are vital, but no mention is made of the role of consumerism as a driver of the economic growth that is considered to be the bedrock of our economy (see Higgs 2014: Chap. 5). The SDGs do not refer to the role of advertising in harnessing human desire to markets and inciting mass shopping and endemic waste, to keep the consumer economy functioning. Some detailed targets are laudable and essential, such as managing chemicals and their wastes (12.4), and reducing waste in general (12.5) but the over-arching target—sustainable management and efficient use of natural resources (12.2)—is not defined. Similarly, the target for fossil fuel subsidies (12.c) is vague:

> Rationalize inefficient fossil-fuel subsidies that encourage wasteful consumption by removing market distortions, in accordance with national circumstances, including by restructuring taxation and phasing out those harmful subsidies … to reflect their environmental impacts.

While this formulation allows room for developing countries to adjust according to their own circumstances, it lacks any categorical rejection of such subsidies in the developed world.

5.7 SDG8: 'Sustainable Economic Growth'?

The most problematic of the goals is SDG8 which calls for "sustained, inclusive and sustainable economic growth," (purportedly generating jobs for all). Apart from the title line, the word 'sustainable' is not included in the targets and indicators; economic growth is the only item that is slated to be sustained here, while the *ecological* basis of economies, usually regarded as intrinsic to the concept of sustainability (Daly 1990), is absent. 'Sustainable growth' is not defined and has, in any case, been recognised as an oxymoron (Daly 1990). The connection between economic growth and job-creation is also simply assumed, despite the fact that the track record of market-oriented, capital-intensive expansion is not encouraging (Hirway and Shah 2011). Where the object is profit-generation under the market model, cost reduction is a routine strategy and leads to mechanisation and job-shedding. Community-based models such as Ashok Khosla's (2015) social enterprise, Development Alternatives, operating in central India, provide a realistic model of business creating jobs and manufacturing items needed by local people. However, this is not the kind of development envisaged in SDG8.

Target 8.1 requires sustained per capita GDP growth, including at least 7% per annum for the least-developed countries. Material extraction is essential in less-developed contexts in order to provide sanitation, water, and services in general, and this will involve some degree of GDP growth, but, as Khosla's enterprise shows, the emphasis can be on meeting urgent needs rather than simply boosting GDP.

Moreover, SDG8 demands ongoing economic growth throughout the world. This is hardly surprising, given that economic growth has been advanced as the panacea for all problems for well over 50 years and continues to be prominent on the lips of politicians, business people and most of the global institutions involved in sustainability policy (Higgs 2014: Part II).[3] GDP growth requires extraction and waste-disposal and has been associated over the past century with mounting carbon emissions, depletion of resources and ongoing transformation of land systems and marine environments (Steffen et al. 2015a). Although claims are made that growth can be decoupled from these negative consequences (OECD 2011; World Bank 2012; Hatfield-Dodds et al. 2015; UNEP 2016a; UNEP 2019), hard evidence of absolute decoupling is scarce (Haberl et al. 2020; Vadén et al. 2020). Instead, the empirical evidence shows both an ever-increasing rate of extraction (Fig. 5.1) and a tight association between GDP and material footprint (Fig. 5.2).

Sources agree fairly closely on the current extent of material extraction, estimated at 87–88 billion tonnes a year in 2015, having tripled over the previous 50 years, and still growing (UNEP 2019). Circle Economy (2020) included recycled materials and estimated material extraction at 100.6 billion tonnes a year. In line with others who quantify material extraction, UNEP (2019) and Circle Economy

[3] GDP measures the flow of monetary exchanges in the market economy based on production. While GDP includes genuine positives, it is indiscriminate and adds together both costs and benefits to society.

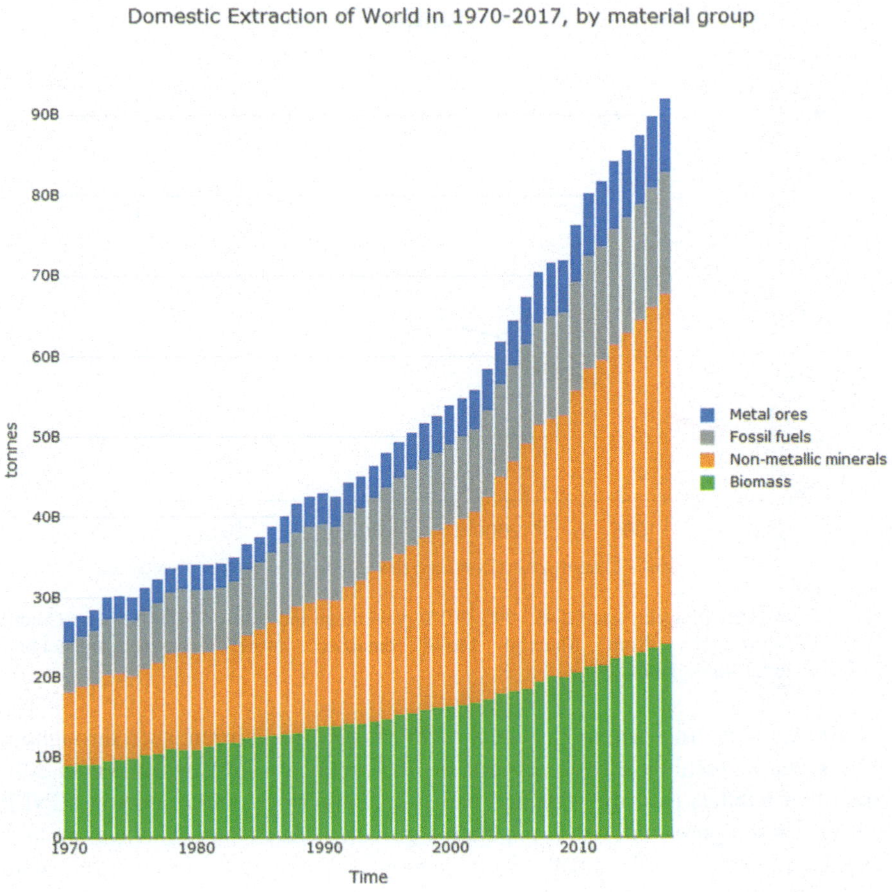

Fig. 5.1 Material Extraction, 1970–2017 in billions of tonnes. (Credit: www.materialflows.net. The Material Flow Analysis Portal. Vienna University of Economics and Business (WU), Austria)

(2020) include in their calculations metals, non-metallic minerals (such as sand), fossil fuels, biomass, water and land conversion. Cooper et al. (2018), in addressing the extent of humans' sediment production in the mineral and construction industries, include all forms of material shifted from its original location (such as overburden, waste rock etc.) and arrive at a figure of around 316 billion tonnes in 2015, much greater than estimates that ignore waste.

Estimates of an approximate target for sustainable resource extraction cluster around 50 billion tonnes a year (UNEP 2014; Hoekstra and Wiedmann 2014; Bringezu 2015; Hickel 2019a). If this figure is regarded as reasonable, we are already extracting double that target and still moving in the wrong direction.

It is also widely accepted that the continuation of historical trends and relationships, which depend on 'a permanent throughput of materials that are extracted, traded and processed into goods and finally disposed of as waste or emissions' will,

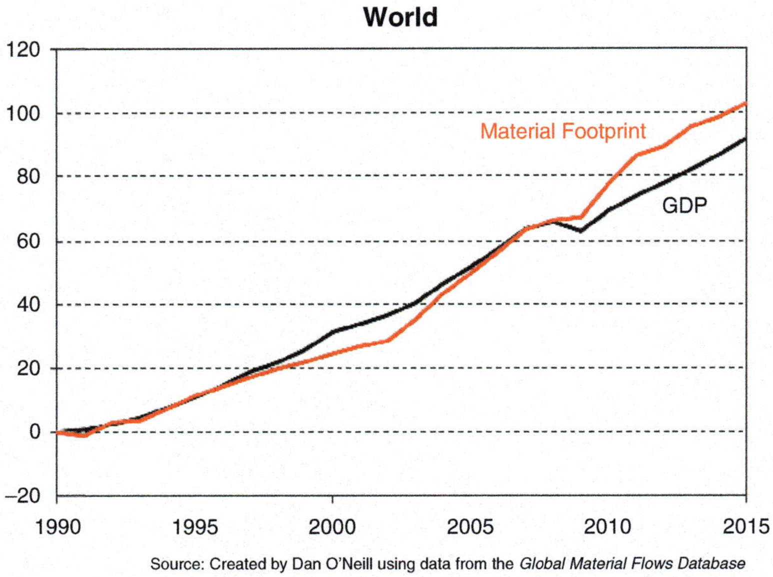

Fig. 5.2 Material footprint and GDP, 1990–2015: percentage increase on 1990 values. (Credit: Dan O'Neill, data from Material Flows database, https://www.resourcepanel.org/global-material-flows-database)

by 2060, inflate gross annual resource use drastically. In addition, such throughput will increase greenhouse gas emissions by 43%, reduce forests by more than 20% and other habitats (mainly grassland and savannahs) by a similar amount (UNEP 2019). These figures will rise if population growth exceeds the UN's medium projection (O'Neill et al. 2018). Such outcomes are clearly inconsistent with SDGs 12–15.[4]

UNEP (2016a, b, 2019) is adamant that the decoupling of material extraction from environmental impact 'will be essential for ensuring future human well-being based on much lower material throughput' (UNEP 2016a). It argues that this can be achieved through resource efficiency, technological advances and circular production with maximum reuse and recycling. By these means, it predicts, the volume of material extraction can be reduced. There is no evidence that such useful (but inadequate) measures will exert sufficient downward pressure on the ever-increasing level of extraction. Instead, over the past decade or so, while UNEP (2011, 2014, 2016a, b, 2019) has backed decoupling, arguing both that it is essential and that it can be done, the trends in the real world run the other way—ever-increasing material flows, more extraction, more consumption and more waste. While maintaining optimism about decoupling, UNEP (2016a, 2019) concedes that efficiencies could

[4] See Ward et al. (2016) on the underlying physical realities that govern non-substitutable resources such as land, water, raw materials and energy; also Hickel and Kallis (2019) on levels of material extraction compatible with Earth's ecological integrity.

cause a rebound effect and trigger accelerated economic growth, thus negating efforts to reduce gross material demand. It suggests that tax measures could counteract this trend (2019).

UNEP (2019: 27) is fully aware that continuation of business-as-usual will make even relative decoupling impossible and argues for worldwide implementation of a 'Towards Sustainability' pathway, which will require 'an urgent and systemic transformation of how we use and manage natural resources' (UNEP 2019: 19, 31). While the policies outlined here are positive, it is unclear how they will be adopted under prevailing political and economic conditions where 'the market' is expected to govern choices. If decisions rely on short-term profit and avoidance of government involvement, such policies are highly unlikely to receive widespread adoption.

While the recommended strategies of efficiencies, circular production and technological advance may mitigate the problems, genuine absolute decoupling is improbable (Higgs 2019) or perhaps impossible (Victor and Jackson 2015; Alexander and Rutherford 2019, Vadén et al. 2020). Absolute decoupling requires ongoing reductions in impacts alongside increases in material flows. Nothing of this kind, especially on a global scale, has yet been observed. The pursuit of decoupling may be a fool's errand in any case. Lenzen et al. (2016) note that 'it is illogical to expect technological progress to stretch to unprecedented limits in order to let populations enjoy unchecked growth in numbers and affluence'. But the assumption that technology can enable growth in extraction and pollution rests on faith in successful decoupling.

According to UNEP (2016a, b) we now use more materials per unit of global GDP than in 2000 and, in the 2019 assessment of progress made towards realising SDG12, the UN (2019) reported that material consumption has continued to increase—from 87–88 billion tonnes in 2015 to 92.1 billion in 2019.[5] It notes that 'worldwide material consumption has expanded rapidly, as has material footprint per capita, seriously jeopardizing the achievement' of SDG12. It warns that 'urgent action is needed to ensure that current material needs do not lead to the over-extraction of resources or to the degradation of environmental resources' (archived by Leibniz Universität 2020). Unless effective action is taken, we will be extracting 190 billion tonnes of materials by 2060, almost four times the amount that numerous researchers consider sustainable (see above). As in 2015, the UN recommends the same solutions: improved resource efficiency, reduced waste, and establishing sustainability practices across all sectors of the economy.

[5] While Circle Economy (2020) includes recycled materials and puts the figure at 100.6 billion tonnes in 2017.

5.8 Problems of Population and Affluence

Population gets no explicit attention in the SDGs. The endorsement of 'universal reproductive health and rights' (SDG3) indicates that all women should be assigned the right to control over their own bodies. But, while the issue of the rights of women is also reflected in SDG4 and SDG5 (education and ending all forms of discrimination against women), population is not addressed. This omission is likely to have resulted from the UN requirement for all countries to endorse the proposed text; as the history of population policy shows, powerful forces deeply opposed to any form of family planning have been influential in international negotiations since at least 1980.

In the early post-World War II decades, reducing population growth was a preferred emphasis of the wealthy consumer economies in dealing with hunger and poverty in poor nations, diverting attention away from our own conspicuous and unsustainable consumption. However, emphasis shifted away from population in the 1970s, when US Evangelicals joined forces with Catholics and conservative Islam (Saudi Arabia) and imposed a natalist ideology. At the 1984 UN conference on population, the US withdrew funding, declaring a new focus on 'family values' and claiming that free market economies were the 'natural' way to slow population growth (Higgs 2014: 36–38).

The global rate of population growth has been slowing since the early 1960s,[6] a process theorised as 'the demographic transition', where high birth and death rates were said to be gradually replaced by low birth and death rates, as countries industrialised and grew wealthy. However, this change applies to a relatively limited sector of these populations—the wealth created during 70 years of economic growth has not been well distributed and vast numbers of people remain trapped in poverty with little prospect of any transition. This outcome suggests that direct action such as explicit provision of family planning may be needed to reduce population growth, rather than entrusting the task to the automatic operation of the 'free market.' Natalist interests may block such a move, however. Even though the average *rate* of population growth has slowed markedly since the1960s, the huge base created then still generates much the same absolute increment as was seen at that time, so we still have more than 80 million extra people to feed, clothe and shelter every year.

The socio-economic SDGs are concerned with development in the global South, where population increase remains at its most rapid. It is vital to remember that, on average, people in the developing world consume relatively little in per capita terms and it is population growth in wealthy countries such as Australia and the United States (and burgeoning middle-classes in China and India) that contributes most to impacts on land, oceans and climate. This is partly due to the high consumption levels of the rich world which 'contribute disproportionately to exhausting and polluting the planet' (Coole 2019: 257; Wiedmann et al. 2020); and partly to the increasing role played by trade, as developed countries reduce their environmental

[6] This rate peaked in 1961–62 at 2.2%.

impact by exporting it to the South. In some countries, such as Malaysia, Madagascar, Honduras and Papua New Guinea, 50–60% of species loss is directly attributable to activities conducted on behalf of the North, for example clearing for crops such as coffee, soy, sisal, cocoa and palm oil, or for timber and grazing. Average species decline attributable to such activities in the developing world was found to be 30% (Lenzen et al. 2012; Wiedmann et al. 2015).

The trend towards increasing affluence in the emerging middle classes of the South involves the adoption of western dietary habits such as increased meat-eating and western travel habits such as personal cars. The idea that everyone can replicate these wasteful practices ignores the levels of extraction and pollution that are inevitably attached. Swelling numbers with such expectations will limit the ability of nations to meet the crucial social goals of the SDGs as well as further undermine the environmental SDGs. Population determines the scale of services that will be required and the scale of the economy that will be required to provide them.

The problems of affluence are also neglected by the SDGs, despite the fact that the affluence enjoyed in the rich world and by growing élite minorities in the South, is a key driver of unsustainable material flows. In 1991, Paul Ekins noted that 'universal opulence' was simply not an option. He calculated that to supply even 20% of Northern affluence across all people in the South would necessitate a total and immediate freeze on Northern consumption. Using oil and paper as examples, Higgs (2014: 103) analysed the immense increase in material production that would be needed to allow China alone to adopt US or even European consumption patterns. Rising inequality (Alvaredo et al. 2018) has multiplied the challenge posed by Ekins (1991). Apart from China, developing countries showed greater internal inequality than Europe, Russia, or North America. Oxfam (2018) estimated that 82% of all wealth created globally in 2017 flowed to the top 1%, with no increase whatsoever for the bottom 50%, most of whom live in the South. Chancel and Piketty (2017) describe the Indian version of these trends. Notwithstanding many decades of extraordinary growth, prosperity is still concentrated among a privileged minority (Higgs 2014: 105–162; Hickel 2017). Clearly, radical steps will be needed to change this situation.

UNEP (2016a: 29–30) confirms that 'the level of well-being achieved in wealthy industrial countries cannot be generalized globally based on the same system of production and consumption'—without further jeopardising environmental thresholds that are already under pressure. Here, UNEP concedes that a different system of production and consumption will be necessary if wellbeing is to be available to all. But the existing world economic system is entrenched and there seems little prospect of replacing it. Indeed, the injunction to pursue economic growth, a mainstay of the accepted solutions put forward for decades, is a key plank in the existing system which requires ongoing profit. A few transnational corporations (TNCs) own and control a majority of the entire global productive apparatus.[7] TNCs also own and control much of the world's mining, energy, transport and manufacturing. They are multifaceted organisations able to contract out elements of their production chains in different countries to take advantage of low wages, permissive regulation, or low taxes. Moreover, they depend on growth of both production and consumption for their continued success at generating their profit.

Control of global wealth is even more concentrated than ownership. Vitali et al. (2011) studied the connections between 43,000 TNCs using topological analysis. This network study revealed that 747 of these corporations controlled 80% of all TNCs worldwide, and that a core 'super-entity' of just 147 tightly interlocking TNCs controlled 40% of all global revenue. The majority of the dominant corporations are financial institutions and this concentrated ownership and control of the world's financial and productive apparatus constitutes a major obstacle to reform. The controlling corporations have no incentive to change since the pursuit of ongoing growth allows them to continue to hold out the promise (however illusory) of future prosperity for all, supposedly including the world's poor.

5.9 Conclusion

The SDGs, while including indispensable environmental goals for the first time, skate over crucial contradictions between these and genuine solutions to the unaddressed deficits still experienced by billions of people in the developing world. The 'growth solution' remains at the heart of the 2015 goals, ignoring the extraction and pollution effects inevitably attached to it. Just as in the early post-World War II years, the rich world continues to avoid the claims of redistributive justice.

To resolve the internal contradictions of the SDGs will require difficult and strategic choices—especially in agriculture, urban development, infrastructure and industry; these are discussed above. It will also demand significant reform, if not transformation, of the economic system. Whether decoupling is feasible or not, we urgently need to abandon all growth that involves increasing material flows, unless it meets actual human needs. We may wish to abandon all economic growth everywhere, but we cannot abandon significant increases in material flows throughout the developing world. We need to accommodate this by cutting growth in the rich world. No solution will emerge without at least minimal redistribution.

Randers et al. (2018: 15–35) modelled four possible pathways to achieve the aim of 'inclusive and prosperous world development within a stable and resilient earth system'. They believe that there is growing acceptance among governments that maximising GDP as a top priority is not the best path to sustainable human

[7] The WCED (1987) noted that 80 to 90% of the trade in each of the world's key commodities—tea, coffee, cocoa, cotton, timber, tobacco, jute, copper, iron ore, and bauxite—was controlled by fewer than six transnationals. Consolidation has continued; soy and palm oil should be added to the key commodities.

wellbeing, and that: 'a conventional market-based growth approach has weak incentives to achieve SDGs and none at all for protecting the commons, social welfare, or any other non-economic values'. Of their four scenarios, 'transformational change' is the only model that delivers on the objective. This model includes five major strands:

- accelerated renewables;
- genuinely sustainable food production;
- new development approaches;
- investment in education for all, gender equality, health and family planning;
- serious reductions in inequality.

On this last question, Randers et al. (2018: p 33) propose that 'the 10% richest take no more than 40% of income,' an aim to be achieved via progressive taxation of income and wealth, a shorter working year, and similar measures—they argue that this would also help to fund necessary social investments. As noted above, the SDGs do not give thorough and explicit endorsement to such measures: family planning is not mentioned; food and development goals conform to 'free-market' solutions rather than sustainable ones; and although reductions in inequality get a mention, there is no hint at how this is to be achieved. In addition, it remains to be seen whether Randers et al.'s relatively moderate proposals could gain acceptance in the world's current political and economic framework.

Clearly, material growth will be essential for the South, but to meet that need, the consumption patterns of the North must contract.[8] It will not suffice to set the market free to do its work. Design and planning must be promoted—and this will require a retreat from the prescriptions of the neoliberal economics that has dominated economic thinking for nearly 50 years.

References

Alexander S, Rutherford J (2019) A critique of techno-optimism. In: Kalfagianni A, Fuchs D, Hayden A (eds) Routledge handbook of global sustainability governance. Routledge, Abingdon and New York, pp 231–241

Alvaredo F, Chancel L, Piketty T et al (2018) World inequality report. http://wir2018.wid.world/. Accessed 2 Dec 2020

Bringezu S (2015) Possible target corridor for sustainable use of global material resources. Resources 4:25–54

Chancel L, Piketty T (2017) Indian income inequality, 1922–2014: from British Raj to Billionaire Raj? WID.world working paper series 2017/11. http://wid.world/document/chancelpiketty2017widworld/. Accessed 30 Jan 2020

Circle Economy (2020) The circularity gap report. https://www.circularity-gap.world/2020/. Accessed 3 Jan 2021

[8] Maxton and Randers (2016: pp 108–174) propose 13 policies that they believe to be politically feasible in the rich world, perhaps opening a pathway to contraction.

Coole D (2019) The population challenge. In: Kalfagianni A, Fuchs D, Hayden A (eds) Routledge handbook of global sustainability governance. Routledge, Abingdon and New York, pp 254–265
Cooper A, Brown T, Price S et al (2018) Humans are the most significant global geomorphological driving force of the 21st century. Anthrop Rev 5(3):222–229. https://doi.org/10.1177/2053019618800234
Cordell D, Drangert J-O, White S (2009) The story of phosphorus: global food security and food for thought. Glob Environ Chang 19:292–305
Crist E, Mora C, Engelman R (2017) The interaction of human population, food production, and biodiversity protection. Science 356(6335):260–264
Daly H (1990) Sustainable growth: an impossibility theorem. In: Daly H, Townsend K (eds) Valuing the earth. MIT Press, Cambridge, MA
Daniel S, Mittal A (2009) The great land grab: rush for world's farmland threatens food security for the poor. Oakland Institute, Oakland. www.oaklandinstitute.org/sites/oaklandinstitute.org/files/LandGrab_final_web.pdf. Accessed 23 Feb 2020
Ekins P (1991) The sustainable consumer society: a contradiction in terms? International Environmental Affairs 3(4):243–258
Elhacham E, Ben-Uri L, Grozovski J et al (2020) Global human-made mass exceeds all living biomass. Nature. https://doi.org/10.1038/s41586-020-3010-5
George S (1992) The debt boomerang: how third world debt harms us all. Westview, Boulder
Global Development Policy Center (2020) China's global energy finance. https://www.bu.edu/cgef/#/all/EnergySource. Accessed 29 Feb 2010
Global Footprint Network (2019). https://www.footprintnetwork.org/our-work/ecological-footprint/. Accessed 23 Feb 2020
GRAIN (2014) Hungry for Land: small farmers feed the world with less than a quarter of all farmland. https://www.grain.org/article/entries/4929. Accessed 23 Feb 2020
Haberl H, Wiedenhofer D, Virág D et al (2020) A systematic review of the evidence on decoupling of GDP, resource use and GHG emissions, part II: synthesizing the insights. Environ Res Lett 15:065003. https://doi.org/10.1088/1748-9326/ab842a
Hatfield-Dodds S, Schandl S, Adams P et al (2015) Australia is 'free to choose' economic growth and falling environmental pressures. Nature 527(7576):49–53
Henry J (2012) The price of offshore revisited. Tax Justice Network. https://taxjustice.net/cms/upload/pdf/Price_of_Offshore_Revisited_120722.pdf. Accessed 12 Mar 2020
Hickel J (2017) The Divide: a brief guide to global inequality and its solutions. William Heinemann, London
Hickel J (2019a) The contradiction of the sustainable development goals: growth versus ecology on a finite planet. Sustain Dev 2019:1–12
Hickel J (2019b) The imperative of redistribution in an age of ecological overshoot: human rights and global inequality. Humanity 10(3):416–428
Hickel J, Kallis G (2019) Is green growth possible? New Polit Econ. https://doi.org/10.1080/13563467.2019.1598964
Higgs K (2014) Collision course: endless growth on a finite planet. MIT Press, Cambridge, MA
Higgs K (2019) Growth and development. In: Kalfagianni A, Fuchs D, Hayden A (eds) Routledge handbook of global sustainability governance. Routledge, Abingdon and New York, pp 180–192
Hirway I, Shah N (2011) Labour and employment under globalisation: the case of Gujarat. Econ Polit Wkly 46(22):57–65
Hoekstra A, Wiedmann T (2014) Humanity's unsustainable environmental footprint. Science 344:1114–1117
Khosla A (2015) To choose our future. Academic Foundation, New Delhi
Krueger A (2004) Letting the future in. 4 Jun http://www.imf.org/external/np/speeches/2004/060404.htm. Accessed 23 Feb 2020
Laurance W (2017) The dark legacy of China's drive for global resources. Yale Environment 360, 28 Mar. http://e360.yale.edu/features/the-dark-legacy-of-chinas-drive-for-global-resources. Accessed 29 Feb 2020

Leibniz Universität (2020) SDG Education. https://sdg-education.net/en/sdg-videos/12-responsible-consumption-and-production.html. Accessed 21 Dec 2020
Lenzen M, Malik A, Foran B (2016) Letter. J Clean Prod 139:796–798
Lenzen M, Moran D, Kanemoto K et al (2012) International trade drives biodiversity threats in developing nations. Nature 486:109–112
Losos E, Pfaff A, Olander L et al (2019) Reducing environmental risks from belt and road initiative Investments in Transportation Infrastructure. World Bank Group
Magdoff F (2013) Twenty-first-century land grabs: accumulation by agricultural dispossession. Mon Rev 65(6):1–18
Maxton G, Randers J (2016) Reinventing prosperity. Greystone Books, Vancouver
OECD (2011) Towards green growth. https://www.oecd.org/greengrowth/48012345.pdf. Accessed 23 Feb 2020
O'Neill D, Fanning A, Lamb W et al (2018) A good life for all within planetary boundaries. Nature Sustainability 1(2):88–95
Oxfam International (2018) Reward work, not wealth. https://www.oxfam.org/en/research/reward-work-not-wealth. Accessed 23 Feb 2020
Randers J, Rockström J, Stoknes P et al (2018) Transformation is feasible. Stockholm Resilience Centre and Norwegian Business School
Ren Peng, Liu Chang, Zhang Liwen (2017) China's involvement in coal-fired power projects along the belt and road. Global Environmental Institute, May
Riddell R (2009) Is aid working? Open Democracy, 20 Nov. http://www.opendemocracy.net/roger-c-riddell/is-aid-working-is-this-right-question-to-be-asking. Accessed 1 Mar 2020
Rockström J, Steffen W, Noone K et al (2009) A safe operating space for humanity. Nature 461:472–475
Rockström J, Gaffney O, Rogelj J et al (2017) A roadmap for rapid decarbonization. Science 355(6331):1269–1272
Shaxson N, Christensen J, Mathiason N (2012) Inequality: you don't know the half of it. Tax Justice Network. http://www.taxjustice.net/cms/upload/pdf/Inequality_120722_You_dont_know_the_half_of_it.pdf. Accessed 13 Mar 2020
Steffen W, Broadgate W, Deutsch L et al (2015a) The trajectory of the Anthropocene: the great acceleration. Anthrop Rev 2(1):81–98
Steffen W, Richardson K, Rockström J et al (2015b) Planetary boundaries: guiding human development on a changing planet. Science 347(6223):736–746
UN (2019) Special edition: report of the Secretary-General on SDG Progress 2019. https://sustainabledevelopment.un.org/content/documents/24978Report_of_the_SG_on_SDG_Progress_2019.pdf. Accessed 23 Feb 2020
UN Conference on Environment and Development (1992) Agenda 21. https://sustainabledevelopment.un.org/content/documents/Agenda21.pdf. Accessed 23 Feb 2020
UNDESA (2015) Sustainable development goals. Knowledge base. UN Department of Economic and Social Affairs. https://sustainabledevelopment.un.org/sdgs. Accessed 28 Apr 2020
UNEP (2011) Decoupling natural resource use and environmental impacts from economic growth, a report of the working group on decoupling to the International Resource Panel. http://wedocs.unep.org/handle/20.500.11822/9816. Accessed 23 Feb 2020
UNEP (2014) Managing and Conserving the Natural Resource Base for Sustained Economic and Social Development. https://www.resourcepanel.org/managing-and-conserving-natural-resource-base-sustained-economic-and-social-development-english. Accessed 23 Feb 2020
UNEP (2016a) Global material flows and resource productivity. Summary report. https://www.resourcepanel.org/global-material-flows-and-resource-productivity-summary-english. Accessed 23 Feb 2020
UNEP (2016b) Resource efficiency: potential and economic implications: a report of the international resource panel. Summary for Policy-makers. Ekins P et al. https://www.resourcepanel.org/reports/resource-efficiency. Accessed 23 Feb 2020

UNEP (2016c) Emissions gap report. http://uneplive.unep.org/theme/index/13#egr. Accessed 23 Feb 2020
UNEP (2019) Global resources outlook 2019. https://www.resourcepanel.org/reports/global-resources-outlook. Accessed 23 Feb 2020
Vadén T, Lähde V, Majava A et al (2020) Decoupling for ecological sustainability: a categorisation and review of research literature. Environ Sci Policy. https://doi.org/10.1016/j.envsci.2020.06.016
Victor P (2019) Managing without growth: slower by design not disaster, 2nd edn. Edward Elgar, Cheltenham
Victor P, Jackson T (2015) The problem with growth. In: Starke L (ed) 2015 state of the world report, confronting hidden threats to sustainability. Worldwatch Institute, Washington
Vitali S, Glattfelder J, Battison S (2011) The network of global corporate control. PLoS One 6(10)
Ward J, Sutton P, Werner A et al (2016) Is decoupling GDP growth from environmental impact possible? PLoS One 11(10)
Wiedmann T, Lenzen M, Keyßer L, Steinberger J (2020) Scientists' warning on affluence. Nat Commun 11:3107
Wiedmann T, Schandl H, Lenzen M et al (2015) The material footprint of nations. PNAS 112(20):6271–6276
Wolfenson K (2013) Coping with the food and agriculture challenge: smallholders' agenda. Preparations and outcomes of the 2012 United Nations Conference on Sustainable Development (Rio+20). www.fao.org/3/a-ar363e.pdf. Accessed 23 Feb 2020
World Bank (2012) Inclusive green growth—the pathway to sustainable development. http://siteresources.worldbank.org/EXTSDNET/Resources/Inclusive_Green_Growth_May_2012.pdf. Accessed 23 Feb 2020
World Commission on Environment and Development (1987) Our common future. Oxford University Press, Oxford

Part II
How We Got Here

Chapter 6
The Evolution of Neoliberalism

John Quiggin

Abstract The term 'neoliberalism' is widely used and often misused and misunderstood. Properly understood, it is the ideological underpinning of the era of financialised capitalism which emerged from the economic crises of the early 1970s and remained dominant for the rest of the twentieth century. Its central theme was the idea that markets, particularly financial markets, generally outperform governments in the allocation of resources and investments. Neoliberalism came in a variety of forms, reflecting the variety of liberalism itself, but was different from preceding forms of liberalism because of the need to respond to the successes and failures of social democracy in the second half of the twentieth century. In one form or another, neoliberalism became the unquestioned basis for the thinking of both centre-right and centre-left parties around the world. This chapter describes the rise of neoliberalism in the twentieth century, its failure in the twenty-first and the prospects for a progressive alternative.

6.1 Introduction

The term 'neoliberalism' is widely used and often misused and misunderstood. Properly understood, it is the ideological underpinning of the era of financialised capitalism which emerged from the economic crises of the early 1970s and remained dominant for the rest of the twentieth century. Its central theme was the idea that markets, particularly financial markets, generally outperform governments in the allocation of resources and investments. Neoliberalism came in a variety of forms, reflecting the variety of liberalism itself, but was different from preceding forms of liberalism because of the need to respond to the successes and failures of social democracy in the second half of the twentieth century. In one form or another,

J. Quiggin (✉)
School of Economics, University of Queensland, Brisbane, Australia
e-mail: j.quiggin@uq.edu.au

S. J. Williams, R. Taylor (eds.), *Sustainability and the New Economics*,
https://doi.org/10.1007/978-3-030-78795-0_6

neoliberalism became the unquestioned basis for the thinking of both centre-right and centre-left parties around the world.

The era of unquestioned neoliberal dominance came to a sudden end on 15 September 2008 when the bank and financial services firm Lehmann Brothers filed for bankruptcy. This collapse signalled the failure of attempts by the US authorities to manage the crisis that had originated in markets for housing loans to "subprime" (that is, low income or high risk) borrowers. Within a week, the global financial system was in crisis, and governments were forced to engage in massive bailouts.

The years since the Global Financial Crisis (GFC) have been characterised by chronic upheaval. The bailouts needed to prevent the collapse of the financial system in turn produced fiscal crises, most notably the 'sovereign debt crisis' in the European Union. These crises were resolved by 'austerity' policies which produced drastic cuts in public services and living standards, while leaving the financial sector essentially unscathed. As a result of the crises, the neoliberal consensus that had prevailed up to the GFC broke down.

The most dramatic illustration of this was the capture of the Republican party and the Presidency of the US by Donald Trump. Trump's rhetoric in the leadup to the election suggested that he would be a new kind of Republican, putting the interests of working-class voters ahead of a globalised system based on free markets and free trade. In practice, he did nothing for workers. His only significant legislative success was a massive tax cut for corporations and high-income earners. In executive actions he helped his corporate supporters, while threatening businesses who opposed him. Trump's crony capitalism is in no way progressive, but it bears little relationship to neoliberal aspirations for a free market economy in which capital is presumed to flow to its most valuable use, and government keeps out of the road.

The final shock came with the Covid-19 pandemic. Faced with a drastic economic shock, governments around the world responded with intervention on a massive scale. Dogmas about balanced budgets and the perils of public debt, already honoured more in the breach than the observance, were discarded overnight. The pandemic required massive intervention in labour markets, housing, financial markets, the health system and international trade. Some of these interventions will be unwound when the pandemic recedes, but not all. The system that emerges from the pandemic will undoubtedly be capitalist, and more or less inequitable, but it will not, in any meaningful sense be neoliberal.

The aim of this chapter is to describe the evolution of neoliberalism, beginning as an oppositional movement in the era of Keynesian social democracy and the Bretton Woods system, attaining dominance from the early 1970s onwards, and collapsing in the face of the crises of the twenty-first century. Prospects for the future will be considered briefly.

The chapter is organised as follows. Section 6.2 presents neoliberalism in relation to liberalism, the intellectual tradition from which social liberalism and social democracy are also primarily derived. Section 6.3 describes the decades of social democracy after 1945, and the rise of neoliberalism during the crisis of the 1970s. Section 6.4 describes the dominant 'hard' form of neoliberalism, embodied in the policies of the Thatcher government in the UK, and emulated by right-wing governments around the world. Section 6.5 focuses on 'soft neoliberalism', the modified version of neoliberalism adopted by social democratic parties in the 1980s and 1990s. Section 6.6 deals with the decline of neoliberalism, from the crises of the late 1990s, to the Global Financial Crisis of 2008. Section 6.7 deals with the political implications of the failure of neoliberalism. A 'three party' model is presented in which the main movements driving contemporary politics are categorised as neoliberal (soft or hard), Trumpist and leftist. In conclusion, the requirements for a successful left response to the crisis are discussed, and some suggestions offered for a way forward.

6.2 What is Neoliberalism?

The evolution of neoliberalism has been a complex process, exacerbated by disputes over the meaning, and even the legitimacy of the term, which has been criticised, mostly by defenders of neoliberal policies, as a meaningless all-purpose pejorative. This criticism is unsurprising. The need for a definition of an ideological movement is felt much more by its critics than by its supporters. Ideology always looks like common sense from the inside, and a dominant ideology just seems like 'what everyone knows'. For this reason, followers of a dominant ideology resist the use of any specific name or label to describe their position.

One result is that labels of this kind, including 'neoliberalism' are almost always used negatively, and as a result, overused to refer to anything the speaker dislikes. In a discussion of broader historical and political trends, however, 'neoliberalism' is too useful a term to forgo, despite the confusion that surrounds it. It is, instead, necessary to address some of the sources of this confusion.

As the name implies, neoliberalism is a descendant of liberalism, a tradition which, in various forms, underlies most contemporary streams of political thought. An obvious starting point is the greatest of nineteenth-century liberals, John Stuart Mill. Mill is best known today for his strong defence of free speech, his advocacy of equality for women and his steadfast opposition to slavery. Over the course of his lifetime he shifted his views on economic issues, from the free-market orthodoxy of classical economics to an abstract form of socialism (Mill 1848).

All of these aspects of Mill's thought are reflected in different streams of liberal thought. In North America, and most of the English-speaking world, the dominant version of liberalism was social liberalism, combining support for civil liberties with advocacy of a range of government policies to limit income inequality and alleviate poverty.

In Europe, from the early twentieth century onwards, social liberalism was largely subsumed by social democratic and labour movements. Here the term 'liberal' was more closely associated with the political right, and with free-market economic analysis, in which Millian concerns about free speech were less prominent. This tendency is sometimes referred to as 'classical liberalism'.

In the European context, the term 'neoliberalism' (and the closely related idea of 'ordoliberalism') was first used as early as 1938 to describe attempts to develop an updated version of classical liberalism capable of responding to the crisis created by the Great Depression and, after 1945, of competing with the social democratic welfare state (Hartwich 2009). The result was a 'hard' form of neoliberalism in which concessions to social democratic ideas were limited as far as possible. In the US context, where the term neoliberalism referred to the 'soft' variety, hard neoliberalism was most commonly described as 'free market conservatism' (Nell 1984).

Because it was based primarily on a critique of Keynesian social democracy, hard neoliberalism placed more weight on economic freedom than on personal freedom or civil liberties, reversing the emphasis of classical liberalism. On matters of personal freedom, hard neoliberalism is basically agnostic, encompassing a range of views from repressive traditionalism to libertarianism.

The term 'neoliberalism' embodies these ambiguities and has been coined at least twice, to apply to different, but closely related, developments within liberalism. I will distinguish these as 'hard' and 'soft' neoliberalism.

In the United States the term 'neoliberalism' was used from the 1990s onwards to describe a shift to the right by social liberals, particularly those termed 'New Democrats'[1]. Here the key ideas were a focus on 'sound' economic management, acceptance of financial deregulation and market-based policies more generally, and avoidance of class rhetoric. Both the Clinton and Obama Administrations were heavily influenced by this version of neoliberalism.

In one form or another 'soft' versions of neoliberalism came to dominate social liberal and social democratic parties throughout the English–speaking world. The term 'Third Way' was often applied to this development (Giddens 1999, 2000), with the suggestion that this approach represented a new development in social democracy rather than a capitulation to its opponents.

In terms of economic policy, neoliberalism is constrained by the need to compete with the achievements of social democracy especially the improvements the working class generally enjoyed in the decades after 1945. Hence, it was inconsistent with the kind of dogmatic libertarianism that would leave the poor to starvation or private charity, and would leave education to parents.

The need to compete with the appeal of social democracy has continued in the period of neoliberal dominance. As the result, despite extensive efforts at privatisation and deregulation, the importance of the public sector, as measured, for example, by the revenue share of national income[2], has remained largely unchanged.

---

[1] 'New Labour' in the UK was part of the same 'Third Way' movement. A cynic might observe, with reference to products like 'New Coke', that the attachment of the term New to an existing movement is rarely a good sign.

[2] It is more common to focus on Gross Domestic Product (GDP). However, the capacity of a society to meet its needs is more properly measured by national income. Quiggin (2014) discusses the distinction.

6.3 The Rise of Neoliberalism in the 1970s

In the decades after 1945, social democratic ideas were dominant throughout the developed world. Social democracy gained ascendency through programs like US president Roosevelt's 'New Deal' to counter the effect of the Great Depression. Whether or not social democratic parties held office, they drove the policy debate, to the extent that terms like 'progressive' inherently incorporated the notion of 'progress in the direction of more social democracy' (Quiggin 2003).

The starting point of twentieth–century social democracy was the combination of the welfare state, Keynesian macroeconomic stabilisation and the mixed economy. Their combined effect was to transform the lived experience of capitalist society.

The risks of falling into destitution as a result of unemployment, illness or old age – previously an ever-present reality for the great majority of workers – were eliminated almost completely by social security systems and, except in the United States, publicly provided healthcare. However, the trends that would permit the emergence of the neoliberal counter-revolution were already evident.

Keynesian macroeconomic policies produced decades of near-full employment. This was partly the result of the direct effects of fiscal policies designed to stimulate economic activity in times of recession and partly a self-fulfilling commitment. As long as governments were committed to maintaining full employment, businesses and workers could make plans consistent with that outcome.

The mixed economy used public ownership to deal with monopoly and other problems that had produced inequitable and unsatisfactory outcomes, while leaving competitive industries to private business. The result was improved performance in both sectors.

As environmental concerns developed, beginning with the 1952 Great Smog of London, and continuing with the emergence of chronic smog problems in Los Angeles and other cities, governments acted to control pollution and regulate industries. These actions were complemented by measures to preserve natural environments and promote clean water. Despite yielding substantial improvements in the quality of life, environmental policies and their advocates ('greenies' or 'enviros') were, and remain, hated by large sections of the political right.

The opponents of the social democratic and Keynesian analysis developed both an intellectual critique and a network of think tanks and sympathetic university departments. At a global level, the most prominent think tank was the Mont Pelerin Society, named for the Swiss resort where the first meeting, convened by Friedrich von Hayek, was held in 1947.

The thinking of the Mont Pelerin Society was most clearly represented by Hayek's (1944) *The Road to Serfdom*, which predicted that the policies of the British Labour Party and other social democratic European parties would lead inexorably to dictatorship. Hayek and his followers were unembarrassed by the failure of this prediction, and treated the neoliberal reaction against social democracy as a validation, even when (as in Chile) it was undertaken by a brutal dictatorship.

The Mont Pelerin Society gave rise to a plethora of think tanks in various countries, notably including the Centre for Independent Studies in Australia (Cahill and Beder 2005). Cockett (1995) gives an account of the role of neoliberal think tanks in the UK, the most notable of which was the Institute of Economic Affairs. The ideas developed by these think tanks formed the basis for the first systematic implementation of a neoliberal policy program, undertaken by the government of Margaret Thatcher in the United Kingdom from 1979 onwards.

The most important academic centres of intellectual resistance to social democracy were the University of Chicago and the University of Virginia. The Chicago critique, originated in the 1930s by Frank Knight, focused primarily on economic policy issues. The monetarist analysis developed by Milton Friedman was the most prominent alternative to Keynesian macroeconomics. Chicago economists also criticised interventions in particular markets, including anti-monopoly legislation, financial regulation and minimum wages. The advocacy of Friedman (1962) and also Friedman and Friedman (1980) played a crucial role in promoting free market conservatism in the United States. The University of Virginia critique focused on 'public choice', a model of political decision-making that emphasised the distorting effects of powerful interest groups. The lead figure here was James Buchanan, a student of Knight.

These groups were relatively marginal during the decades after 1945. But when the social democratic model ran into trouble, they were ready with both a critique and an alternative policy program. Both were welcomed by large sections of business, which, despite prospering during the long post-World War II boom, had been forced to make accommodations with governments and trade unions. The financial sector was particularly keen to break the shackles of regulation introduced after the Great Depression.

The last years of the social democratic era saw a struggle over income distribution that virtually guaranteed an inflationary outburst. Union militancy, fuelled in many countries by Marxist rhetoric, came into sharp conflict with an emerging speculative capitalism, driven by revived global financial markets. Firms raised prices to meet wage demands, spurring yet further wage demands to compensate for higher prices and maintain living standards (Brenner et al. 2010).

Previous episodes of inflation had been brought under control quite rapidly through Keynesian contractionary policies. Unfortunately, these policies were becoming less effective as inflationary expectations became embedded and as the social restraint generated by memories of the Great Depression broke down. The result was that fiscal contraction stifled economic growth and produced higher unemployment without substantially reducing inflation, producing the ugly portmanteau term 'stagflation'.

For a while it appeared that wage and price controls might work where Keynesian macroeconomics had failed. The Nixon Administration imposed a wage and price freeze which lasted for 90 days, followed by a system of wage and price controls.

The freeze and controls were imposed in response to the breakdown, in 1971, of the Bretton Woods system of fixed exchange rates which had been the basis of the international financial system within which Keynesian macroeconomics operated.

The key event was the abandonment of the convertibility of the US dollar into gold, at a price which had been fixed since 1944. Underlying this was the gradual development of international financial markets such as the Eurodollar market, operating outside the control of national monetary authorities[3]. The growth of international markets ensured that any attempt to impose national policies seen as too radical could be met with capital flight.

The coup de grace came with the oil shock of 1973, which was both a reflection of the inflationary outburst that was already under way and the cause of a further upsurge. The price and wage controls were abandoned for all commodities except gasoline. The resulting years of queues and shortages produced a false impression of the central importance of oil which has distorted US foreign policy ever since.[4]

Within a couple of years the entire edifice of post-war prosperity had collapsed and the Keynesian 'Golden Age' came to a painful and chaotic end. Meanwhile, the collapse of the Bretton Woods system led to a self-perpetuating cycle in which the rapid growth of international financial flows led to the breakdown of both domestic and international financial regulations, allowing yet further expansion of the financial sector.

In the decades following the breakdown of the Bretton Woods system, financial sector activity exploded, and was almost completely decoupled from any connection to real economic activity. By 2014, global financial market transactions totalled $US4 trillion each day — more than 20 times real economic activity. Trade in derivatives was even larger, with notional outstanding volumes of at least $500 trillion (Bank for International Settlements 2015).

With this growth in activity came hugely increased power and profitability. As early as 1987, financial firms were being described as the 'Masters of the Universe', a phrase coined in Tom Wolfe's (1987) novel, *The Bonfire of the Vanities*. The rise of the financial sector, and the concomitant decline of the trade union movement tipped the balance of political power in favour of neoliberalism.

6.4 Hard Neoliberalism

Although the rising political power of the financial markets facilitated the general shift towards neoliberalism, this development also depended on a resurgence of neoliberal policy ideas. In sharp contrast with social democrats, the advocates of neoliberalism were ready with answers to the stagflation crisis of the 1970s.

---

[3] Ironically, the first mover in the creation of the Eurodollar market (by most accounts) was the Moscow Narodny Bank, an arm of the government of the (then) Soviet Union.

[4] The ostensible cause of the oil crisis was an embargo imposed by oil-exporting countries in response to US support for Israel in the 1973 war with Arab states. However, the inflationary surge predated the embargo, and continued well after it ended. Price controls were doomed to fail regardless.

In macroeconomics, the monetarist critique of Keynesianism, developed by Friedman (1968) was rapidly accepted. Friedman opposed attempts to manage the economy through active fiscal policy but nonetheless conceded a significant role for government in stabilising the economy. In the course of the 1970s, Friedman's approach was superseded by more radical critiques of Keynesianism, including Rational Expectations, New Classical Macroeconomics and Real Business Cycle Theory. Although varying in details, and lacking in significant empirical support, these critiques had in common the view that recessions were either natural and desirable or the product of mistaken government policy, and that in neither case was policy intervention justified.

In policy terms, the immediate alternative to Keynesian macroeconomics was Friedman's model of targeting the growth rate of the money supply. This proved unworkable and eventually evolved into the 'inflation targeting' regime that remained in effect until the Global Financial Crisis.

In fiscal policy, the 'tax revolts' of the 1970s (exemplified by Proposition 13 in California) led to a consensus on the need to restrain the growth of government. The perceived success of airline deregulation in the United States led to a broader movement in favour of deregulation and privatisation. To emphasise the distinction from macroeconomic policy, these policies were frequently referred to as 'microeconomic reform' (Quiggin 1996). This phrase captured the positive connotations of 'reform', a term long used to describe liberal and social democratic policy innovations that assisted the disadvantaged, while reversing the substantive content.

In the UK, the core of the hard neoliberal program was:

> (i) to abandon Keynesian macroeconomic stabilisation, based on active fiscal policy, in favour of an independent central bank with a directive to control inflation at all costs; (ii) to remove the state altogether from 'non-core' functions through privatisation of government business enterprises and the sale of public housing; (iii) to reject redistribution of income except for a basic 'safety net'; (iv) to minimise the role of the state in core functions such as health, education and income security through contracting out, voucher schemes and so on.

Thatcher was most successful in the first and second of these goals. Although policies of monetary contraction implemented from 1979 produced a deep recession, with millions left unemployed, the government persisted, and succeeded in bringing an end to the inflationary upsurge of the 1970s. Most of the publicly owned infrastructure sector (electricity, water, telecommunications, airports and railways) was privatised.

Rather less progress was made on the third and fourth objectives. There were also substantial reductions in the progressivity of the tax system, but Thatcher's most ambitious move in this direction, the replacement of council rates by a poll tax, was a disaster, leading eventually to her downfall. The attempt to wind back public involvement in health and education was similarly limited, meeting particular resistance in the case of the National Health Service. The result was that, although the scope of public sector activity was wound back through privatisation, the size of government was not. The century-old trend of growth in the share of national income going to government to fund public services and income redistribution was halted in the 1970s, but not reversed.

Thatcher's policies formed the core of a 'hard' neoliberalism which rapidly became dominant throughout the English-speaking world, especially with the election of Ronald Reagan as US president in 1981. It was embodied in such documents as the *Fightback!* plan put forward by the Liberal and National parties in Australia in 1993 and the Contract with America proposed by the US Republican Party led by Newt Gingrich in 1994. Although neither of these programs produced immediate electoral success, most of the policies they proposed were eventually implemented.

Beyond the English-speaking world, the global spread of neoliberalism was driven less by political advocacy and more by international institutions. The debt crises of the 1990s produced what Williamson (1990) described as the Washington Consensus, a term that reflected the shared views of the US Treasury, the International Monetary Fund and the World Bank, all based in Washington DC.

Less remarked upon, but equally significant, was the Europe-based consensus of the OECD, the European Commission and the European Central Bank (ECB). The ECB, created in the 1990s as part of the political project of unifying Europe around a common currency, represents a particularly pure institutional embodiment of hard neoliberalism (Palley 2013).

The rise of neoliberal ideas reinforced, and was reinforced by, the resurgence of faith in the financial sector. The breakdown of social democracy in the 1970s saw the rise of financialised capitalism, dominated by global banks like Citibank and Wall Street investment banks such as Goldman Sachs.

The key idea here was the 'efficient (financial) markets hypothesis', In its strong form, put forward by Fama (1970) the hypothesis states that financial markets provide the best possible estimate of the value of any investment. Although there was never any good supporting evidence for this claim, it became part of the 'common sense' of the neoliberal era. One result was the trepidation with which governments awaited the verdict of 'the markets' on budgets and other policy decisions.

6.5 Soft Neoliberalism

The resurgence of a financialised form of global capitalism from the 1970s onwards came as a shock to the left. There were some attempts at resistance, notably by the Mitterrand government which came to office in France in 1981, but all such attempts failed in the face of the power of global capital markets[5]. By the 1990s, the triumphalist decade that followed the collapse of the Soviet Union, the dominance of

[5] The Mitterrand government experienced the flight of capital from France after 1981, and imposed a series of capital controls to attempt to halt the flight. But it was high US interest rates, in combination with the franc being pegged to the German mark (a precursor to the European common currency) that led Mitterrand to abandon his socialist agenda in 1983 (Mitchell and Fazi 2017, pp76–83).

market liberalism was clearly re-established. Over the course of the 1980s and 1990s, most social democratic parties accommodated to what were seen as inescapable realities.

As noted above, this accommodation was often presented as a new 'Third Way', allegedly transcending the dispute between social democrats and market liberals. In reality, however, the 'Third Way' amounted to little more than a 'soft' version of neoliberalism (Callinicos 2001). Soft neoliberalism involved acceptance of most of the core elements of the neoliberal program, including privatisation, attacks on trade unions, uncritical acceptance of the dominant role of the financial sector, and attempts to halt or reverse the growth of the public sector. On the other hand, soft neoliberals generally supported measures to protect the environment, and (within the limits of a market economy) to promote gender equality.

Over the course of the 1980s and 1990s, most social democratic and liberal parties in the English-speaking world adopted soft neoliberalism in one form or other. Some examples, significant in themselves, but also symbolic of the shift in economic thinking include:

- The decisions by the Hawke–Keating Labor government in 1983 and 1984 to float the Australian dollar and deregulate[6] the financial system.
- In the United States, the Clinton Administration's support for 'the end of welfare as we know it', demanded by the Gingrich-led Republican party in 1994. This measure was initially seen as successful, because its adverse effects were masked by the strong growth of the 1990s. After growth slowed in the 2000s, however, the absence of welfare support contributed substantially to the growth of poverty. Even more striking was increasing mortality rates among significant groups, such as middle-aged white Americans.
- In the United Kingdom, Tony Blair's creation of a 'New Labour' party, in which the socialist objective formerly stated in Clause IV of its Constitution was abandoned (Gani 2015). The crucial change was from:

> common ownership of the means of production, distribution, and exchange, and the best obtainable system of popular administration and control of each industry or service.

To:

> a dynamic economy, serving the public interest, in which the enterprise of the market and the rigour of competition are joined with the forces of partnership and co-operation to produce the wealth the nation needs.

The rejection of public ownership embodied in the New Labour platform was reflected in policies of privatisation and private ownership of public infrastructure. Along with faith in the efficiency of financial markets and acquiescence in, or support of, anti-union policies, support for privatisation represented one of the most

---

[6] In this, as in other contexts, 'deregulation' is something of a misnomer. Australian governments continued to be closely involved with the financial system, which was effectively bailed out in both 1991 and 2008. An extensive system of regulation remained in replace, but regulations restricting interest rates, the creation of new financial instruments and bank profits in general were removed.

consistent areas of agreement between soft and hard neoliberals. This support persisted despite outcomes that ranged from mediocre to disastrous. The Private Finance Initiative (FPI), pursued with vigour by the Blair government in the United Kingdom, was one of the most notable examples of failure. The PFI was an attempt to reduce public debt by allowing private consortiums to build and own public infrastructure such as schools and hospitals. It resulted in higher costs and lower quality services (Quiggin 2019).

Despite accepting the core elements of hard neoliberalism, soft neoliberals attempted, in various ways, to mitigate the growing inequality that inevitably resulted from the implementation of the neoliberal program. In particular, soft neoliberal governments attempted to improve the functioning of the social welfare system rather than eliminating it or stripping it down to a minimal 'safety net'. Nevertheless, the egalitarianism of traditional social democracy was abandoned with arguments about the distribution of income and access to community services being replaced by discussion of 'safety nets' or the efficient provision of services to 'customers'.

This shift was frequently expressed in terms of older debates about 'equality of opportunity' as opposed to 'equality of outcomes'. However, although the advocates of soft neoliberalism are particularly friendly to the upwardly mobile, the hostility to inherited privilege that characterised earlier advocates of equality of opportunity largely disappeared. Some supporters of soft neoliberalism took the argument to its logical conclusion, rejecting even the idea of equality of opportunity (Cavanagh 2003). This is at least consistent. In the presence of serious inequality of outcomes it is impossible to prevent parents from passing their advantages on to their children. Under these circumstances it is therefore impossible to achieve equality of opportunity.

More marked divisions between hard and soft neoliberalism arose in relation to social issues, particularly those related to multiculturalism, feminism and environmentalism. In the United States and Australia, the sharpness of these divisions, commonly referred to as 'culture wars', masked a substantial convergence on economic policy (Frank 2007).

The most successful implementation of soft neoliberalism was probably that of the Hawke–Keating Labor government in Australia between 1983 and 1996. The platform on which Labor was elected was an interventionist one, centred on the idea of an 'Accord' on prices and wages. The Accord was negotiated between the Labor party and the Australian Council of Trade Unions, of which Hawke had been a successful president, though it was hoped that a bargained consensus, incorporating business groups, could be achieved (Gruen and Grattan 1993).

The combination of financial market pressure and the spread of neoliberal ideas ensured that the government took a different direction, beginning with the decisions, in 1983 and 1984, to float the dollar and undertake substantial deregulation of financial markets. However, the success of the Accord in constraining wage growth and allowing a non-inflationary recovery from the recession of the early 1980s was an important countervailing force. A notable example was the failure of the hard neoliberal reform of the tax system favoured by the then Treasurer Paul Keating, based on the idea of using a goods and services tax (GST) to finance cuts in the top marginal rate of income tax. In the face of resistance from the union movement, and a lack of support from business, Prime Minister Bob Hawke rejected Keating's preferred option. Instead, the government implemented a reform program with substantial progressive elements such as a capital gains tax. The Hawke–Keating government also redesigned the social welfare system, integrating it with the tax system and maintaining or improving its progressive redistributive effects (Gruen and Grattan 1993).

The relative success of soft neoliberalism under Hawke and Keating was not sufficient to prevent growth in inequality over time, or the development of a bloated and dangerously unstable financial system. Nevertheless, a measure of their success can be gained by looking at the disastrous performance of New Zealand, where both the Labour government elected in 1983 and the National Party government that succeeded it from 1990 to 1999 embraced hard neoliberalism in a particularly doctrinaire form, and with substantially worse economic outcomes (Hazledine and Quiggin 2006).

So, while not as significant as claimed by advocates of the Third Way, the differences between 'hard' and 'soft' neoliberalism were more than cosmetic and cultural.

Neoliberalism reached its peak of political and economic success in the 1990s. Neoliberal globalisation was seen as an unstoppable force, both by enthusiasts (Friedman 1999) and by critics (Martin and Schumann 1997). Social democratic parties were in retreat throughout the world. Financial markets were booming, in developed and developing countries alike. The collapse of the Soviet bloc had finally discredited the alternative offered by communism.

6.6 The Crises of the Twenty-First Century

Amid the triumphalism of the 1990s, many predicted that the twenty-first century would be one of hyper-capitalist prosperity. The results have turned out very differently.

The century began with a spectacular crash: the bursting of the US stock market bubble, focused on 'dotcom' Internet stocks, in 2000. The 'dotcom' bubble and bust was notable because it occurred at the centre of global capitalism, rather than in peripheral countries where financial capitalism was a recent arrival, as had been the case with the Asian financial crisis of 1997 and earlier emerging market crises.

Expansionary monetary policy allowed a reasonably rapid recovery from the dotcom bust, but fuelled a speculative boom, focused on real estate, that culminated in the Global Financial Crisis of 2008. Far from being the omniscient masters of the universe, the financial sector was shown up as corrupt and incompetent. As well as bringing an end to the widely shared prosperity of the 1990s, the stock market collapse undermined the central tenet of neoliberalism, namely, the efficient (financial) markets hypothesis (Quiggin 2010).

The GFC was followed by years of austerity, in which the costs of bailing out the financial sector were borne by the population as a whole, in the form of wage stagnation and reduced public services. Bankers and the financial system were bailed out, while ordinary people were made to pay the price. The situation was worst in the Eurozone, where the design of the European Central Bank (ECB) made it virtually impossible to adopt any policy except ‘austerity’, a counterproductive focus on cutting budget deficits and controlling the non-existent threat of inflation. The result has been a decade of depression in most of the developed world. Even in the United States and United Kingdom, which have, on some measures, recovered, living standards have never returned to the previous growth path, and the inequality of wealth and income has been more evident.

Despite its death as a credible theory of economics and politics, however, neoliberalism has stumbled on in zombie form for nearly a decade, maintaining its hold over major political parties and over organisations such as the OECD, the International Monetary Fund and the European Commission. The economics profession as a whole has learned almost nothing from the Global Financial Crisis (Quiggin 2013). Ideas like austerity that should have been decently buried long ago continue to wreak havoc throughout the world, and most notably in Europe (Blyth 2012).

Finally, in early 2020, just as the long and grinding recovery from the GFC seemed to be nearing completion, the Covid-19 pandemic smashed the global economy. The management of the pandemic varied from competent to disastrous, but it depended almost entirely on governments. The financial sector played virtually no role on the management of the pandemic, but came off unscathed as usual.

6.7 Political Implications

Just as the economic ideology of neoliberalism lumbers on in zombie form, so, until recently has the political system it supported. Insurgents of various kinds have gained support nearly everywhere, but the alternation between different versions of neoliberalism continued for nearly a decade after the GFC, but arguably came to an end with the Brexit referendum of 2016 and the election of US President Trump in the same year.

During the decades of neoliberalism that began in the 1970s, the political system, nearly everywhere, was based on electoral competition between the hard and soft versions of neoliberalism, typically represented by (nominally) conservative and social democratic parties respectively. Within the political class, and among business leaders and policymakers, there was a near-universal consensus in support of neoliberal ideas. To take any position outside the spectrum defined by the soft and hard variants of neoliberalism guaranteed marginalisation and exclusion from serious political debate.

Yet, despite its dominance, neoliberalism hardly ever achieved broad support among the public at large. Rather, the seeming electoral success of neoliberalism concealed the continued strength of currents that remained submerged for decades, becoming politically significant only in occasional eruptions. The failure of neoliberalism has brought these submerged currents to the surface. The result is a 'three party' form of politics in which neoliberals (soft and hard) must compete with a revived political left and an authoritarian populist movement which may be called Trumpism, after its most successful exponent (so far).

Many of the elements of Trumpism are revivals of past forms of right-wing authoritarianism, such as fascism and related movements in the first half of the twentieth century. Notable examples are uncritical leader-worship and xenophobic nationalism. Nevertheless, because it has arisen in a different context, Trumpism has unique features.

The core of Trumpism is a form of negative or default identity politics best understood using the idea of the 'unmarked category', from linguistics and sociology. Campos (2019) illustrates:

> An unmarked category is present when the category is considered so normal or ordinary in a particular context that it goes unnoticed. The category is the default setting in regard to social expectations, and it in a sense remains invisible precisely because it's so dominant ... For example, if you had asked a lawyer in 1960 to name three characteristics that every current Supreme Court justice shared, it's very likely the lawyer would not have mentioned either race or gender.

Campos observes:

> What "identity politics," so-called, has done is to slowly and painfully and partially transform being a white man in America into a marked category. And makes a lot of the people who have become white men rather than members of society's invisible default category very uncomfortable.

We may usefully add the unmarked categories 'Christian' where 'Christian' is interpreted in a sense of cultural identification rather than any specific religious belief, and heterosexual, among others.

The most politically potent form of default identity politics and the relevant one here, is that of a formerly unchallenged dominant group facing the real or perceived prospect of becoming a politically weak and economically declining minority.

Trumpist default identity politics is closely linked to support for 'strongman' rule, exemplified by current practitioners such as Erdogan, Orban, Putin and, of course, Trump himself. This may be due to the fact that the one universal feature of default identity politics is a worship of masculinity. Theatrical displays of masculinity are a standard feature of Trumpist politics (Trump himself is absurd enough but Putin's bare-chested horseback rides take these displays to the point of parody and beyond).

During the period of neoliberal dominance, hard neoliberals pandered to Trumpists in rhetorical terms, but ignored them where their views conflicted with the neoliberal policy agenda, most obviously in relation to globalisation. The neoliberal world was one without borders, where the wealthy and powerful moved freely, regardless of race, gender sexuality or (to the extent that any of them took it seriously) religion. But they relied on the votes of people who could be mobilised by the claim that their dominant racial or religious identity, and the gender roles underpinned by this identity, was under threat from cosmopolitan 'elites'.

Now the balance has been reversed. The Trumpists are in charge of right-wing parties, most notably the US Republican party, but use tax cuts, deregulation and handouts to favoured cronies to keep business support.

Opposed to the Trumpists in critical ways, but similar in others, is a disparate group that may be called, for want of a better term 'the left'. As well as a small group who adhere to Marxist or other radical critiques of capitalism, the 'left' in this sense includes environmentalists, feminists, unionists, old-style liberals and social democrats, and a wide variety of groups whose personal or cultural identity is threatened by Trumpism.

Although the left has not been as successful as the Trumpists in political terms, the issues raised by the left, including inequality, racial injustice, gender roles and global heating have now replaced the concerns of neoliberalism, such as markets and competition, at the centre of the policy agenda for centre-left parties. Whereas they formerly claimed to offer a more humane version of neoliberalism, the leaders of the centre-left are now seeking a compromise between an unsatisfactory status quo and the policy program of the left.

An indication of how things have changed can be gained by comparing the policy program of the incoming Biden Administration with that of Barack Obama, in which Biden served as vice-president. In healthcare for example, Obama's Affordable Care Act was based on policy ideas developed by Mitt Romney when he as the Republican governor of Massachusetts, and designed in the futile hope of attracting bipartisan support. Obama rejected the idea of a 'public option' as an alternative to private health insurance.

Leftist candidates for the Democratic presidential nomination, such as Sanders and Warren, proposed a comprehensive single-payer scheme, Medicare for All. Biden has resisted this, but adopted a strong version of the public option in which those without insurance would be automatically enrolled.

Similarly, Obama proposed to raise the US federal minimum wage from $7.25/hour to $9.00/hour, but did little to push the idea. Campaigns at state and local level have pushed the demand for a minimum wage of $15/hour with considerable success. Biden has adopted the same policy at a national level and seems likely to pursue it.

Biden has likewise threaded the needle on the left demand for a Green New Deal, saying it "is a crucial framework for meeting the climate challenges we face", while rejecting or vacillating on the more radical components.

On the other side of the ledger, soft neoliberal proposals that were central to Obama's thinking have disappeared. Whereas Obama and his education secretary Arne Duncan backed charter schools, Biden has promised a complete ban on funding of for-profit charters, and has said nothing positive about the kinds of school reform favoured by Obama. Obama's talk of 'reinventing government' has no echo in Biden's electoral rhetoric.

The other big success of global Trumpism has been Brexit, driven by the demand for restrictions on immigration. Hard neoliberals have long been willing to exploit hostility to migration to attract the electoral support they need to govern in support of financialised capitalism. Having distracted their supporters with threats of being swamped by foreigners, they have been free to remove all obstacles to the operations of global financial institutions, and trade in goods and services.

Brexit turned the tables. In order to control migration from the EU, the advocates of Brexit have been willing to abandon capital market integration and unrestricted trade in goods and services with the EU in order to impose restrictions on immigration. Indeed, the final deal created trade barriers within the UK, between Northern Ireland and mainland Britain. One of the last issues in dispute was the UK demand to be able to offer state aid to corporations without complying with EU restrictions. The same developments can be seen in purely domestic policies. Archetypal neoliberal policies such as the Private Finance Initiative have been abandoned,

On the Labour side of politics, the left made big gains under Jeremy Corbyn, but has been in retreat since the 2019 election defeat. Nevertheless, while new Labour leader Keir Starmer has abandoned much of Corbyn's socialist rhetoric, he has maintained support for a substantial program of renationalisation and floated the idea of a once-off wealth tax to offset the fiscal cost of the Covid-19 pandemic. Even the right wing of the Labour party is eager to dissociate itself from the soft neoliberalism of Tony Blair, with the result that the term 'Blairite', which once had positive connotations, is now used solely as a pejorative.

6.8 Concluding Comments: The Way Forward

The failure of neoliberalism poses both challenges and opportunities for the left. The greatest challenge is the need to confront Trumpism as a powerful political force in itself, rather than as a source of political support for hard neoliberalism. Given the dangers posed by Trumpism this is an urgent task. One part of this task is that of articulating an explanation of the failure of neoliberalism and explaining why the simplistic policy responses of Trumpist politicians will do nothing to resolve the problems. The other is to present a progressive alternative to the accommodations of soft neoliberalism.

The core of such an alternative must be a revival of the egalitarian and activist politics of the post-World War II social democratic moment, updated to take account of the radically different technological and social structures of the twenty-first century, and of the urgent challenges posed by the global crisis of the Covid-19 pandemic and the even greater crisis of global heating. The demand for a Green New Deal is an important step in this direction. It links the urgent need for a decarbonised and sustainable economy, with a renewed demand for economic justice, drawing on the best traditions of the left.

In technological terms, the most important development is undoubtedly the rise of the Internet. Thinking about the relationship between the Internet economy and public policy remains embryonic at best. But as a massive public good created, in very large measure, by the public sector, the Internet ought to present opportunities for a radically remodelled progressive policy agenda.

In political terms, the breakdown of neoliberalism implies the need for a political realignment. This has now taken place on the right, as Trumpists assert their dominance over hard neoliberals. The most promising strategy for the left is to achieve a similar shift in power within the centre-left coalition of leftists and soft neoliberals.

The era of unchallenged neoliberal dominance is clearly over. Hopefully, it will prove to have been a relatively brief interruption in a long-term trend towards a more humane, egalitarian and sustainable society. Whether that is true depends on the success of the left in putting forward a positive alternative.

Acknowledgements I thank Nancy Wallace for helpful comments and criticism.

References

Bank for International Settlements (BIS) (2015) Semiannual Over-The-Counter (OTC) derivatives markets statistics, http://www.bis.org/statistics/derstats.htm

Blyth M (2012) Austerity: the history of a dangerous idea. Oxford University Press, Oxford

Brenner A, Brenner R, Winslow C (2010) Rebel rank and file: labor militancy and the revolt from Below during the long 1970s. Verso, London

Cahill D, Beder S (2005) Neo-liberal think tanks and neo-liberal restructuring: learning the lessons from project Victoria and the privatisation of Victoria's electricity industry. Soc Altern 24:43–48

Callinicos A (2001) Against the third way: an anti-capitalist critique. Polity, Cambridge

Campos P (2019) Marking the unmarked category, Lawyers, Guns and Money, https://www.lawyersgunsmoneyblog.com/2019/03/marking-unmarked-category

Cavanagh M (2003) Against equality of opportunity, Oxford Philosophical Monographs. Oxford University Press, Oxford

Cockett R (1995) Thinking the unthinkable: think-tanks and the economic counter-revolution 1931–1983. HarperCollins, New York

Fama E (1970) Efficient capital markets: a review of theory and empirical work. J Financ 25:383–417

Frank T (2007) What's the matter with Kansas?: how conservatives won the heart of America. Macmillan

Friedman M (1962) Capitalism and freedom. University of Chicago Press, Chicago

Friedman M (1968) The role of monetary policy. Am Econ Rev 58:1–17

Friedman T (1999) The Lexus and the Olive Tree: understanding globalization. Farrar Strauss Giroux, New York

Friedman M, Friedman R (1980) Free to choose. Harcourt Brace Jovanovich, New York

Gani A (2015) Clause IV: a brief history. The Guardian. https://www.theguardian.com/politics/2015/aug/09/clause-iv-of-labour-party-constitution-what-is-all-the-fuss-about-reinstating-it. Accessed 18 Jan 2021

Giddens A (1999) The third way: the renewal of social democracy. Blackwell, London

Giddens A (2000) The third way and its critics. Polity Press, Cambridge

Gruen F, Grattan M (1993) Managing government: labor's achievements and failures. Longman Cheshire, Melbourne

Hartwich O (2009) Neoliberalism: the genesis of a political Swearword, CIS Occasional Paper 114

Hayek F (1944) The road to Serfdom. University of Chicago Press, Chicago
Hazledine T, Quiggin J (2006) No more free beer tomorrow? Economic policy and outcomes in Australia and New Zealand since 1984. Aust J Polit Sci 41:145–159
Martin H-P, Schumann H (1997) The global trap: globalization and the assault on prosperity and democracy. St Martin's Press, New York
Mill JS (1848) Principles of political economy. John W. Parker, London
Mitchell W, Fazi T (2017) Reclaiming the State: a progressive vision of sovereignty for a post-neoliberal world. Pluto, London
Nell E (1984, republished 2009) Free market conservatism (Routledge Revivals): a critique of theory & practice. Routledge, Abingdon
Palley TI (2013) Europe's crisis without end: the consequences of neoliberalism. Contrib Polit Econ 32:29–50
Quiggin J (1996) Great expectations: microeconomic reform and australia. Allen & Unwin, St Leonards
Quiggin J (2003) Word for Wednesday: progressive, http://johnquiggin.com/2003/01/29/word-for-wednesday-2/
Quiggin J (2010) Zombie economics: how dead ideas still walk among us. Princeton University Press, Princeton, NJ
Quiggin J (2013) The state of economics in 2012: complacency amid crisis. Econ Rec 89:23–30
Quiggin J (2014) National accounting and the digital economy. Econ Anal Policy 44:136–142
Quiggin J (2019) The diffusion of public private partnerships: a world systems analysis. Globalizations 16:838–856
Williamson J (1990) What Washington means by policy reform. In: Williamson J (ed) Latin American adjustment: how much has happened? Institute for International Economics, Washington, DC
Wolfe T (1987) The bonfire of the vanities. Farrar, New York

Chapter 7
Population Growth

Ian Lowe

Abstract The first human populations were, like those of all other species, forced to live within the limits of the planet's ecological systems. Since the Industrial Revolution, technologies have allowed a dramatic increase in the population. The demands of humans are now measurably altering planetary systems, most obviously by changing the global climate and by driving a catastrophic loss of biodiversity. While the United Nations has embraced Sustainable Development Goals and many countries have stated commitments to development patterns that could be sustainable, progress toward those goals has been impeded by a widespread failure to recognise the crucial importance of stabilising our population. Unless we develop socially acceptable strategies for limiting population growth, human civilisation faces a bleak future.

7.1 Introduction

Throughout most of the history of humans as an identifiably separate species, populations were limited by the availability of resources. Diamond (2005) argues that societies tended to expand until they reached limits; those that recognised the limits and adapted their practices survived as stable societies, while some persisted with their previous approach and collapsed. A famous example is the settlement established on Greenland by the Vikings during the medieval warm period that collapsed during the so-called Little Ice Age (about 1500–1850). We know that the first Australians had not just customs but a system of laws that allowed them to live within the limits of their harsh and unforgiving continent. It is an issue that is still in dispute, but some think these laws may have been developed in response to the loss of the megafauna which were on the Australian continent when the first humans arrived (Slatre et al. 2019; Hocknull et al. 2020). While the proximate cause of the

I. Lowe (✉)
Griffith University, Brisbane, QLD, Australia
e-mail: i.lowe@griffith.edu.au

S. J. Williams, R. Taylor (eds.), *Sustainability and the New Economics*,
https://doi.org/10.1007/978-3-030-78795-0_7

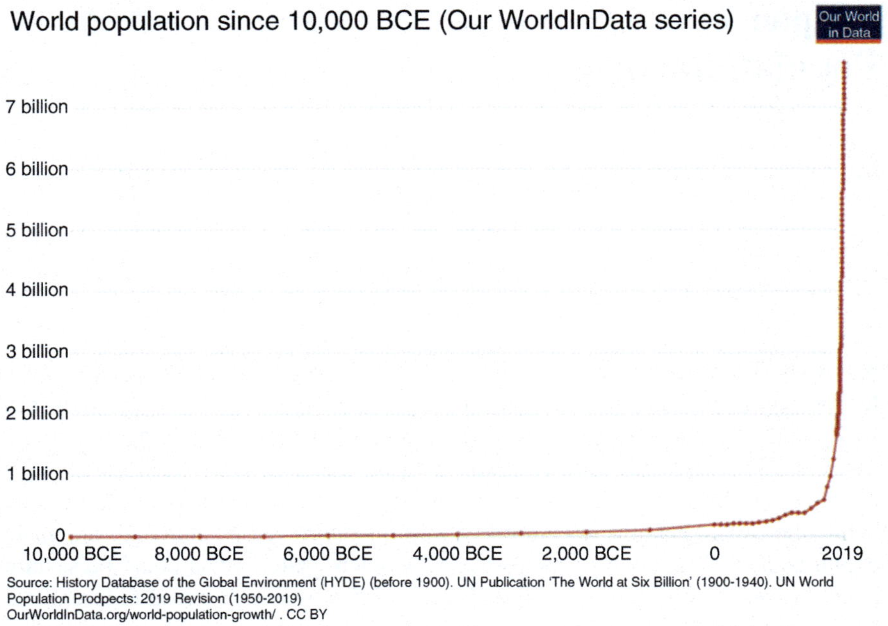

Fig. 7.1 World population since 10,000 BCE to 2019

loss of those amazing species was climate change, the humans would have been seriously affected by their disappearance, so it would have been logical for the thoughtful ones to have proposed measures to protect the remining food supply.

Until the Industrial Revolution, global settlements that survived did so by using what could be sustainably produced locally or traded from neighbouring societies. When the Little Ice Age reduced their capacity to grow crops and feed animals, the settlement on Iceland turned to obtaining food from the sea and trading dried fish for the vegetables and fruit they could no longer grow (Zorich 2019). The development of technology to utilise coal and steam engines provided extra resources and began the process of freeing societies from limitations of local capacity to produce food and water or manage waste. This began the exponential increase in human population, a process that now threatens the very survival of civilisation (Fig. 7.1).

7.2 Expanding the Resource Base

In 1798 (when the global population was about 1 billion) English economist Thomas Malthus famously speculated that the human population was growing faster than it would be possible to expand food production, arguing that the inevitable result would be famine and disease in turn limiting population growth. It was the first discussion of the possible limitations of population growth, which had traditionally been encouraged in Jewish and Christian societies by the Biblical injunction to "be fruitful and multiply", with its implicit assurance that natural systems and other living things were at our disposal. The 250 years since the beginning of the Industrial Revolution has seen the human population grow rapidly, decoupled from the traditional ecological limits by using fossil resources. In particular, the entire contemporary system for producing, processing and distributing food uses massive quantities of petroleum fuels. As a direct result, food prices have for decades risen and fallen in lock-step with the price of oil (Esmaeli and Shokooi 2011). It is no exaggeration to say that the huge oil deposits of the Middle East have enabled food production to increase greatly since 1945 – the so-called Green Revolution – in turn allowing a much larger global population to be supported.

The two world wars and the 1918 influenza pandemic caused significant losses of population, but the overall human numbers continued to grow. Most societies saw growth as desirable, since historically wars had tended to be won by more populous states that could mobilise larger armies. In the Australia of the 1940s and 1950s, there was a widespread belief that a larger population would be needed to defend the continent: 'populate or perish' was the slogan. Some perceptive critics drew attention to the fallacy underlying the argument that Australia needed to populate its 'empty north' to prevent it being over-run by invading hordes. Cyril Pearl pointed out in a 1961 ABC radio talk that the north of the continent had been relatively empty, and islands like Java densely populated, for centuries before Europeans invaded Australia. He argued that there were sound geographical reasons why Java, with its rich, deep volcanic soils, supported a large population and Arnhem Land in Australia's north, with its weathered nutrient-poor land, did not. He went on to say that believing northern Australia needed a bigger population to prevent it being invaded was a bit like Algeria worrying that the Sahara Desert would be over-run if they didn't spread into it.

7.3 Early Environmental Concerns

When it became apparent in the 1960s that the ever-expanding demands of the human population was damaging natural systems, environmental scientists and activists started to raise their concerns. Ehrlich (1968) warned about a 'population bomb', spurring a political movement advocating zero population growth: ZPG. Since contraception is the most effective way of slowing growth, the argument was resisted strongly by the Catholic Church and more moderately by the other Abrahamic religions, Judaism and Islam. In Western societies, the development and subsequent broad availability of the contraceptive pill had a dramatic impact on birth-rates, despite the views of religious leaders. Even solidly Catholic countries like Italy have seen the average number of children per adult woman drop rapidly from about 4 to 1.29 in 2018 (World Bank 2020), revealing that the Church's teachings are being widely ignored. Population in most affluent countries has essentially stabilised, while it is slowly declining in several including Italy and

Japan (Engelman 2012). The small number of OECD nations which still have rapidly growing population are those receiving large numbers of immigrants: the USA, Canada, the UK, New Zealand, and Australia. Much of the slower population growth in western Europe is also a result of migration, often unwanted, from Africa and the Middle East. It is worth noting that populations can still be growing even if the birth rate is below the replacement level; in the specific case of Australia, with a fertility rate of about 1.75 births per women, the annual "natural increase" – births minus deaths – has been about 150,000 in recent years, because the number of women in the reproductive age range is still increasing and the population enjoys long life expectancy (Lowe 2012). This is usually referred to as demographic momentum.

Since the overall size of a country's economy is influenced by the number of people and their production-consumption patterns, economists, usually intent on economic expansion, tend to be in favour of population growth. All other things being equal, if the number of people living in a country increases by 5%, there will be a corresponding increase in the amount of clothing needed, the amount of food consumed and so on, increasing the amount of economic activity by about 5%[1]. Steady levels of migration had been supported by successive Australian governments since World War II, but in the late 1990s the Howard government dramatically increased immigration levels as well as introducing a 'baby bonus' – a financial incentive for Australian women to have more children. These measures significantly increased the rate of overall population growth in Australia, from about 250,000 a year to almost double that level (van Onselen 2014). Governments have supported rapid population growth because it produces more economic activity and gives an illusion of progress, a cynical exercise that has been criticised by political insiders (van Onselen 2018). Surveys of community opinion have shown a marked reduction in support for large-scale immigration, probably because those living in Australian cities correctly perceive that infrastructure is not keeping pace with the increases in population and consequent demand for public transport, road space, recreational facilities etc (O'Sullivan 2012). As this chapter was being written, there was a rare break from the usual mindless support of population growth by economic analysts. Leading finance presenter Alan Kohler showed a graph in his regular spot on the national evening news in June 2020. It was in the context of comments that the 2020 recession, attributed to the catastrophic 2019–20 bushfires and the subsequent COVID-19 pandemic, had ended 29 years of continuous economic growth in Australia. Plotting GDP per capita for the 24 years from 1995–96 to 2019–20 and smoothing annual fluctuations showed a consistent downward trend. Long-term comparisons are complicated by changing exchange rates, but there is no doubt that real GDP per capita in Australia has declined in recent years (CEIC 2020). While the rapid population growth resulting from the Australian government's

[1] In practice, Gross Domestic Product tends to increase more than population growth, giving an increase in GDP per capita. Globally, Gross World Product since 1960 has typically grown between 5% and 2% per year; global population growth peaked in the early 1960s at 2.2% per year, and is now just above 1%.

immigration targets had caused the economy to expand, it had not grown as fast as the population. In so far as GDP per capita is a measure of prosperity, wealth per person was slowly but surely shrinking. That means the decline in the quality of life experienced in Australian cities was not being offset by increasing wealth, but compounded by declining average incomes. The social problem has been exacerbated by growing income inequality, which reached a peak during the Global Financial Crisis after steadily increasing since 1980 and has since stabilised at the 2007 level (ACOSS 2020), so many people would have experienced a lowering of their material standards.

7.4 The Global Problem

At the global level, the issue of the demands of the growing population was put squarely on the agenda in 1972 by the first report to the Club of Rome, *The Limits to Growth* (Meadows et al. 1972, see Chap. 8). The title could have been designed to antagonise the politicians and economists who do not believe there are any limits to growth. The report provoked a storm of criticism, much of it attacking a straw man. In a dishonestly simplified version of the argument, opponents claimed that the report said we would run out of resources or be choked by waste. The report was enabled by the rapid development of computer systems, allowing the first ever global systems model to analyse the long-term consequences of the sort of growth which had happened since the end of World War II. It showed that if the trends of growth in population, resource use, industrial output, agricultural production and pollution were all to continue, the world would reach limits within a hundred years, i.e. by 2070. The most likely result of those trends continuing, it said, would be a collapse of environmental, social and economic systems starting about 60 years from the date of the report, so by about 2030. That was far enough away for it to be possible for elected politicians to ignore the uncomfortable conclusions. They totally ignored the report's second conclusion, that these trends were not inevitable and it would be possible to redirect human development onto a path that would be sustainable into the distant future, allowing everyone on Earth to live satisfying lives. This idea was studiously ignored because the third conclusion was that such an approach would require governments to take deliberate policy action to steer the path of development. The 1970s was a period of economic upheaval resulting from the OPEC oil crisis when prices rose from less than $US2 a barrel to over $US35 (Statista 2020). The wave of inflation that spread through the economies of affluent countries was used by the so-called Chicago School of economists to spread their neoliberal ideology: in hindsight, a great leap backwards to pre-Keynesian economics with a naïve faith in the magic of market forces (see Chap. 6). With that ideology dominant in most advanced economies, there was no chance for the sort of purposive action that could have enabled a sustainable future. As a result, recent analysis shows the growth trends that were analysed 50 years ago have steadily continued, leaving the world on track for social, economic and ecological collapse in the near future (Turner 2014). As this chapter was being written, climate change was accelerating with disastrous potential consequences for the supply of food and water, while the COVID-19 pandemic was being interpreted as a direct consequence of the population stretching the capacity of natural systems to meet our needs (IPBS 2020).

Consideration of population growth as a problem was arguably shunted aside by the 1987 report of the World Commission on Environment and Development. The former Prime Minister of Norway and later head of the World Health Organisation, Gro Harlem Brundtland, put the notion of sustainable development into the conversation (WCED 1987). Her report argued that it would still be possible to shape a trajectory of economic development that would lift poor countries out of poverty within the limits of natural systems.[2] The Australian government led by prime minister Bob Hawke subsequently set up a process of nine working groups that culminated in the Council of Australian Governments adopting a National Strategy for Ecologically Sustainable Development (CoAG 1992). I was a member of one of the working groups and have on my bookshelf copies of both the national strategy and the compendium of recommendations. Both still read as forward-looking plans for a better future. The national strategy was meant to be a commitment by all Australian governments to a path of economic progress that would not reduce opportunities for future generations; that would strive for equity within and between generations; and that would recognise the global dimensions of their actions. It also included specific principles of protecting Australia's unique biodiversity and maintaining the integrity of natural systems. It would be difficult for even a generous observer to detect any sign that recent Australian governments, either progressive or conservative, even know there was – and still could be – a national strategy for sustainable development. It is certainly not seen as a framework for the day-to-day decisions about population, consumption or the provision of infrastructure, even though Australia is a party to the UN Sustainable Development Goals; reports document the country's systematic failure to approach those goals (National Sustainable Development Council 2020).

In 1992, a group of researchers published The World Scientists' Warning to Humanity (Ripple et al. 1992), drawing attention to the increasingly serious environmental consequences of the increasing demands of the growing human population. The warning had no discernible impact on decision-makers. Twenty-five years later, the same group coordinated a second warning (Ripple et al. 2017) which was signed by more than 21,000 researchers around the world. It identified one success story: the reduction in the release of ozone-depleting chemicals had stopped the destruction of the ozone layer, which has now stabilised at a lower level than in pre-industrial times. All the other environmental problems identified in the 1992 report, it observed, had got significantly worse. In the period from 1992 to 2017, the world lost more than 100 million hectares of forest; the fish catch fell

[2] Such thinking is still reflected in the United Nations Sustainable Development Goals, which champion economic growth for all (see Chap. 5).

about 20%; fresh water per person declined about 25%; and releases of greenhouse gases continued to increase, as a result of which climate change accelerated. Most worryingly, species abundance, which in 1992 was only 60% of the 1970 level, was down to 40% of the 1970 level by 2017. The report was clear that all of these problems were the direct result of an expanding human population and steadily increasing per capita consumption, putting compounding pressure on natural systems. While optimists have drawn attention to a slowing of the annual percentage increase in the human population, the annual increase is still more than 80 million a year (Lianos 2020) because of the high base population (7.8 billion in late 2020). The report's stark warning was:

> Humanity is now being given a second notice, as illustrated by these alarming trends. We are jeopardising our future by not reining in our intense but uneven material consumption and by not perceiving continued rapid population growth as a primary driver behind many ecological and even societal threats.

It continued to argue that time is running out to redirect the trajectory of human development away from the disastrous path we are still following, which will inevitably result in "widespread human misery and catastrophic biodiversity loss". This is restating a warning that has been given for decades (Ehrlich et al. 1977; Crist and Cafaro 2012). The UN's population projections – with low, medium and high variants – reflect the fact that the amount of future population growth is not inevitable, but largely dependent on policy decisions (UN 2019). The medium variant predicts 9.7 billion people by 2050 and 10.9 billion by 2100. As a specific example of the impact of policy decisions, some 270 million women still have an unmet need for modern contraception (WHO 2020). Indeed, many poor countries with high population growth have policies to ameliorate the increase, but need assistance from wealthy countries (UN 2017).

7.5 Population or Consumption?

Any questioning of population growth inevitably raises the issue of consumption as a driver of ecological harm. It has been argued for decades that the impact of humans on natural systems is the product of our number and our individual demands. Ehrlich and Holdren (1971) expressed the relationship as the equation I = PAT: **I**mpact equals **P**opulation times **A**ffluence times **T**echnology. That is undoubtedly true. But focusing on overconsumption as the main driver of ecological harm deflects political attention from the issue of overpopulation. In the final analysis, most elected politicians in the Western world are equally unwilling to canvass serious action to actively reduce consumption, even when the methods being proposed could be argued to have obvious benefits to individuals. One concrete example was the National Framework for Energy Efficiency (Australian Government 2003). It analysed energy use and concluded that energy-related greenhouse gas emissions could be reduced by 30%, simply by applying existing technology more generally, with payback times of less than 4 years. Measures such as introducing the minimum appliance efficiency standards of the European Union would save consumers money as well as reducing the environmental impacts of running refrigerators, freezers and washing machines, but there has been no systematic move in that direction (Harrington and Wilkenfeld 1997). Australia is one of the few OECD nations with no vehicle-efficiency standards, as a result of which the average fuel efficiency of the fleet is no better than it was when figures were first collected in 1963 (PTUA 2019). At a time when the size of the average household is shrinking, with the majority now consisting of one or two people, the scale of new dwellings steadily increased until the average size of new houses in Australia peaked in 2009 at more than 240 square metres (Buildsearch 2020). The increasing size of houses means that increasing amounts of resources are needed for their construction and operation.

In the age of neoliberal economics, with governments paralysed by their ideological reluctance to regulate in the community's interest, there seems little prospect of serious measures to restrain consumption. That means that the demands of the population will increase in direct proportion to population growth. The fundamental problem is, as discussed earlier, the present demands of the human population are already exceeding the limits of natural systems, so almost all the significant environmental indicators are in decline. A systematic comparison of present human demands with planetary limits confirms the seriousness of the situation (Rockstrom et al. 2009; Wijkman and Rockstrom 2012). The loss of biodiversity is the worst consequence, because there is no practical way to reverse extinction of species. Our understanding of complex ecological systems does not allow us to predict with any confidence the results when a species disappears, but there are obviously consequences up and down the food chain. As we stand by and watch accelerating loss of species, we are effectively pulling random bricks out of the wall of life, with no understanding of the risks that whole sections will collapse. In the short term, however, climate change is the urgent priority. It is not just a serious problem in itself; it is also an additional driving force adding to those already causing the loss of species: destruction of habitat, introduced species and chemical pollution. Of course, those driving forces are all directly proportional to the scale of the human population. It follows that any Green New Deal that seeks to move society towards sustainability must address the unsustainable nature of endless growth – of both populations and resource 'throughput' (Daly 1990; von Weizsäcker et al. 2009). The fundamental reason most politicians favour both growing population and increasing production-consumption is that they think the choice is between growth and recession. It needs to be recognised that the idea that total scale of measured economic activity (GDP) is an indicator of prosperity has been widely discredited; it does not measure much productive economic activity such as household work and unpaid care, while adding the costs of accidents, violence and crime. The emerging ideas of degrowth and steady state economics offer the chance of futures that could be sustainable (Quiggin 2014; Victor 2008; Jackson 2009; Czech 2013). The fundamental point is that all countries, whether rich or poor, should be aiming to develop along paths that could be sustainable. To pursue a trajectory that is unsustainable is essentially stealing from future generations (Inayatullah 1997).

7.6 Climate Change as 'The Great Educator'?

US political analyst Lester Milbrath (1989) argued that climate change could prove to be the global wake-up call that would be needed to educate decision-makers about the fundamental limits to growth. If civilisation survives, he proposed, it will be because the crisis of climate change forced us off the growth trajectory. Scientists warned that climate change was a serious issue more than 30 years ago, in the statement issued by the 1985 conference in the Austrian town of Villach. The concern was then reflected in the Framework Convention on Climate Change, adopted at the 1992 Rio Earth Summit, and the establishment of the Inter-governmental Panel on Climate Change (IPCC) to advise UN agencies. The early reports of the IPCC led to the Kyoto Protocol being adopted in 1997, committing most developed countries to reduce emissions. The actions of a few rogue states, especially the USA and Australia, delayed the ratification of the treaty and there was comparatively little action to slow climate change in the early years of this century. The IPCC Fifth Assessment Report warned that climate change was accelerating and observed that greenhouse gas emissions from all areas were increasing in direct proportion to population growth (Blanco et al. 2014). That palpable acceleration of the impacts of climate change caused the largest ever gathering of world leaders to adopt the Paris agreement, setting a target of keeping the increase in average global temperature below 2.0 °C with an ambition of attempting to keep the increase below 1.5°.

At the time of writing, it was almost impossible to find a credible climate scientist who thought the more ambitious target of 1.5° could be achieved. Until the COVID-19 pandemic caused a dramatic reduction in energy use, especially for international travel, global annual emissions were still increasing. Emissions for 2020 were about 7% lower than the 2019 figure, but most governments around the world saw the reduction in economic activity as a problem that needed to be rectified as a matter of urgency and analysts predict a rebound in 2021 (Newburger 2021). There has been a rapid move away from investing in fossil fuel systems for generating electricity, driven as much by spectacular reductions in the cost of solar panels and wind turbines as by concerns about climate change (see Chap. 12). 2015 was the first year in which the global new generating capacity from renewable systems – solar, wind, hydro and biomass – became equal to that from coal, oil and gas. By 2019, the new capacity from renewables installed globally was 170 GigaWatts, compared with about 40 GW from gas and about 30 GW from coal (IRENA 2019). However, in many countries the clean electricity supply systems are supplementing the existing fossil fuel electricity, rather than displacing it, while little has been done to reduce greenhouse gas emissions from other sectors like transport, so that globally in 2019, renewables only account for about 11% of all primary energy use (Ritchie and Roser 2020). The fundamental problem is that the human population is still growing, as is the global economy, and so demand for energy is still increasing. While the cleaning up of electricity supply is welcome – although it is happening far too slowly – it needs to be put in the context of the increasing global population and consequent proportional increase in demand. There has been little discussion of the potential role of slowing population growth, although it is explicitly mentioned in one recent study (Ripple et al. 2019). One comparative study of affluent countries found that having fewer children would do far more to slow climate change than other approaches that are often proposed in those nations (Wynes and Nicholas 2017); having one fewer child was calculated to reduce emissions by 25 times as much as living car-free and 70 times as much as adopting a plant-based diet.

Globally, the pace of change has been glacial and most relevant scientists think we are on track for a much warmer world, with alarming consequences for our ability to provide food and water for the still increasing global population. The first report to the Club of Rome in 1972, *The Limits to Growth*, as discussed earlier, projected that these problems of food and water would lead to a rapid decline in the human population in the middle decades of this century. That was a stark reminder that failure to stabilise the population by socially acceptable measures will probably lead to it being reduced by starvation, disease and fighting (Cribb 2020).

7.7 Australian Environmental Organisations

Tragically, most environmental groups have been silent about the issue of population in recent decades. They have campaigned vigorously to protect outstanding natural assets and to prevent some of the driving forces of climate change like coal mining, but they have rarely been critical about either population growth or increasing consumption (economic growth). One of Australia's main environmental organisations – the Australian Conservation Foundation (ACF) – has criticised consumption levels and their impact on the natural world, but it has mostly been unwilling in recent times to engage in debate about the effects of population numbers on the natural environment. That has not always been the case. As O'Connor and Lines (2008) noted, in the 1960s and 1970s ACF "fearlessly spoke out about population and questioned Australia's immigration policies". The debate became more difficult after the historian Geoffrey Blainey in 1984 gave a widely reported speech to the unlikely audience of a Rotary Club in the small country town of Warrnambool, asking how long the Australian public would continue to support the increasing levels of immigration from Asian countries. For some time after that event, anyone questioning the level of net migration was likely to be seen as speaking in code to support a racist argument. This became even more of a problem after the election of ultra-nationalist Pauline Hanson to the Australian parliament in 1996. Birrell et al. (1994) had written that the main attack on immigration levels had been from "opponents arguing from a 'White Australia' or racist perspective", so those arguing about environmental impacts or the effects of population growth on the quality of life "found it difficult to avoid being tagged as morally suspect". ACF did make a submission to the 1994 Jones Inquiry,

(Australia's most-recent official inquiry into population[3]), arguing that net immigration should be reduced to 30,000 a year to achieve future stabilisation. Its publication *Better Than Growth* (ACF 2000) called for the population to be stabilised at a level that could be sustainably supported and called for the immigration quota to be administered without discrimination, explicitly rejecting the racist agenda. When I was president of ACF (2004–14), I explicitly called for a long-term goal of stabilising the population in two separate addresses to the National Press Club. All that being said, it is reasonable to criticise ACF, and more strongly the other national environmental groups, for not having campaigned strongly on the obvious threat that the growing population poses to the integrity of natural systems and our unique biodiversity.

It might have been expected that the emergence of a Green political party would have put the issue of population on the political agenda, but that has not happened. The Australian Greens had its origin in purely environmental issues, beginning in 1972 with the United Tasmania Group, arguably the world's first green party. Then Green independents were elected to the Tasmanian State parliament, at one stage holding the balance of power and supporting a minority government with agreement on important environmental issues such as forest protection. The Greens eventually became a significant minor party in the Australian federal parliament; after the 2013 national election in Australia, the centre-left Labor party needed the support of Green members to form a minority government and the result was a comprehensive package of measures to address climate change. However, the Australian Greens are no longer simply a political party concerned with environmental issues, but have embraced a range of social-justice issues. Their population policy was abruptly changed at their national conference in 1998, at least partly as a result of the toxic political climate discussed above (Davis 1998). A background paper presented to the conference showed that the migration level accepted at that time by the Australian government – about 60,000 a year – would have seen the population stabilise by about 2020. The Greens decided to accept a range of migration policies concerning family reunion, refugees, international students, and skilled workers that have added up to a continuing rapid population increase of about 1.6% a year (pre-COVID-19). Concern about this issue has led to the emergence of a new political party, the Sustainable Australia Party, which has an explicit policy of stabilising the population in the range from 26 to 30 million. This would represent a very significant slowing from the rate of growth supported by the major parties. However, it has made little impact on the political scene, possibly demonstrating the difficulty of challenging what Patricia Kelly (2019) called "the toxic meme of growth".

---

[3] By one count there have been six official inquiries since the 1960s: Vernon 1965; Borrie 1975; FitzGerald 1988; National Population Council 1991; Withers 1992; and Jones 1994. In 2016 the Productivity Commission produced a report 'Migrant Intake into Australia'.

7.8 The Harsh Reality of 2020 Meets Delusions of Limitless Futures

In the troubled world of the COVID-19 pandemic, many countries have suffered terrible levels of premature death. By mid-January 2021, more than 2 million people had officially died from the infection, although the true number is unknowable but certainly even larger. It was a stark reminder that there is a price to pay for pushing the limits of natural systems. Sadly, many decision-makers are so obsessed with short-term economic issues that they see the highest priority as trying to return to the growth trajectory which has caused the problem. The Boston-based Tellus Institute has explored alternative possible futures (Raskin 2006). Until the pandemic, most nations were following the *market forces* (neoliberal) approach, relying on the 'self-correcting' logic of competitive markets to respond to problems, with governments playing only a supporting role. Their study concluded that this approach cannot even in principle lead to a sustainable future, as it inevitably produces environmental degradation and widening inequality, forces which undermine the social stability and international cohesion needed for the working of effective markets. The most likely result of this approach would be either *fortress world* – a future in which a minority of countries try to use armed forces to maintain order and protect their privileged lifestyles – or the equally plausible outcome of *breakdown* – a collapse of the systems that we see as vital to a civilised future: supply of food, water and energy to a mostly urbanised population; the treatment and safe disposal of wastes; adequate health care; and so on. In principle, the report argues, a *policy reform* approach could reduce resource use and its inevitable environmental consequences, but the obstacle is the lack of political will for such a solution, given what the report called "the resistance of special interests, the myopia of narrow outlooks and the inertia of complacency". This problem was earlier analysed by Kennedy (1993), who argued that elected politicians are very unlikely to take concerted action now in the interests of the future as long as they can argue that experts are divided and more research is needed. With the inherent complexity of the issues we now face, it will always be the case that experts are divided and so it will always be defensible to argue that more research is needed. As a concrete case illustrating the scale of the problem, it has taken 50 years to persuade some decision-makers in some countries to implement modest steps to curb smoking tobacco, despite the hard evidence that it shortens the lives of half its users; in many other countries, inaction still prevails. So Raskin argued that the only possibility of a civilised future requires a *great transition*, with market-led wealth generation and government-guided technological change supplemented by a values shift. The values which have driven progress for at least the last 100 years have been profit, production-consumption, individualism and domination of nature. Unless they are replaced by a new set of values - quality of life, global citizenship and recognition of ecological limits - we are likely to see the bleak futures caused by failure to live within the limits of natural systems. The optimistic view that consumption can be curbed depends explicitly (Victor 2008; Kopnina et al. 2020) or implicitly (Czech 2013) on stabilising the human population as a prerequisite to limiting resource demands.

References

Australian Conservation Foundation (2000) Better than growth. ACF, Carlton
Australian Council of Social Services (2020) Inequality in Australia. Accessed at http://povertyan-dinequality.acoss.org.au/inequality/
Australian Government (2003) Towards a national framework for energy efficiency. Energy Efficiency and Greenhouse Working Group, Melbourne
Birrell R, Hill D, Neville J (1994) Populate and perish. Fontana, Sydney
Blanco G et al (2014) Drivers, trends and mitigation. In: Edenhofer O et al (eds) Climate change 2014: mitigation of climate change. Contribution of Working Group III to the Fifth Assessment Report of the Intergovernmental Panel on Climate Change. Cambridge University Press, Cambridge/New York
Buildsearch (2020) Average house size in Australia. Accessed at https://buildsearch.com.au/house-size
CEIC (2020) Australia GDP per capita. Accessed at https://www.ceicdata.com/en/indicator/australia/gdp-per-capita
Cribb J (2020) Food or war. Cambridge University Press, Cambridge
Crist E, Cafaro P (2012) Life on the brink: environmentalists confront overpopulation. University of Georgia Press, Georgia, pp 3–15
Czech B (2013) Supply shock: economic growth at the crossroads and the steady state solution. New Society Publishers, Gabriola Island
Daly H (1990) Toward some operational principles of sustainable development. Ecol Econ 2:1–6
Davis F (1998) Greens change their immigration policy. Green Left Weekly Issue 333, September 16. Accessed at https://www.greenleft.org.au/content/greens-change-their-immigration-policy
Diamond J (2005) Collapse: why societies choose to fail or succeed. Viking Press, New York
Ehrlich P (1968) The population bomb. Buccaneer Books, New York
Ehrlich PR, Holdren JP (1971) Impact of population growth. Science 171:1212–1217
Ehrlich P, Ehrlich A, Holdren J (1977) Ecoscience: population, resources, environment. W.H. Freeman and Co, New York
Engelman R (2012) Nine population strategies to stop short of 9 billion. In: Starke L (ed) State of the world 2012: moving toward sustainable prosperity. Island Press, Washington, DC
Esmaeli A, Shokooi Z (2011) Assessing the effect of oil price on world food prices: application of principal component analysis. Energy Policy 39(2):1022–1025
Harrington F, Wilkenfeld G (1997) Appliance efficiency programs in Australia: labelling and standards. Energy Buildings 26(1):81–88
Hocknull SC et al (2020) Extinction of eastern Sahul megafauna coincides with environmental degradation. Nature Commun 11:Article number 2250. Accessed at https://www.nature.com/articles/s41467-020-15785-w
Inayatullah S (1997) Future generations thinking. Futures 29:701–706
IPBS (2020) IPBS pandemics report: Escaping the ear of pandemics. Accessed at https://ipbes.net/pandemics
IRENA (2019) Renewable energy now accounts for a third of global power capacity. https://www.irena.org/newsroom/pressreleases/2019/Apr/Renewable-Energy-Now-Accounts-for-a-Third-of-Global-Power-Capacity
Jackson T (2009) Prosperity without growth: economics for a finite planet. Earthscan, London
Kelly P (2019), quoted in Nelson CA (2019) Providentia's prospectus 2019, top 30 trends futuristas are watching. The Futures Forum. Accessed at https://issuu.com/thefuturesforum/docs
Kennedy PM (1993) Preparing for the twenty-first century. Harper-Collins, London
Kopnina H, Washington H, Lowe I (2020) Why do society and academia ignore the 'scientists' warning to humanity' on population? J Futures Stud 25(1):93–106
Lianos T (2020) World population: the elephant in the living room. Real-World Econ Rev 91:70–82. Accessed at https://www.academia.edu/42846281/World_population_the_elephant_in_the_living_room?email_work_card=abstract-read-more

Lowe I (2012) Bigger or better?: Australia's population debate. University of Queensland Press, Brisbane

Malthus TR (1798) An essay on the principle of population. Online at the US Library of Economics and Liberty. http://www.econlib.org/library/Malthus/malPop.html

Meadows D, Meadows D, Randers J, Behrens W (1972) The limits to growth. Universe Books, Washington

Milbrath L (1989) Envisioning a sustainable society. State University of New York Press, Buffalo

National Sustainable Development Council (2020) Transforming Australia: progress toward SDG goals. Accessed at https://www.monash.edu/__data/assets/pdf_file/0005/1471991/transforming-australia-sdg-progress-report-summary-dashboard-05092018.pdf

Newburger E (2021) Covid pandemic drove a record drop in global carbon emissions in 2020. CNBC. Accessed at https://www.cnbc.com/2020/12/11/covid-record-drop-global-carbon-emissions-2020.html

O'Connor M, Lines W (2008) Overloading Australia: how governments and media dither and deny on population. Envirobook, Sydney

O'Sullivan JN (2012) The burden of durable asset acquisition in growing populations. Economic Affairs 32(**1**):31–37

Our World in Data (2020) World population growth since 10,000 BCE. Accessed at https://ourworldindata.org/world-population-growth

Public Transport Users' Association (2019) Myth: cars are becoming more efficient. Accessed at https://www.ptua.org.au/myths/efficient/

Quiggin J (2014) Zombie economics: how dead ideas still walk amongst us. Black Inc, Melbourne

Raskin P (2006) The great transition today: A report from the future, GTI Paper no. 2. Great Transition Initiative, Boston

Ripple WJ et al (1992) World scientists warning to humanity. Accessed at https://www.ucsusa.org/resources/1992-world-scientists-warning-humanity

Ripple WJ et al (2017) World scientists' warning to humanity: a second notice. Bioscience 67(12):1026–1028

Ripple WJ et al (2019) World scientists' warning of a climate emergency. Bioscience 70:8–12

Ritchie H, Roser R (2020) Renewable energy. Published online at OurWorldInData.org. Accessed at https://ourworldindata.org/renewable-energy

Rockstrom J et al (2009) Planetary boundaries: exploring the safe operating space for humanity. Ecol Soc 14(2):32. see: http://www.ecologyandsociety.org/vol14/iss2/art32/. Accessed 21 Jan 2021

Slatre F et al (2019) Climate-human interaction associated with south-east Australia megafauna extinction patterns. Nature Commun 10:Article number 5311. Accessed at https://phys.org/news/2017-01-humans-climate-australian-megafauna.html

Statista (2020) Average annual OPEC crude oil price from 1960 to 2020. Accessed at https://www.statista.com/statistics/262858/change-in-opec-crude-oil-prices-since-1960/

Turner G (2014) Is global collapse imminent? An updated comparison of *The limits to growth* with historical data, Research Paper no. 4. Melbourne Sustainable Society Institute, Melbourne

UN (2017) Population facts. Accessed at https://www.un.org/development/desa/pd/sites/www.un.org.development.desa.pd/files/files/documents/2020/Jan/un_2017_factsheet10.pdf

UN (2019) World Population Prospects 2019. Accessed at https://population.un.org/wpp/Publications/Files/WPP2019_Highlights.pdf

van Onselen L (2014) Howard explains the population Ponzi. MacroBusiness, 31 Jan. Accessed at https://www.macrobusiness.com.au/2014/01/howard-explains-australias-population-ponzi/

van Onselen L (2018) Credlin blows lid on how fraudulent migrant intake is set. MacroBusiness, 13 March. Accessed at https://www.macrobusiness.com.au/2018/03/credlin-blows-lid-on-how-fraudulent-migrant-intake-is-set/

Victor P (2008) Managing without growth: slower by design, not disaster. Edward Elgar, Cheltenham von Weizsäcker E, Hargroves K, Smith M, Desha C, Stasinopoulos P (2009) Factor 5: transforming the global economy through 80% increase in resource productivity. Earthscan, London

Wijkman A, Rockstrom J (2012) Bankrupting nature: denying our planetary boundaries. Routledge, London

World Bank (2020) Fertility rates in Italy. Accessed at https://data.worldbank.org/indicator/SP.DYN.TFRT.IN?locations=IT

World Commission on Environment and Development (1987) Our common future. Oxford University Press, Oxford

World Health Organisation (2020) Family planning/contraception methods. Accessed at https://www.who.int/en/news-room/fact-sheets/detail/family-planning-contraception

Wynes S, Nicholas KA (2017) The climate mitigation gap: education and government recommendations miss the most effective individual actions. Environ Res Lett 12:7. Accessed at http://iopscience.iop.org/article/10.1088/1748-9326/aa7541

Zorich Z (2019) Letter from Iceland: surviving the Little Ice Age. Archaeol Archive 65:5. Accessed at https://archive.archaeology.org/1209/letter/iceland_hjalmarvik_irminger_east_greenland_current.html

Chapter 8
A Brief History of The Limits to Growth Debate

Kerryn Higgs

Abstract Commissioned by the Club of Rome, the Meadows team at the Massachusetts Institute of Technology (MIT) produced *The Limits to Growth* in 1972. Using system dynamics modelling, the team found that economic growth will reach physical limits by the middle of the twenty-first century. Economists in particular rejected the notion that human economies have any limits and treated the work with derision. Scientists took a more supportive approach and have written independent reviews and developed the limits ideas. A number of them have tested the modelling against real-world outcomes 30, 40 or 50 years later, and have found that the *Limits* projections are reliable. While it is often claimed that collapse can be avoided by large technological advances, it is likely that a paradigm shift in human, social and economic expectations is also necessary.

8.1 Introduction

The first 25 years after World War II were characterised by unprecedented economic growth[1] as Europe was rebuilt and consumer culture expanded to majorities in the global North and elites in the South. Global population grew from a little more than 2.5 billion in 1950 to nearly 4 billion in 1975, while world GDP more than doubled. By the end of this 25-year period, both the economic system and human numbers were large, doubling times were short, and economic growth had been adopted as the fundamental objective of economic policy (Higgs 2014, Chap. 10). Growth had been the implicit objective of businesses for several centuries but, after that war, the pursuit of growth became an explicit and central aim of both business and governments. It remains so today.

[1] 'Growth' or 'economic growth' refers throughout to GDP growth.

K. Higgs (✉)
University of Tasmania, Hobart, Australia

Club of Rome, Sydney, Australia

S. J. Williams, R. Taylor (eds.), *Sustainability and the New Economics*,
https://doi.org/10.1007/978-3-030-78795-0_8

The growth ideology embodied several key assumptions. Economic growth was naturalised as an inevitable stage of human civilisation, an element in humanity's innate and desirable progression from more 'primitive' social forms to modernity; in this version of history, the achievements of Europeans were assumed to provide a universal template for all (Higgs 2014). In the second half of the twentieth century, GDP growth 'came to be presented as the common good, thus justifying the particular interests of those who benefited most from the expansion of market transactions as beneficial for all' (Schmelzer 2017: p 351). Not just inevitable and beneficial, economic growth was depicted as indefinite, a process unlimited in space or time.

8.2 The Club of Rome and *The Limits to Growth*

This was the economic and ideological climate in which the Club of Rome emerged. The Meadows team (Donella Meadows, Dennis Meadows, Jørgen Randers and William Behrens) was commissioned by the founding members of the Club of Rome to explore the physical impacts of the indefinite expansion of the human economic system. The Meadows team was based at MIT, where Jay Forester was developing the new discipline of system dynamics. The outcome of their project was condensed into a short volume, *The Limits to Growth* (Meadows et al. 1972). This book was not the first to warn about such concerns, but it attracted considerable attention and brought the limits ideas into mainstream thinking.[2]

The Club of Rome, founded by Italian industrialist Aurelio Peccei and Scottish scientist Alexander King, had gathered a select group of prominent, thoughtful and often wealthy individuals who wanted to address what they called the *problematique*, or 'the predicament of mankind'. For Peccei, the astonishing economic and industrial growth that followed World War II was not an unmixed blessing. Despite its benefits, Peccei was 'perplexed and worried by the orderless torrential character of this precipitous human progress' (Barney 1982: p 607). Soon after it was established in 1968, the Club commissioned the MIT project.

The predicament as defined by the MIT team was characterised by five major features: accelerating industrialisation, rapid population growth, extensive malnutrition, depletion of non-renewable resources, and environmental decline. The team asked if growing populations, engaged in ever-expanding industrialisation, could avoid immense environmental decay, depletion of the resources on which the entire project depends, and the social chaos that would result from decline or collapse. To tackle this question, they developed World3, a computer program based on Forester's pioneering work. World3 brought together extensive data on the many interacting

---

[2]At the end of the eighteenth century, Thomas Malthus hypothesised that food production would not keep up with population growth. He pre-dated modern industrial production, so did not deal with issues of the overall sustainability of modern society (Higgs 2014: pp 40–43).

aspects of society, economy and environment.[3] A non-linear model, it incorporated feedback loops between the elements, and included the delays that can apply to such feedbacks, delays that may generate unexpected results and jeopardise effective solutions. Fertile land may be gradually eroded by economic exploitation; minerals may become much more difficult to locate and extract; climate feedbacks may cause runaway warming. Overall, World3 attempted to integrate the dynamic relationships between a vast array of factors, sectors and variables that affect each other and the outcomes in the physical world. 'The real challenge then becomes solving issues in multiple sectors concurrently' (Turner 2008).

World3 ran different scenarios for various changes that might (or might not) be made. These options included business as usual (BAU) or the standard run, at one end of the spectrum—BAU assumed continuation of the trajectory Peccei described as 'precipitous human progress;' there were several combinations that assumed extremely advanced technology; and, at the other end of the spectrum, both physical throughput and population were stabilised. The standard run yielded collapse in the middle of the twenty-first century. Massive technological advance delayed but could not avert this outcome. There were scenarios that could, however: those that stabilised population and wound back the scale and rate of material extraction and waste.

Although *The Limits to Growth* became the best-selling environmental book ever published, its reception was mixed. Both US President Carter and Prime Minister Trudeau of Canada embraced its concerns and launched national studies along similar lines (Barney et al. 1981; Barney 1982; Voyer and Murphy 1984).

8.3 Reactions from Economics and Science

From the outset, however, most economists poured scorn on the work, especially the idea that physical limits might apply to human economic systems (Beckerman 1972; Economist 1972; Nordhaus 1973; Solow 1973). Within the discipline of economics, this view has prevailed. Although modelling is an essential tool in economics, economists rejected the non-linear modelling of system dynamics. One critic condemned the World3 computer program as a case of 'Garbage in, garbage out' (Kiernan 1972); John Koehler (1973), of the Rand Corporation, described the work as 'fraud,' 'fantasy,' and 'fudge;' while sociologist Robert McGinnis (1973) accused it of fostering 'that faddish hysteria which passes as concern for The Ecology.' The 'doomsday' label was ubiquitous.

Neither did economists canvas redistributive options, but relied on continuing GDP growth to deal with poverty and hunger. According to Robert Solow (1973), 'The *only* prospect for a decent life for Asia, Africa and Latin America is in more

[3] They used five basic sectors: population, capital, agriculture, non-renewable resources, and persistent pollution and included variables such as population, capital, pollution, and arable land (Turner 2012).

total output.' Beckerman (1974) agreed: 'A failure to maintain economic growth means continued poverty, deprivation, disease, squalor, degradation and slavery to soul-destroying toil for countless millions of the world's population.' Robert Gillette (1972) held that the 'assumption of inevitable economic growth' constitutes 'the very foundation' of the economics profession, a focus that may help to explain the intensity of the assault from economists.

The assumption of endless economic growth is captured in the way mainstream economics models economic processes: economic growth is understood to be the result of two factors, capital and labour, depicted as a circular flow between production and consumption. This circular diagram is conceptualised without a physical context from which resources must be drawn and where wastes can be discarded; it was developed while energy and resources were plentiful and economists could ignore the physical basis of economic activity, including the role of energy. By contrast, most physical scientists and all ecological economists (see Chap. 11 of this volume) regard the physical world as the indispensable foundation of human economies, both to provide resources (including energy) and to absorb wastes (Fig. 8.1). In this framework, depletion and pollution are inescapable.

Physical scientists also regard energy as the 'master resource', since no other commodity can be produced without it (Cleveland 1991; Zencey 2013). When Robert Solow (1957) examined the empirical evidence, he found that capital and labour could explain only a small fraction of observed economic growth in the United States. Solow thought that 'technical change' could account for the unexplained 87.5%, a very large 'residual.' More recently, Robert Ayres and Benjamin Warr (Ayres and Ayres 2009: pp 9–18) argued that the missing factor is actually energy or, more accurately, exergy: the increasing efficiency with which energy and resources can be made into useful work. Thus, technical innovation has a role, but 'labour and capital extract energy; they don't make it.' Energy is a prerequisite for,

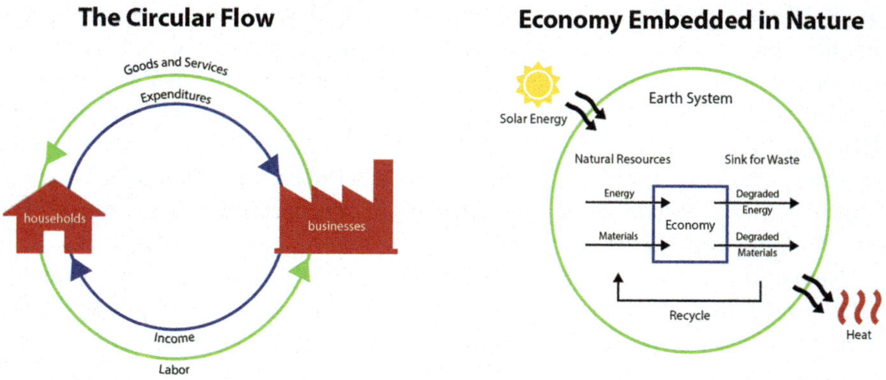

Fig. 8.1 The circular flow of economics in contrast with the embedded economy of ecological economics. (Prepared by Rhianna Topschij using data from Bureau of Economic Analysis (2015): p 2 and Daly (1991). Courtesy of Sage Handbook of Nature, 2018, Vol 3 (ed Terry Marsden))

not a product of, economic activity. And energy, like all material resources, is ignored in the standard economic paradigm.

Few scientists condemned *The Limits to Growth* as a doomsday tract or took part in the attack on it. However, the prominent English physicist John Maddox (1972) wrote a polemic called *The Doomsday Syndrome* and the Science Policy Research Unit at the University of Sussex published a book of critical essays, called *Models of Doom* in the US (Cole et al. 1973); this collection included numerous economists as well as scientists—'doom' and related concepts appeared frequently. The editors allowed the Meadows team to include a detailed response in the same book. Their response noted numerous places where the Sussex critics had distorted or misunderstood their work. For example, they pointed out that, although models are simplifications, they are based on explicit assumptions, allowing critics to inspect them—in striking contrast with the unexamined preconceptions on which policy decisions are typically based. They also noted that their critics simply ignored the fact that extraction of new natural resources was doubling every 15 years, while waste material was 'being dumped into the environment' at much the same rate. Rates of extraction and waste are much greater today than was the case in the early 1970s when the research was carried out, and are still climbing (see Higgs, Chap. 5 of this volume for a detailed examination of rates of material extraction and their uninterrupted escalation). As for the claim that the poor can only be assisted via ongoing GDP growth, Meadows et al. (1973) argued that '[h]istorically at least, growth of population and of capital has been correlated with the concentration of wealth and with rising gaps in the absolute income of the rich and the poor.' 45 years later, this remains the case (Oxfam 2018; Alvaredo et al. 2018).

8.4 Attack from Activist Economists

The MIT team published sequels in 1992 and 2004, where they reviewed their findings and found that the conclusions of the original text remained valid. They concluded that

> A sustainable society would be interested in qualitative development, not physical expansion. It would use material growth as a considered tool, not a perpetual mandate. It would be neither for nor against growth. … Before this society would decide on any specific growth proposal, it would ask what the growth was for, and who would benefit, and what it would cost, and how long it would last, and whether it could be accommodated by the sources and sinks of the planet (Meadows et al. 1992: p 210).

This judgement echoes that of Kuznets, the man who devised the concept of GDP during the latter stages of the Great Depression. Kuznets told the US Congress that social wellbeing could not be inferred from GDP income, and the question 'growth of what and for whom?' would always need to be asked (Gadrey 2005). Among proponents of GDP growth, Kuznets's caveat is not much remembered.

While scientists' warnings of ecological disaster have often been brushed aside or entirely ignored, mainstream economists have been outrageously successful in propagating their worldview, largely as a consequence of their connections to and funding by big business. In the US, business began forming organisations to lobby for their interests in the late nineteenth century, as the consolidation of small businesses into large corporations began. Public relations (PR) sections were soon formed inside corporate structures and PR emerged as a profession by the 1920s. From the mid-1940s, the first tranche of private enterprise think tanks was founded. These were a small foretaste of the avalanche of 'free market' think tanks that mounted the neoliberal takeover of economic, social and political analysis through the 1970s and 1980s (Higgs 2014: Part 3). Funding came from corporations that opposed regulation, such as those associated with tobacco, asbestos, chemicals, fossil fuels, mining and automotive manufacturing; and from the family foundations of their CEOs. Activist economists set out to 'litter the world' with 'free enterprise' think tanks (Cockett 1995: p 307; Higgs 2014: pp 90, 191–193), which have disseminated industry propaganda as 'independent research' ever since. Most claim charity status and are tax-exempt. Their employees exert immense influence in the media of the English-speaking world, where they are depicted as scholars on an equal footing with peer-reviewed academics.

In their second major review, the Meadows team acknowledged that perception often outweighs material reality and perception is influenced—or sometimes created—by

> those groups (largely comprised of economists) who have spent the past 30 years pushing the concept of free trade. Unlike us, they have been able to make their concept a household word. Unlike us, they have convinced numerous politicians to fight for free trade. … Ecological overshoot seems to us to be a much more important concept in the twenty-first century than free trade. But it is far behind in the fight for public attention and respect (Meadows et al. 2004: p xx).

In addition to these two comprehensive sequels, Dennis Meadows (2007) took issue with the fudge and distortion propagated by many of their critics. For example, although the team's scenarios included the stabilising scenarios mentioned above, the influential *New York Times Book Review* simply turned a blind eye to this aspect and claimed that the Meadows team 'jigger the assumptions just enough to eliminate non-catastrophic possibilities' and the world economy 'obligingly' collapses (Passell et al. 1972). Meadows (2007) noted that this review's portrayal of a 'false inevitability of doom' is based on ignoring the scenarios that do not lead to collapse (the subject of the entire fifth chapter of the original book) and speculated that the reviewers did not actually finish reading the book. He also pointed to several other false and misleading assertions, such as the claim that the MIT team had declared that reserves of vital materials would be exhausted within 40 years. This error was much-repeated through the following decades, by reputable commentators as well as ideological warriors (Higgs 2014: pp 259–262). The MIT team were exploring the tendencies of the global economic system, its process and trends, and they stressed this explicitly, while Passell et al. (1972), like others, insisted on treating it as an exact forecast. Meadows (2007: pp. 408–413) concluded that these reviewers were engaged in deliberate misrepresentation.[4]

8.5 New Approaches to Limits

While economists continued to attack the idea of limits to economic growth, the actual growth has indeed been 'torrential,' as Peccei observed in the 1970s. Although some economists have claimed that human exploitation of resources has been a linear process since the Stone Age (Solow 1974; Beckerman 1972), there has never been anything like the twentieth century, when physical production increased 20-fold and human population quadrupled. Although the roots of this process lie deep in our history, it was not until about 1950 that the scale of the human project began to outgrow planet Earth decisively, with consequences that continue to cascade through the natural and human world, literally liquidating life on earth. Steffen et al. (2015b) have shown just how unusual it was, in the sets of exponential graphs known as 'The Great Acceleration.' These graphs run from the year 1750 through to 2010; the profound disjunction around 1950 is a feature of all of them (see Chap. 2 of this volume).

Physical scientists have warned of impending calamity since the early 1990s (UCS 1992; Ripple et al. 2017) and some have advanced defences of the limits ideas. Among the most important of these are the ecological footprint and the planetary boundaries work. In 1990, William Rees and Mathis Wackernagel pioneered the idea of the 'ecological footprint,' which 'measures the ecological assets that a given population or product requires to produce the natural resources it consumes,' including disposal of the wastes of production. The footprint measures this impact in hectares or acres of biologically productive land and water. In the footprint framework, a country such as the UK, with limited land, appropriates about three times its productive capacity by importing food and other materials from abroad. Estimates of the global human footprint suggest humans are consuming resources which would take 1.6 planets to supply sustainably (Global Footprint Network 2021; Dasgupta Review 2021). Although it has been criticised, it nonetheless informs our ability to get some sense of the carrying capacity of the planet (Meadows 2019) and to grasp the fact that we are remorselessly devouring our 'natural capital' when we should be living off 'income'. The loss of natural capital diminishes the productive capacity of the planet, reducing income in a vicious circle.

A large team led by climate scientists Johan Rockström and Will Steffen has been developing the planetary boundaries concept since 2009. This attempts to quantify which trends in human economic activity constitute threats to the Earth system. In the most recent boundaries paper, Steffen et al. (2015a) suggest that loss

[4]There have also been numerous independent studies of the history of *The Limits to Growth* controversy and debates about it, most since 2000. Among these are Simmons 2000; Nørgård et al. 2010; Bardi 2011; Higgs 2014; Ragnarsdottir and Sverdrup 2015; Jackson and Webster 2016.

of biological diversity, biogeochemical processes (especially nitrogen and phosphorous cycles), land use change and climate change are either exceeding planetary boundaries, or are close to doing so. Steffen et al. (2018) have warned more recently that failure to reduce carbon emissions will lead to what they call 'hothouse earth,' where many species, including humans, would struggle to survive (see also Steffen, Chap. 2, this volume).

8.6 Assessments of *The Limits to Growth* vs. Real World Outcomes

Graham Turner (2008), then at the Sustainable Ecosystems Unit of Australia's Commonwealth Scientific and Industrial Research Organisation (CSIRO), compared the output from several of the limits scenarios with well-accepted public statistics for the period 1970–2000 to establish how far the projections of World3 tally with outcomes in the real world. Turner chose three of these scenarios, representing the full range: the 'standard run' (BAU), where rates of GDP growth, resource use and industrialisation continued on their 1970 trajectory; the 'comprehensive technology' run, where massive improvements were achieved in agricultural productivity, resource extraction techniques, pollution mitigation, and family planning;[5] and the 'stabilised world' run, in which population and capital investment were stabilised alongside significant technological advance.[6]

Turner (2008) found there was a significant correlation between real-world data and the standard run, with small divergences that did not contradict the basic trends (Fig. 8.2). The stabilised world showed little similarity to the real world as measured in 2000. This is unsurprising, since warnings about biophysical limits have had little or no impact on policymaking since *The Limits to Growth* was published. Neither did 'comprehensive technology' lead to a sustainable world: 'These efforts delay the collapse of the global system to the latter part of the 21st century, when the growth in economic activity has outstripped the gains in efficiency and pollution control.'

Turner (2008) also deals with the 'rebound effect' which explains why significant gains in efficiency facilitate expansion rather than moderating consumption.

[5] 'The comprehensive technology scenario … incorporates levels of resources that are effectively unlimited, 75% of materials are recycled, pollution generation is reduced to 25% of its 1970 value, agricultural land yields are doubled, and birth control is available world-wide' (Turner 2014).

[6] 'For the "stabilized world" scenario, both technological solutions and deliberate social policies are implemented to achieve equilibrium states for key factors including population, material wealth, food and services per capita. Examples of actions implemented in the World3 model include: perfect birth control and desired family size of two children; preference for consumption of services and health facilities and less toward material goods; pollution control technology; maintenance of agricultural land through diversion of capital from industrial use; and increased lifetime of industrial capital' (Turner 2014).

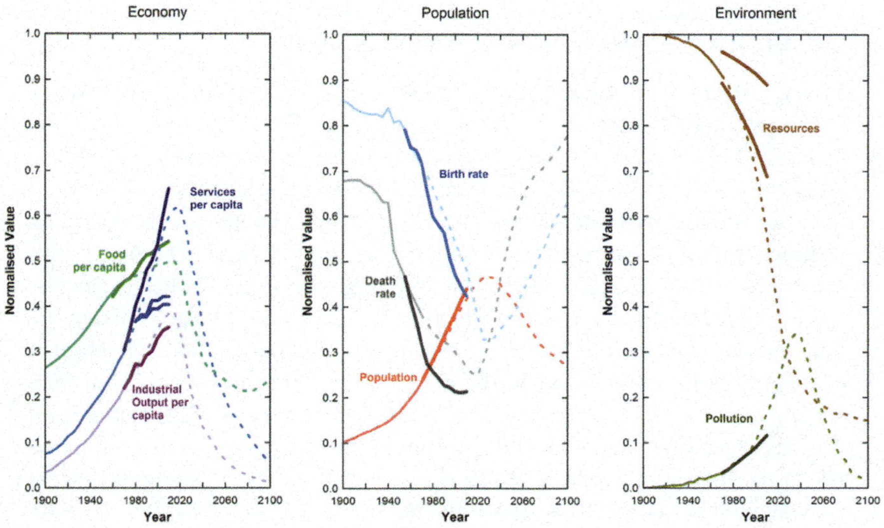

Fig. 8.2 Standard Run (BAU) scenario (dotted lines) compared with historical data from 1970 to 2010 (solid lines). For environmental variables: global persistent pollution, fraction of non-renewable resources remaining (upper curve uses a limit of 150,000 EJ for ultimate energy resources, lower curve uses a lower limit of 60,000 EJ); for economic output variables: industrial output per capita, food per capita, services per capita (upper curve: electricity per capita; lower curves: literacy rates for adults and youths); for demographic variables: population, crude birth rate, crude death rate. (Courtesy of Graham Turner 2014: 8)

Theoretically, efficiencies should reduce impacts, but this does not happen in practice. For example, the carbon intensity of industrial production declined throughout the twentieth century, but the rate of carbon emissions continued to grow exponentially. Engineer Michael Huesemann (2003) argued that there is no real paradox here: 'technological innovation has never been used to stabilize the size of the economy; [its] main role has always been exactly the opposite, namely the enhancement of productivity, consumption, and economic growth.'

Turner (2014) pointed out that capital investment must increase as resources 'become harder to extract and process … [and] continues to increase as the resource base is further depleted' (see Box 8.1). What is sometimes called the 'low-hanging fruit' has already been exploited. For oil and gas, resources are deeper, under the ocean, less accessible, or require hydraulic fracturing (known as fracking), which demands numerous wells across the deposit. For minerals, the ores are lower grade and at greater depth, requiring removal of more surface material and finer grinding which takes more energy and water. All of this involves greater environmental impact (Michaux 2016).

Huesemann (2003) raises related issues, arguing that the root causes of humanity's unsustainable behaviour must be tackled. These he suggests are 'our society's obsession with economic growth … driven by an excessive desire for affluence (A) and a lack of limits on population (P).' Since these are not technological, but rather

Box 8.1: It Takes Energy to Make Energy Available

Energy Return on Investment (EROI) or Energy Return on Energy Invested (EROEI)

Fossil fuels remain the major sources of energy worldwide (about 70%). They supply transport, heating and electricity, and serve as chemical stocks to manufacture fertiliser and plastics. These fuels must be extracted, largely by drilling (including hydraulic fracturing) or robotic machinery. These activities use energy—an increasing amount as the target becomes harder to access. Once recovered, the fuel has to be distributed and, in some cases, refined; the infrastructure that enables distribution must be built and maintained. EROI is expressed as the ratio between the energy made available for use and the energy expended in the process. In the early years of fossil fuel capture, oil was easy to extract, and its EROI was high; some estimates for new oil in the US in the nineteenth century, when oil was still literally gushing out of the ground, indicate an EROI ratio of 100:1 but this had declined to 40:1 by 1970, and 14:1 in 2000 (Hall and Day 2009).

Renewable energy systems also require significant inputs of energy to mine and transport the many essential minerals, to manufacture equipment such as turbines and panels, and to install them. These systems also require decommissioning at the end of their useful lives, with presumably some recycling (which also requires energy). In all cases, life-cycle assessment permits calculation of the energy that has to be invested to deliver useful energy to users. Estimates of EROI for solar power range from as low as 3:1 up to 30:1 or more (Bardi and Hall 2016); wind, whether offshore or onshore, is widely agreed to be more efficient. Ratios at the upper end of this range would enable adequate energy to be produced so that, after some of it is used to produce the next generation of renewable infrastructure, there is also sufficient for ongoing social and economic use. Bardi and Sgouridis (2017) warn, however, that the window for a successful transition to renewable energy is narrow. In their view, a very large deployment of currently available energy is urgently needed to produce renewable infrastructure, even though adequate energy will still have to be devoted to ongoing social uses. Moreover, market forces are unlikely to allocate the required resources in a viable timeframe; capital investment in the renewable transition is only about 10% of what is needed, despite broad appreciation of the urgency.

social and ethical issues, they will not be solved by eco-efficiency. Similarly, Lenzen et al. (2016) argue that it is illogical to expect technological progress to 'stretch to unprecedented limits in order to let populations enjoy unchecked growth in numbers and affluence.' Wiedmann et al. (2020) identify consumption as 'by far the strongest determinant of global impacts.' They note that 'the world's top 10% of income earners are responsible for between 25 and 43% of environmental impact.' They also cite overwhelming evidence that 'burgeoning consumption has diminished or cancelled out any gains brought about by technological change.' For these researchers, the key problem is overconsumption. They are among a growing group of scholars who work on recognising the drivers of overconsumption, the role of trade in allowing the rich world to export many of its ecological impacts, the impact of the demands of the North on people and their ecologies worldwide and the consequent imperative for redistribution (for example Lenzen et al. 2012; Weinzettel et al. 2013; O'Neill et al. 2018; Hickel 2019).

Systems ecologists Charles Hall and John Day (2009) also compared the standard run with actual 2008 data. Like Turner, they found that the standard run (BAU) matched population and industrial output per capita in 2008. They looked at a range of resources (copper, oil, soil, and fish) and, again, the actual data were very close to the BAU output. Using CO_2 and nitrogen, they found that pollution levels were close to, though somewhat less than, the standard run. 'We are not aware,' they wrote, 'of any model made by economists that is accurate over such a long timespan.'

Jørgen Randers (2018), one of the original authors of *Limits*, has suggested that the adjustment of the standard run to allow 'doubled resources' is a better fit for what has happened than the standard run itself. This scenario leads to a many-faceted pollution crisis, which includes carbon emissions and the resultant global warming. It varies from the standard run only in that it postpones any collapse from resource constraints until after the pollution crisis has triggered decline (Fig. 8.3).

Randers stresses that the core message of *Limits*, which has often been misinterpreted, was that there are limits to humanity's material footprint on a finite planet. This may not always involve limits to GDP growth although it is clear that GDP and material extraction are closely correlated (see Figs. 5.1 and 5.2 in Higgs, Chap. 5, this volume). Dennis Meadows (2019) another of the original authors, also confirmed the validity of the projections in his lecture at Ulm University.

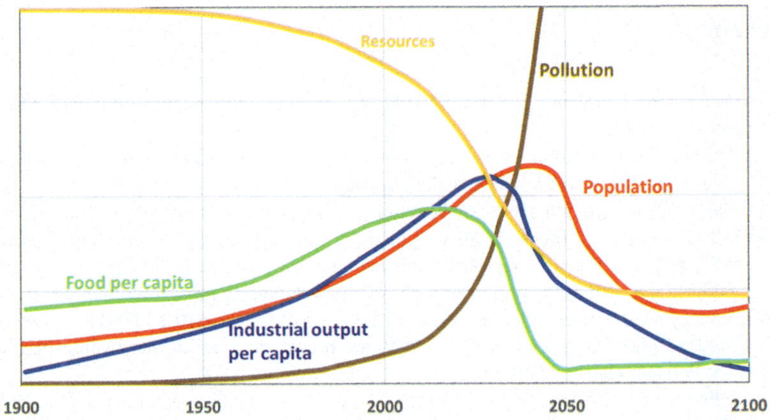

Fig. 8.3 World model with natural resource reserves doubled (pollution crisis). (Courtesy of Jørgen Randers, address to the Club of Rome, 2018. Based on Meadows et al. 1972, Fig. 36)

8.7 Conclusion

For nearly 50 years, questions about physical limits have met systematic derision from business interests and mainstream economists. Most of the vocal critics of *Limits* have ruled out any notion of slowing or restraining growth and ‘Club of Rome’ became easy shorthand for dismissive comments about prophets of doom.

In part because of this barrage of scorn, questions about the continuation of economic growth in a finite world remain, largely, excluded from public discourse and policymaking, though are now more often seen than was the case before 2010 (see for instance Dasgupta Review 2021). Economist JK Galbraith foresaw in 1958 that the ‘gargantuan and growing appetite’ he perceived in 1950s America would, in the long run, need to be curtailed. Instead, the appetite for ever more consumer goods has deepened in the global North, and its extension to sections of the new middle classes in China, India, and Latin America is celebrated as a boon for all. We continue, however, to face extreme and worsening versions of the problems analysed nearly 50 years ago by the Meadows team.

The so-called ‘bigger pie’ has been promoted by economists as the obvious solution to all social problems—debt, unemployment, poverty, and even the environmental damage involved in making and baking the pie. Enlarging the economy still remains the primary strategy for most of the institutions of the OECD world, whether businesses, national governments, or the international bodies allied to business. Some of the UN bodies attempt to contradict this discourse and stress the necessity for a sustainable future (see Higgs, Chap. 5, this volume). If we wish to avoid extreme crises and possible collapse, the fundamental problems with the notion of an ever-expanding pie on a finite planet must be addressed. A paradigm shift is mandatory.

References

Alvaredo F, Chancel L, Piketty T et al (2018) World inequality report. http://wir2018.wid.world/. Accessed 13 Feb 2021

Ayres R, Ayres E (2009) Crossing the energy divide. Pearson Education, Upper Saddle River

Bardi U (2011) The limits to growth revisited. Springer, New York

Bardi U, Hall C (2016) But what’s the real energy return of Photovoltaic energy? The real EROI of photovoltaic systems: Professor Hall weighs in. A comment by Ugo Bardi. https://cassandralegacy.blogspot.com/search?q=eroi+of+photovoltaic. Accessed 13 Feb 2021

Bardi U, Sgouridis S (2017) In support of a physics-based energy transition planning: sowing our future energy needs. Biophys Econ Resour Q 2:14. https://doi.org/10.1007/s41247-017-0031-2

Barney G (ed) (1982) The global 2000 report to the president: entering the twenty-first century, vols 1–3. Pelican, Harmondsworth

Barney G, Freeman P, Ulinsky C (eds) (1981) Global 2000: implications for Canada. Pergamon Press, Toronto

Beckerman W (1972) Economists, scientists, and environmental catastrophe. Oxf Econ Pap 24(3):327–344

Beckerman W (1974) Two cheers for the affluent society: a spirited defense of economic growth. St. Martin's Press, New York

Cleveland C (1991) Natural resource scarcity and economic growth revisited: economic and biophysical perspectives. In: Costanza R (ed) Ecological economics: the science and management of sustainability. Columbia University Press, New York, pp 289–317

Cockett R (1995) Thinking the unthinkable: think-tanks and the economic counter-revolution, 1931–1983. Fontana, London

Cole HSD, Jahoda M, Pavitt KLR et al (eds) (1973) Models of doom: a critique of *The Limits to Growth*. Universe, New York

Daly H (1991) Steady-state economics: second edition with new essays. Island Press, Washington, DC

Dasgupta Review (2021) The economics of biodiversity: the Dasgupta Review (headline messages). https://assets.publishing.service.gov.uk/government/uploads/system/uploads/attachment_data/file/957629/Dasgupta_Review_-_Headline_Messages.pdf. Accessed 19 Feb 2021

Economist (1972) Limits to misconception. Economist, 11 March, pp 21–22

Gadrey J (2005) What's wrong with GDP and growth? The need for alternative indicators. In: Fullbrook E (ed) A guide to what's wrong with economics. Anthem Press, London, pp 262–278, 318–319

Galbraith JK (1958) How much should a country consume? In: Jarrett H (ed) Perspectives on conservation. Johns Hopkins University Press, Baltimore, pp 89–99

Gillette R (1972) The limits to growth: hard sell for a computer view of Doomsday. Science 175(4026):1088–1092

Global Footprint Network (2021). https://www.footprintnetwork.org/our-work/ecological-footprint/. Accessed 13 Feb 2021

Hall C, Day J (2009) Revisiting the limits to growth after peak oil. Am Sci 97(May–June):230–237

Hickel J (2019) The imperative of redistribution in an age of ecological overshoot: human rights and global inequality. Humanity 10(3):416–428

Higgs K (2014) Collision course: endless growth on a finite planet. MIT Press, Cambridge, MA

Huesemann M (2003) The limits of technological solutions to sustainable development. Clean Techn Environ Policy 5(1):21–34

Jackson T, Webster R (2016) Limits revisited: a review of the limits to growth debate. http://limits2growth.org.uk/revisited. Accessed 13 Feb 2021

Kiernan M (1972) The coming Apocalypse. Mother Earth News 15 (May–June)

Koehler J (1973) Untitled review of The Limits to Growth: a report for the Club of Rome's project on the predicament of mankind. J Polit 35(2):513–514

Lenzen M, Moran D, Kanemoto K et al (2012) International trade drives biodiversity threats in developing nations. Nature 486:109–112

Lenzen M, Malik A, Foran B (2016) Letter. J Clean Prod 139:796–798

Maddox J (1972) The doomsday syndrome. McGraw-Hill, New York

McGinnis R (1973) Untitled review of The Limits to Growth: a report for the Club of Rome's project on the predicament of mankind. Demography 10(2):295–299

Meadows D (2007) Evaluating past forecasts: reflections on one critique of *The Limits to Growth*. In: Costanza R et al (eds) Sustainability or collapse? MIT Press, Cambridge, MA, pp 399–415

Meadows D (2019) *Limits to Growth* after forty-five years. https://www.youtube.com/watch?v=aRXb4bJhSSw&ab_channel=FAWnUlm. Accessed 16 Feb 2021

Meadows D, Meadows D, Randers J, Behrens W III (1972) The Limits to Growth: a report for the Club of Rome's project on the predicament of mankind. Signet, New York

Meadows, D. Meadows D, Randers J, Behrens III W (1973) A response to Sussex. In: Cole HSD, Jahoda M, Pavitt KLR et al (eds) Models of doom: a critique of The Limits to Growth. Universe, New York, p 217–240

Meadows D, Meadows D, Randers J (1992) Beyond the limits: confronting global collapse; envisioning a sustainable future. Chelsea Green, Post Mills

Meadows D, Randers J, Meadows D (2004) The Limits to Growth: the 30-year update. Chelsea Green, Post Mills

Michaux S (2016) The coming radical change in mining practice. Minor Metals Trade Association, UK. https://mmta.co.uk/2016/02/14/radical-changes-in-mining/. Accessed 13 Feb 2021

Nordhaus W (1973) World dynamics: measurement without data. Econ J 83(332):156–1183

Nørgård J, Peet J, Ragnarsdottir K (2010) The history of *The Limits to Growth*. Solutions 1(2):59–63, Feb–March. https://thesolutionsjournal.com/2016/02/22/the-history-of-the-limits-to-growth/. Accessed 13 Feb 2021

O'Neill D, Fanning A, Lamb W et al (2018) A good life for all within planetary boundaries. Nat Sustain 1(2):88–95

Oxfam International (2018) Reward work, not wealth. https://www-cdn.oxfam.org/s3fs-public/file_attachments/bp-reward-work-not-wealth-220118-summ-en.pdf. Accessed 13 Feb 2021

Passell P, Roberts M, Ross L (1972) Review of *Limits to Growth*. New York Times Book Review, April 2: 1, 10, 12–13

Ragnarsdottir K, Sverdrup H (2015) *Limits to Growth* revisited. The Geological Society website. https://www.geolsoc.org.uk/Geoscientist/Archive/October-2015/Limits-to-growth-revisited. Accessed 13 Feb 2021

Randers J (2018) *The Limits to Growth* (1972) in a fifty year perspective https://www.youtube.com/watch?v=ykFXvUDnw9Y. Accessed 13 Feb 2021

Ripple W, Wolf C, Newsome T et al (2017) World scientists' warning to humanity: a second notice. Bioscience 57:1026–1028. https://doi.org/10.1093/biosci/bix125

Schmelzer M (2017) The hegemony of growth: the OECD and the making of the economic growth paradigm. Cambridge University Press, Cambridge

Simmons M (2000) Revisiting *The Limits to Growth*: could the Club of Rome have been correct, after all? An energy white paper. http://www.greatchange.org/ov-simmons,club_of_rome_revisted.html

Solow R (1957) Technical change and the aggregate production function. Rev Econ Stat 39(3):312–320

Solow R (1973) Is the end of the world at hand? Challenge (March–April):39–50

Solow R (1974) The economics of resources or the resources of economics. Am Econ Rev 64(2):1–14

Steffen W, Richardson K, Rockström J et al (2015a) Planetary boundaries: guiding human development on a changing planet. Science 347:736–746

Steffen W, Broadgate W, Deutsch L, Gaffney O, Ludwig C (2015b) The trajectory of the Anthropocene: the great acceleration. Anthropocene Rev 2:81–98. https://doi.org/10.1177/2053019614564785

Steffen W, Rockström J, Richardson K et al (2018) Trajectories of the earth system in the anthropocene. Proc Natl Acad Sci U S A 115:8252–8259. https://doi.org/10.1073/pnas.1810141115

Turner G (2008) A comparison of *The Limits to Growth* with 30 years of reality. Glob Environ Chang 18:397–411

Turner G (2012) On the cusp of global collapse? Updated comparison of The Limits to Growth with historical data. GAIA 21/2:116–124

Turner G (2014) Is global collapse imminent? Melbourne Sustainable Society Initiative, research paper no.4: Melbourne University, Melbourne

Union of Concerned Scientists (1992) World scientists' warning to humanity. Union of Concerned Scientists, Cambridge, MA

Voyer R, Murphy M (1984) Global 2000: Canada, a view of Canadian economic development prospects, resources and the environment. Pergamon, Toronto

Weinzettel J, Hertwich E, Peters G et al (2013) Affluence drives the global displacement of land use. Glob Environ Chang. https://doi.org/10.1016/j.gloenvcha.2012.12.010

Wiedmann T, Lenzen M, Keyßer L, Steinberger J (2020) Scientists' warning on affluence. Nat Commun 11:3107. https://doi.org/10.1038/s41467-020-16941-y

Zencey E (2013) Energy as master resource. In: Worldwatch state of the world 2013: is sustainability still possible? Island Press, Washington, DC, pp 73–83

Chapter 9
The Role of the Fossil Fuel Industry

Ian Dunlop

Abstract The global fossil fuel industry has played a pivotal role in the quadrupling of population since World War II and in the commensurate exponential increase in economic growth with its associated wealth creation. The foundation stone was massive expansion in the production and consumption of fossil fuels, globally and in Australia, which led to a corresponding accrual of power and influence on the part of fossil fuel industry players. The implications of increasing fossil fuel use, in terms of global warming as a result of increased atmospheric carbon concentrations, have been understood scientifically for decades. Yet the fossil fuel industry chose to ignore the science, and in its own perceived self-interest, deliberately prevented the development of serious policy to avoid the potentially catastrophic outcomes of this warming. The result is that the world must now take emergency action to rapidly reduce atmospheric carbon concentrations, in large part due to the intransigence of the industry itself in facing up to this reality, and its refusal to constructively manage its own decline. This is particularly the case in Australia. The industry should either proactively co-operate in handling the emergency, or it will be rapidly phased out.

9.1 Fossil Fuels: The Foundation of Exponential Population and Economic Growth

Solar energy has been the prime determinant of the rise and fall of civilisations for centuries. Prior to the Industrial Revolution in the 1700s, that amounted to immediately-available, plant-based solar energy in the form of trees and other biomass. Consistency of supply was problematic – for example, as trees disappeared in the

I. Dunlop (✉)
Australian Coal Association, Sydney, Australia

The Club of Rome, Winterthur, Switzerland

S. J. Williams, R. Taylor (eds.), *Sustainability and the New Economics*,
https://doi.org/10.1007/978-3-030-78795-0_9

face of competing demands for fuel, timber, land for agriculture, and climatic variations.

Thomas Homer-Dixon (2006) catalogued in detail the rise and fall of the Roman Empire in terms of its energy balance and its ‘energy return on investment’ (EROI), arguably the most important parameter determining our future, yet a concept largely unknown by most politicians, the economics profession and corporate leaders. That is the amount of energy a project produces, divided by the amount of energy it consumes in delivering that result.

As an example, take an isolated agrarian society where the farmers do all the work without the help of animals. If they consume the same amount as they produce, the EROI would be 1. If their basic needs require more than produced, the EROI falls below 1 and they would gradually starve if that continued. On the other hand, producing more than they consume pushes the EROI above 1, providing a surplus with which to improve their security and lifestyle, possibly by acquiring animals to further increase the ratio. A ratio much greater than 1 is essential for running even a modestly complex society.

Before the Industrial Revolution, EROI constraints meant that the global population never rose above half a billion, with a growth rate over 10,000 years of around 0.04%pa (Fig. 9.1).

When the Romans occupied Britain, they found outcrops of a velvety black mineral, which could be polished and carved into beautiful jewellery (Freeze 2003). It was also flammable, a fossil fuel solidified and stored over millennia – coal. (In fact it was another form of solar energy as coal is ancient plant material.) It was used on a limited scale for heating, but it was not until the 1600s that its use expanded, to

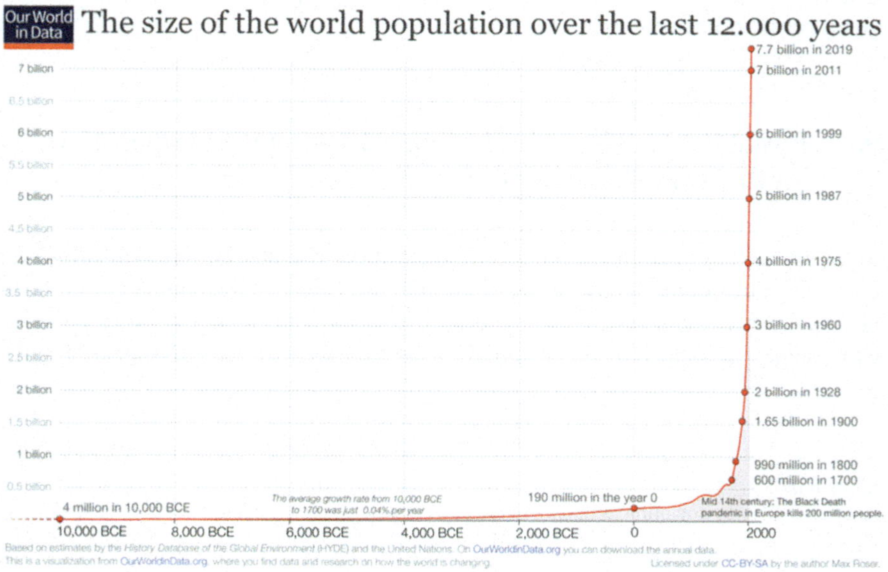

Fig. 9.1 World population growth: 10,000 BCE to 2000

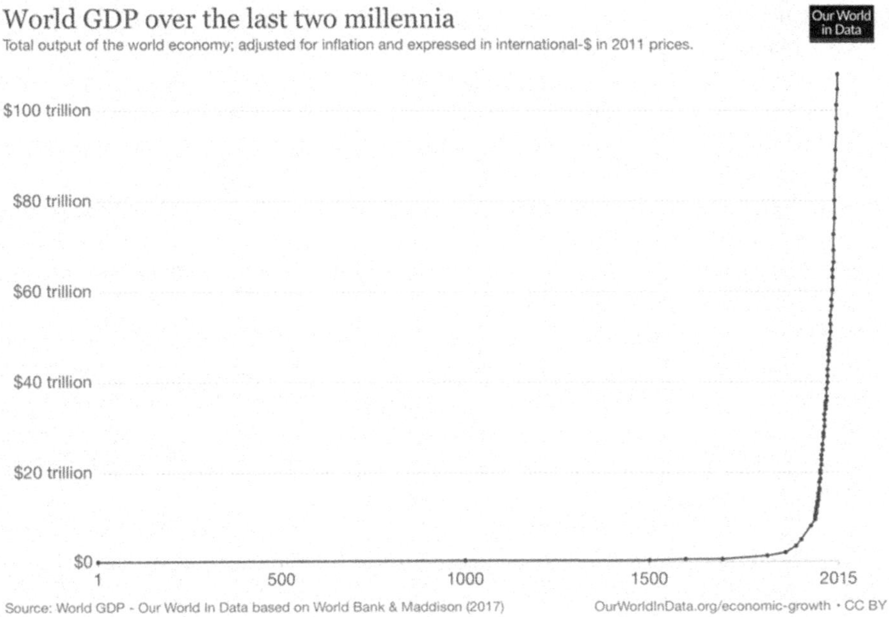

Fig. 9.2 World GDP for the past 2000 years

the point that pollution from coal smoke caused it to be banned periodically in Britain, until the exigencies of severe winter cold forced its re-emergence.

The steam engine, transforming coal energy into motion, was the trigger for the Industrial Revolution, as ingenious innovations compounded one after another, releasing the potential of this energy in new ways. The result was the beginning of an explosion in both population and economic activity which has continued to this day, starting in Britain then spreading globally (Fig. 9.2).

9.2 Global Fossil Fuel Evolution

The use of coal transformed global societies through the nineteenth century in both good and bad ways. EROIs of 20–40 were achievable, albeit coal mining was extremely dangerous and unhealthy work. From this coal base, fossil fuel technology progressed, with the next energy breakthrough being the first natural gas well dug in Fredonia, New York, in 1821, supplying the Fredonia Gas Light Company. By 1836, the city of Philadelphia had created the first municipally-owned natural gas distribution company. These initiatives were major steps forward, but of localised effect due to the difficulties of transporting gas over long distances in gaseous form. The large-scale exploitation of oil started not long afterwards with Edwin Drake's first commercial oil well in Titusville, Pennsylvania, in 1859.

Oil overcame the transportation difficulties of gas and expanded rapidly as oil technology spread around the world. Demand accelerated with the invention of the motor car in 1885 – dramatically so when Henry Ford introduced the Model T in 1908 – making motorised transportation available to the wider public.

The first oil was also discovered in the Middle East in 1908, at Masjid-i-Suleiman in Persia (today's Iran), and developed by the Anglo-Persian Oil Company, which eventually morphed into BP. The productivity of these wells proved far superior to anything previously discovered.

The benefits and geopolitical implications of oil were obvious, prompting a race, as World War I loomed, to secure oil supplies and convert military hardware, particularly navies, to oil-based operation where possible. Access to oil was also pivotal to the outcome of World War II.

The transport limitations of gas were not solved until after World War II, when liquefaction technology enabled natural gas to be compressed and shipped in specially designed vessels as Liquified Natural Gas (LNG). The first purpose-built LNG vessel, the *Methane Princess*, entered service in 1964, shipping gas from Algeria to Europe. Thereafter the gas industry developed apace, both with ocean shipment and the extension of pipeline supply.

The cumulative effect of these developments has been an exponential increase in global fossil fuel consumption since the Industrial Revolution, but particularly after 1950, initiated by reconstruction post-World War II (Fig. 9.3).

US oil production had been a major source of supply growth prior to, and during, World War II. As the post-war period progressed, US production peaked, due to the limitations of oil recovery techniques and low oil prices. With the US unable to meet

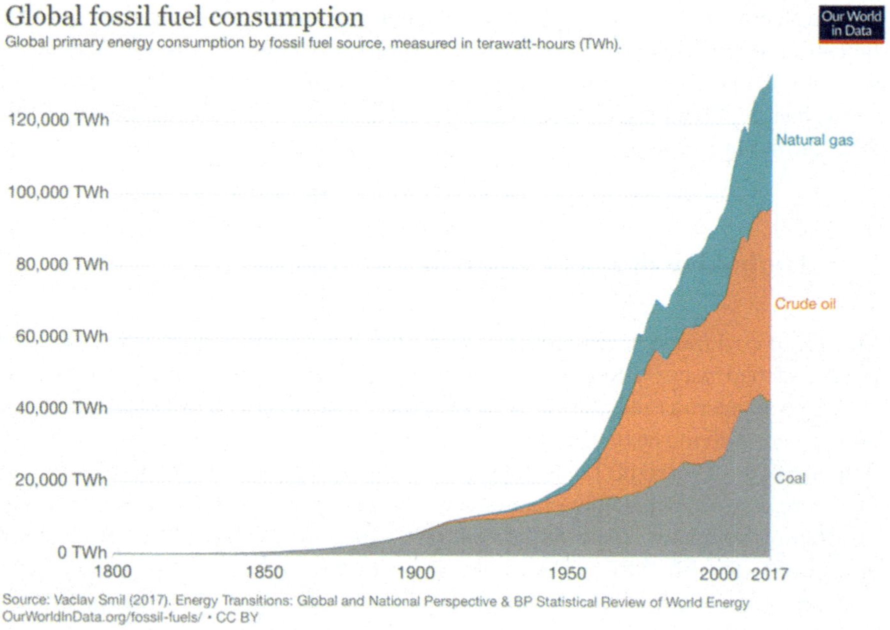

Fig. 9.3 Global fossil fuel consumption: 1800–2017

burgeoning demand, increasing reliance was placed on alternative sources, particularly the Middle East.

In EROI terms, the transformation was remarkable. Middle East oil, being readily accessible, in its heyday had EROIs above 100, an extremely attractive investment. This was obvious to the host Middle East countries, who in 1960 had formed the Organisation of Petroleum Exporting Countries (OPEC) as a result of their increasing dissatisfaction with the low oil prices they were receiving from the major 'Seven Sisters' oil companies[1]. This led to escalating geopolitical confrontation, culminating in the oil shocks from 1969 to 1979, which drove oil prices higher, creating great concern in Western countries over energy security (Fig. 9.4).

The response was a rush for alternative sources of supply, both geographically and technically, as the higher prices justified exploration and development in increasingly difficult environments. The oil industry had always been global, but it was now given added impetus to rely far less on the Middle East. Offshore exploration developed apace, overcoming increasingly hazardous operating conditions with technological innovation in areas such as the North Sea and the Gulf of Mexico. Oil companies also diversified into other energy sources such as nuclear, coal and the embryonic renewables industry, seeing them as bridges to the

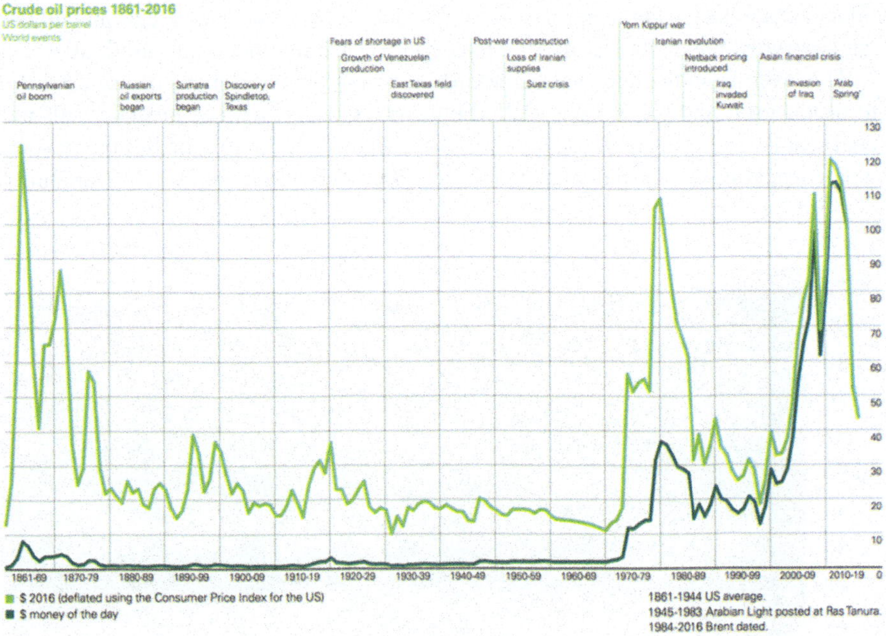

Fig. 9.4 Oil prices and key geopolitical events: 1861–2019. (BP 2019)

[1]The "Seven Sisters" were the multinational oil companies who dominated the global petroleum industry from the mid-1940s to the mid-1970s: BP, Shell, Exxon, Chevron, Texaco, Gulf, Mobil.

energy future. The result was a glut in oil supply which lasted until the late 1980s. Oil prices collapsed, reaching a low point in 1986.

The 1990s were characterised by further geopolitical upheaval initiated by the first Gulf war in 1991. Oil prices remained low through the decade, then began climbing after the second Gulf war in 2003, when Iraqi President Saddam Hussein was toppled and Iraq oil exports halted. At the same time, the cost of extracting oil and gas was increasing, with EROI decreasing, as cheaper sources were depleted and replaced with more expensive options, particularly outside the Middle East (for example deep-water exploration and tar sands) (Fig. 9.5).

As oil prices spiked in 2008 to their highest level since oil was first produced commercially in the 1860s, concern began to emerge about the potentially finite nature, or peaking, of global oil and gas production (International Energy Agency 2008). In hindsight, it is clear that conventional oil production did in fact peak in 2005. However, total oil and gas availability continued to rise beyond that point due to the expansion of the unconventional oil and gas industry, particularly in the US, along with the re-establishment of Iraq oil exports (Fig. 9.6).

The former arose primarily from the development of the fracking industry. Fracking, or more accurately hydraulic fracturing, is the process of opening up fractures in tight subterranean geological formations by injecting fluid at high pressure to release entrapped hydrocarbons. Fracking was developed initially by the oil and gas industry in the late 1940s to improve recovery from conventional oil and gas reservoirs and is widely used in that context (Montgomery and Smith 2010).

However, from the late 1990s fracking gained some notoriety as its use extended to produce oil and gas from so-called unconventional resources in shale beds and coal seams. These deposits are closer to the surface than conventional oil reservoirs, where fracking has the potential to interfere with agriculture and water availability,

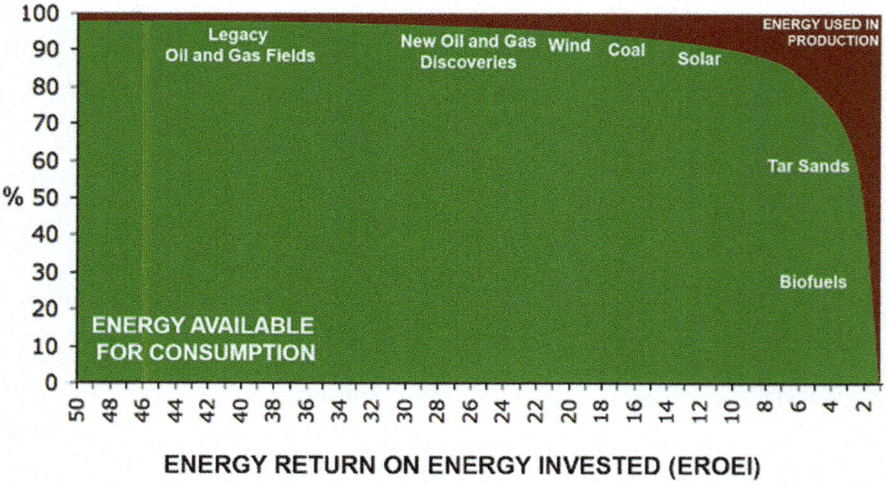

Fig. 9.5 Energy Return on Investment. (Bardi 2020)

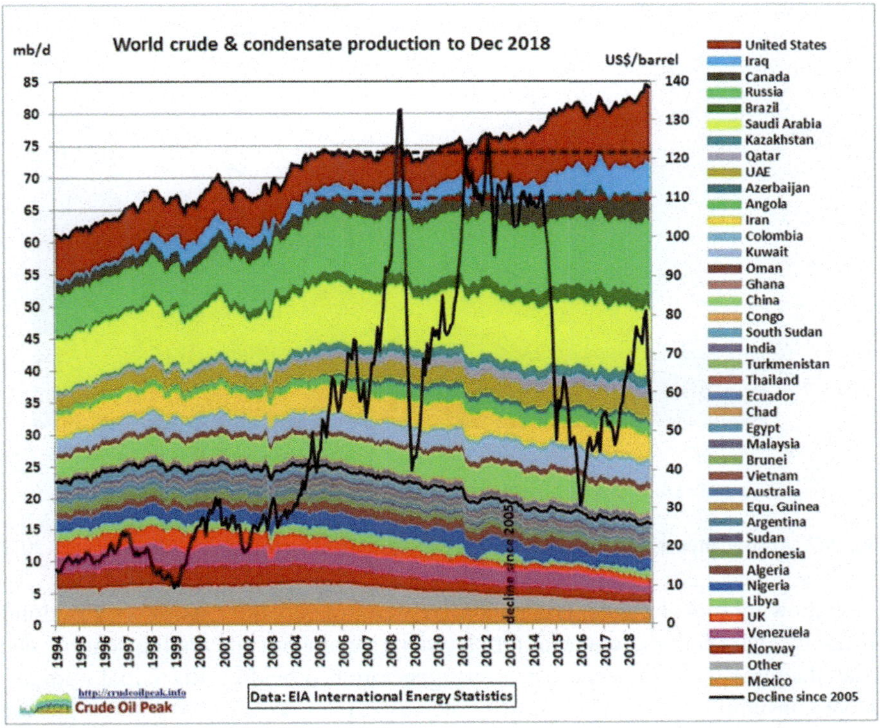

Fig. 9.6 World crude oil and condensate production, 1994–2018. (Mushalik 2019)

create minor earthquakes and other geological disturbances, generate substantial pollution, and cause health risks.

Despite these concerns, fracking of unconventional resources expanded rapidly in the US, driven by higher oil and gas prices and, post the 2008 Global Financial Crisis, the ready availability of cheap loans and/or subsidies provided through quantitative easing policies. This led to a significant increase in unconventional oil and gas production, even though the US industry overall has never been profitable and is now in deep financial trouble as a result of the pandemic-related oil price collapse (Kelly 2020).

9.3 Australian Fossil Fuel Evolution

Australia is particularly well-endowed with fossil fuels, notably coal and gas, less so oil. As a result, our economy is heavily dependent upon those fuels, both for domestic energy supply and in generating export income – far more so than most other nations (Fig. 9.7).

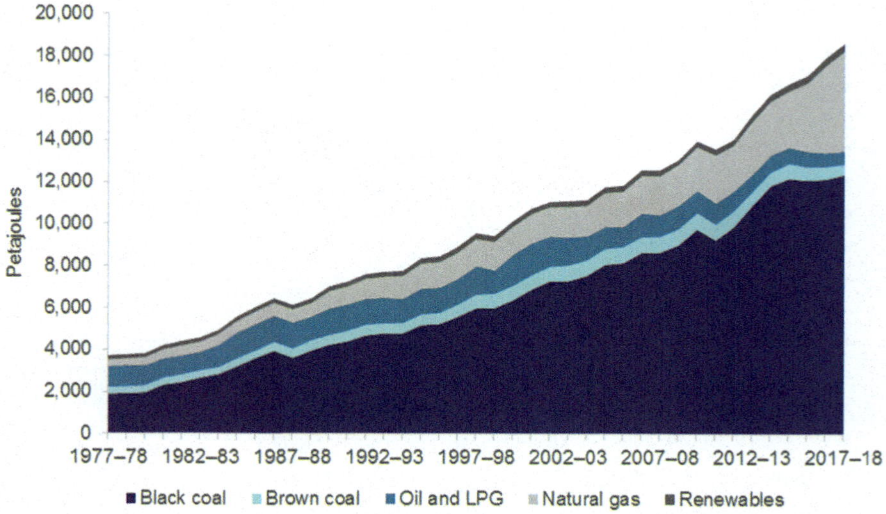

Fig. 9.7 Australian energy production 1977–2018. (Australian Dept of Energy & Environment 2019)

Following global trends, production grew continually after World War II, dominated by coal (both thermal coal for electricity generation and coking coal for steel manufacture) with domestic electricity generation providing the initial stimulus. Coal consumption thereafter steadily increased until around 2008, when gas and then renewables began to make gradual inroads.

In export markets, demand for high-quality coking coal from the rapidly expanding Japanese steel industry triggered growth from the 1970s. This was followed by demand from similar steel expansion in other Asian countries, such as South Korea and Taiwan, continuing through the 1980s and '90s, then China and India as their economies took off. Thermal coal export was slower, but similarly expanded supplying Asian markets from the late 1980s.

LNG export was initiated by the North West Shelf project in Western Australia in 1989, using gas extracted from conventional offshore reservoirs. This was boosted in 2015 with the first shipment of LNG from Gladstone in Queensland, extracted from coal seams by fracking. An unfortunate side effect was a major increase in the price of gas on the eastern seaboard as producers insisted upon international price parity for domestic supply.

A weak point is oil, where Australian production peaked in 2000 as Bass Strait oilfields depleted. The country is now less than 50% self-sufficient, increasingly reliant upon overseas refineries and long global supply lines (Fig. 9.8).

Hydro power has been another significant energy source, for example in Tasmania and the Snowy Mountains of New South Wales. As concerns over the global impact of fossil fuel consumption in triggering climate change have grown, and technological innovation dramatically reduced the cost of solar and wind renewable energy, these

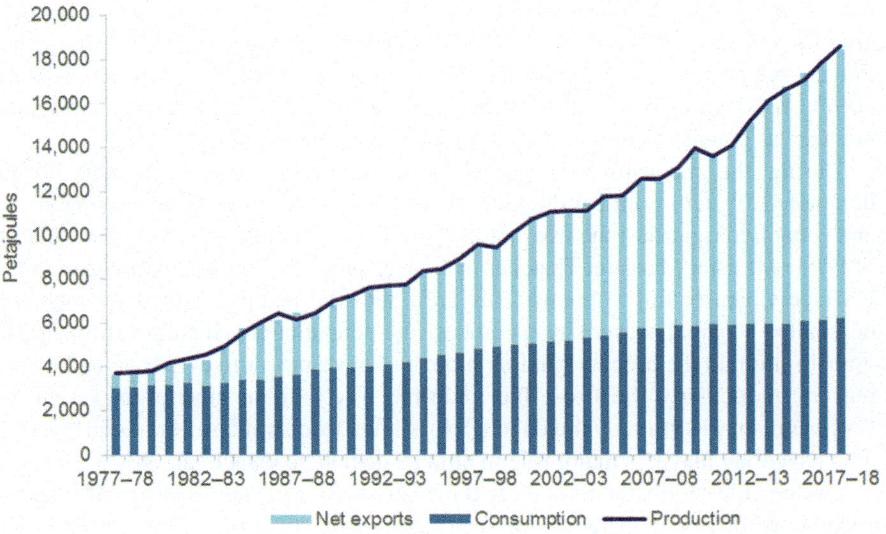

Fig. 9.8 Australian energy balance 1977–2018. (Australian Dept of Energy & Environment 2019)

alternatives are now making rapid inroads into national energy supply, but from a low base.[2]

However, in 2018, fossil fuels still supplied 81% of world primary energy[3] demand (International Energy Agency 2019), with coal and LNG comprising around 23% of Australia's export income (Australian Dept of Foreign Affairs & Trade 2019), and fossil fuels supplying around 94% of Australia's domestic energy needs.

9.4 Power and Political Influence

Since the Industrial Revolution, fossil fuels have played a central role in the development of human civilisation as we know it. Without them, the exponential explosion in population, economic activity and wealth creation would never have occurred. Not surprisingly, those in control of the industry have gained enormous influence over the direction of global and national affairs.

Access to oil in particular has been the rationale for many global conflicts since the early 1900s, and repeatedly triggered geopolitical disruption. Oil was pivotal to

[2] For instance, the state of Tasmania, Australia, is now powered by 100% renewable energy, as is the Australian Capital Territory.

[3] Primary energy is the energy available from the source in its natural state, before any human conversion. Final energy is the amount consumed by the end-user. The difference is the amount consumed in energy transformation and distribution.

the outcome of World War I (Black 2017), and to Middle East territorial disputes after the war following the secret 1916 Sykes-Picot Agreement between Britain and France, which laid the groundwork for conflict that continues to this day (Paul 2002). After World War II, the strategic importance of oil became ever more pronounced as countries competed to accelerate economic growth.

During the reconstruction period prior to 1970, social-democratic ideals dominated Western political thinking. After 1980, in response to increasing social and economic rigidities, the free-market paradigm took hold following the election of Prime Minister Margaret Thatcher in the UK and President Ronald Reagan in the US (see Chap. 6, this volume). This grew into the neoliberal model which has dominated economies ever since, emphasising deregulation, small government and globalisation, in the process handing more and more power to corporate and media players, particularly the fossil fuel industry. The extreme libertarian variant of neoliberalism was strongly linked to US fossil fuel leaders such as the Koch brothers who have long played a major role in climate denialism (MacLean 2017).

One notable neoliberal initiative was the introduction of supposedly performance-enhancing bonus remuneration for senior corporate executives. This started in the US in the late 1980s, subsequently spreading through Anglo-Saxon economies in the 1990s. It is rarely discussed, but the large, often obscene bonuses paid for short-term performance fundamentally changed the ethical basis for business, *inter alia,* away from serious consideration of longer term, systemic issues such as genuine sustainable development. The resulting short-termism became one of the greatest barriers to the development of sensible climate and energy policy.

The influence of fossil fuel leaders was further enhanced through industry advisory bodies in key international institutions such as the International Energy Agency (IEA). The IEA was formed in 1974 at the time of the OPEC oil embargo. Its initial role was to "re-establish oil security", which transformed latterly into "steering the world toward secure and sustainable energy transitions"(International Energy Agency 2021). An IEA Coal Industry Advisory Board was formed in 1979, an Energy Business Council in 2009, and a Renewable Industry Advisory Board in 2011, comprising leaders from these industries. There is nothing inherently wrong with such arrangements, provided their input is balanced, given the importance placed upon IEA analysis by government decision-makers. However, for many years the IEA demonstrated a strong bias in favour of fossil fuels and underestimated the potential of renewable energy (Watts 2018), a position which continues to this day (Farand 2020), undoubtedly influenced by fossil fuel industry pressure (Oil Change International 2018).

In a similar vein, from the time international climate negotiations were initiated under the 1992 UN Framework Convention on Climate Change (UNFCCC), at subsequent annual Conference of the Parties (COP) meetings, fossil fuel industry representatives played a major role in influencing outcomes in favour of continued fossil fuel use (Galey 2019).

The culmination of this activity was the 2015 Paris Climate Agreement in which, despite the supposed success of the agreement in uniting multiple parties behind a common objective, the words 'fossil fuels', 'coal', 'oil or 'gas' do not appear in the entire document, even though reducing their related carbon emissions is the *raison d'etre* of the agreement (Abreu 2019; UNFCCC 2015).

Fossil fuel companies have specialised in using industry associations and think tanks to mask their delaying tactics in the climate and energy arena. It started in the US with the Global Climate Coalition, founded in 1989 and disbanded in 2002, credited with being one of the most successful early climate denialist organisations, contributing to US President George W. Bush's 2001 decision not to sign the Kyoto Protocol (Desmog 2019). Numerous organisations, ranging from the slightly more nuanced American Petroleum Institute and the Competitive Enterprise Institute (CEI), to the outright denialist Heartland Institute, continue to press the fossil fuel industry case to this day, amply funded by the industry itself.

Such practices were rapidly picked up by right-wing figures in Australia, facilitated by extensive US links and substantial foreign ownership of the Australian fossil fuel industry. Western Mining Corporation's Hugh Morgan and Ray Evans were particularly active in furthering climate denialism, in close co-ordination with CEI, initially intent upon derailing the Kyoto Protocol, and establishing the climate denialist Lavoisier Group in 2000 (Hamilton 2007; Wilkinson 2020).

The workings of various industry bodies, such as the Business Council of Australia, the Minerals Council of Australia, the Australian Institute of Petroleum, the Australian Petroleum Production and Exploration Association, the Australian Aluminium Council, the Australian Chamber of Commerce and Industry and the Australian Industry Group, in exerting influence on climate and energy policy have been extensively documented. (Hamilton 2007; Pearse 2007; Pearse et al. 2013; Ritter 2018; Brett 2020, Wilkinson 2020). From the late 1990s, their efforts were coordinated under the umbrella of the Australian Industry Greenhouse Network, which continues today.

In the main, chief executives of the member companies found it convenient to shelter behind these lowest-common-denominator industry bodies, rather than confront contentious climate and energy issues publicly.

One unfortunate outcome was the early politicisation of climate change in fossil-fuel-rich countries such as the US, Canada and Australia, where conservatism irrationally became identified with climate denialism. Close alliances developed between corporate leaders, industry bodies, right-wing think tanks (such as the Institute of Public Affairs, the Centre for Independent Studies and the Sydney Institute in Australia) and the right-wing media (notably the Murdoch press globally), intent upon subverting any serious climate-policy development. Political donations from the fossil fuel industry also played a major role in inclining mainstream politics toward the industry's preferences.

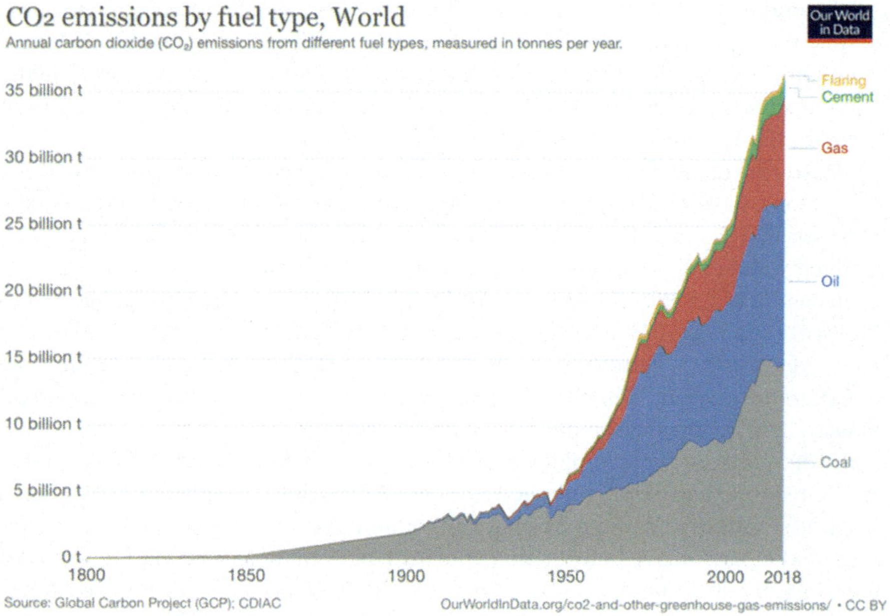

Fig. 9.9 Global CO_2 emissions by fuel type: 1800–2018

9.5 The Climate Change Nemesis

Notwithstanding industry power, the exponential rise in fossil fuel consumption has brought the industry's own nemesis in the form of a commensurate exponential increase in carbon emissions into the atmosphere, and their accelerating impact on the global climate (Fig. 9.9).

Modern climate science can be traced back to Jean Baptiste Fourier in France in the 1820s, progressing through Foote, Tyndall, Arrhenius, Revelle, Keeling, Hansen, to the present day (Mason 2020). US president Lyndon B. Johnson released a report in 1965 which highlighted the potential impacts of climate change on society and the economy (PSAC 1965). At that time it was seen as a long-term issue, much as reflected in the Club of Rome's initial *Limits to Growth* report (Meadows et al. 1972) (Fig. 9.10).

In June 1988, James Hansen, then Director of the NASA Goddard Institute for Space Studies, testified to the US Congress that: "The greenhouse effect has been detected, and it is changing our climate now". The Intergovernmental Panel on Climate Change (IPCC) was formed the same year, issuing its First Assessment Report in 1990. As a result, the United Nations Framework Convention on Climate Change (UNFCCC 1992) was then opened for signature at the UN Conference on Environment and Development in Rio de Janeiro in 1992, ultimately with 197 nations as signatories, including Australia.

The original UNFCCC objective was:

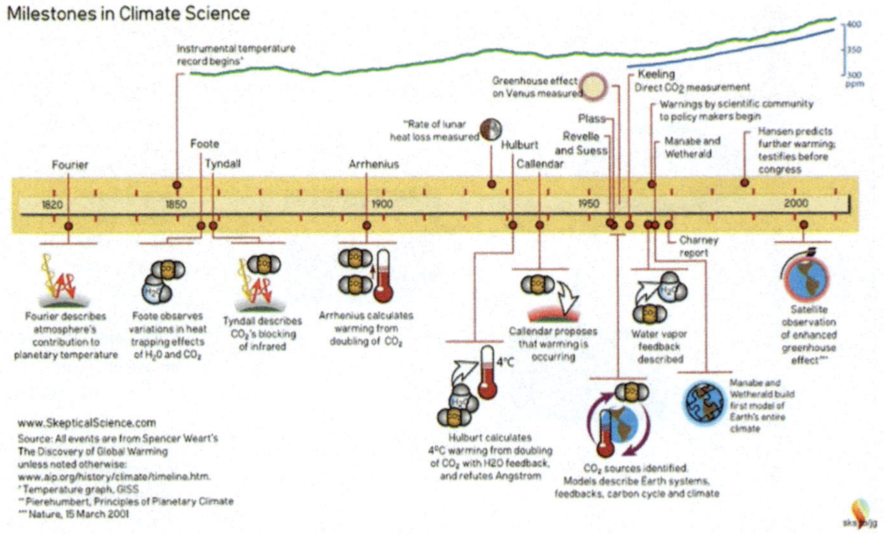

Fig. 9.10 Milestones in climate science. (Mason 2020)

> stabilization of greenhouse gas concentrations in the atmosphere at a level that would prevent dangerous anthropogenic interference with the climate system. Such a level should be achieved within a time frame sufficient to allow ecosystems to adapt naturally to climate change, to ensure that food production is not threatened and to enable economic development to proceed in a sustainable manner. (UNFCCC 1992)

Further:

> The Parties should take precautionary measures to anticipate, prevent or minimize the causes of climate change and mitigate its adverse effects. Where there are threats of serious or irreversible damage, lack of full scientific certainty should not be used as a reason for postponing such measures. (UNFCCC 1992)

The signatories have failed totally to meet those laudable objectives. Dangerous climate change is occurring at the ~1.3 °C global average surface temperature increase, relative to pre-industrial levels, already experienced. Ecosystems cannot adapt to this naturally, food production is threatened, economic development is not proceeding in a sustainable manner and far from precautionary steps being taken, massive expansion of fossil fuel use is being contemplated which will only accelerate adverse climate impacts.

The interminable UNFCCC Conference of the Parties (COP) climate negotiations progressed through the 1997 Kyoto Protocol to the 2015 Paris Climate Agreement. While the latter was hailed as a diplomatic triumph, in terms of serious climate action it is a disaster (Fig. 9.11).

Historically, CO_2 emissions have risen at worst-case levels. For three decades, despite all the climate negotiations and meetings, nothing has been achieved to arrest that rise, although it has now temporarily slowed due to Covid19. The path being followed pre-Covid would lead to a temperature increase of around 4 °C

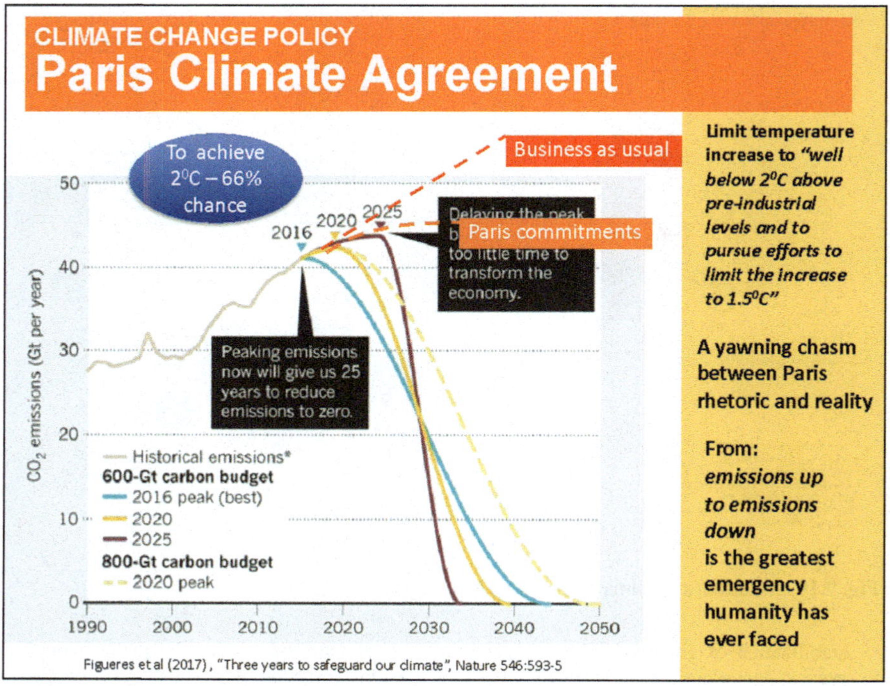

Fig. 9.11 Paris climate agreement

before 2100 – a world incompatible with an organised global society (Gulledge et al. 2007): in short, collapse leading to a substantial reduction in global population (Anderson 2009). The Paris voluntary commitments, if implemented, would only level-off emissions, not reduce them, leading to a temperature increase above 3 °C, quite possibly by 2050 – a world in social chaos (Gulledge et al. 2007). To meet even the upper 2 °C Paris limit requires emissions to peak immediately and rapidly decline thereafter. There is no sign that will happen, given current political and business attitudes, as the fossil fuel industry and political leaders have quite deliberately refused to face up to the implications of the Paris Agreement (Fig. 9.12).

So what would it take to stay below the Paris limits? In the above diagram, global fossil fuels are characterised in the three concentric ovals as: *resources* – what are known to be available; *reserves* – what proportion of resources are recoverable economically today; and *developed reserves* – what proportion of reserves are currently being exploited in existing mines, oil and gas fields. If only the developed reserves are used up, the emissions will be as shown in the left-hand bar. According to the IPCC, the amount of carbon that can henceforth be emitted to the atmosphere (the carbon budget) and have a 66% chance of staying below 2 °C (right-hand bar), is roughly equal to emissions from the developed reserves. For a 50% chance of staying below 1.5 °C, only around 55% of that amount can be emitted (central bar).

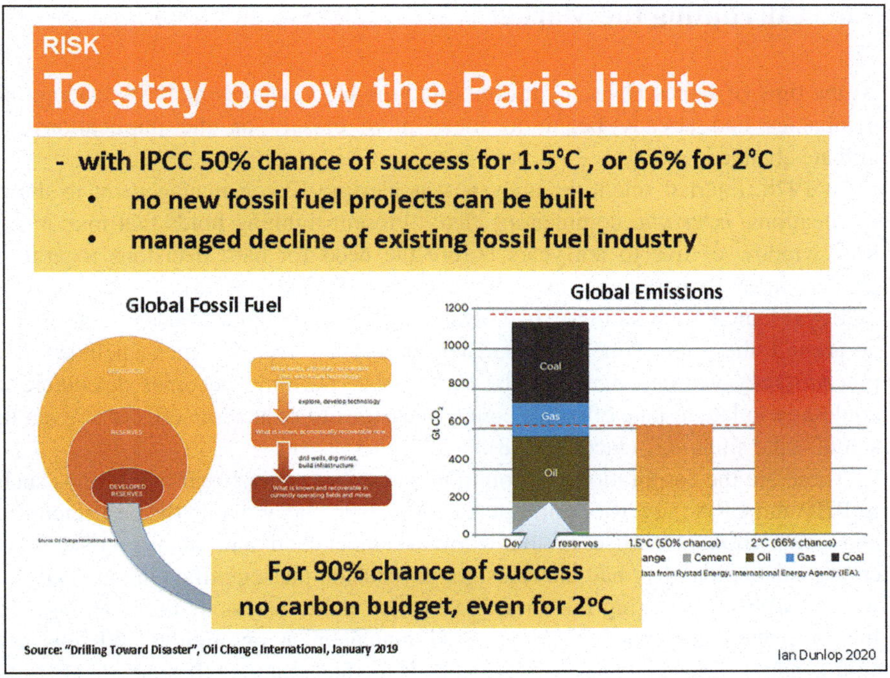

Fig. 9.12 Staying below the Paris climate agreement limits

Thus for 2 °C, no new fossil fuel projects can be developed, and to get *well below 2 °C, toward 1.5 °C,* the existing industry also needs to be wound down rapidly.

But 50% or 66% chances of success are not good odds to preserve the future of human civilisation, which is what we are talking about. If the chance of success is increased to 90%, there is no carbon budget left now to stay below 2 °C, let alone 1.5 °C. This means that the existing fossil fuel industry should be shut down immediately, which is obviously not going to happen given our addiction to those fuels.

Unfortunately this analysis does not take into account the potential non-linear, irreversible climate tipping points referred to in Chap. 2, this volume, which makes current climate risks even more acute, underlining the need for urgent precautionary measures to reduce carbon emissions far more rapidly that currently contemplated.

In short, the world has now reached the point where the fossil fuel industry must be rapidly dismantled if human civilisation in its current form is to survive. That emergency action is now required is largely due to the inaction of the industry itself, and of the politicians beholden to its leaders.

How did matters reach this point?

9.6 Corrupting the Market

At the time of James Hansen's testimony to the US Congress in 1988, leading oil companies had already begun to focus more closely on the implications of climate change.

In 1978, internal scientific analysis was alerting Exxon management to these implications, *inter alia* commenting that: "Present thinking holds that man has a time window of five to ten years before the need for hard decisions regarding changes in energy strategies might become critical" (Black 1978).

Royal Dutch Shell similarly had been commissioning greenhouse effect reports at around the same time, summarised in a 1988 report which included the precautionary warning that: "by the time global warming becomes detectable it could be too late to take effective countermeasures to reduce the effects or even to stabilise the situation" (Jacobs 1988).

These are the corporations, in common with other major fossil fuel groups such as BHP and CRA (now Rio Tinto) in Australia, who had arguably the best scientific and commercial expertise in the world to understand and manage risk. It is commendable that they had the foresight to focus on the scientific climate analysis which had been emerging for some years prior, and to assess its implications – in line with their corporate governance risk-management obligations. What is not commendable, indeed utterly irresponsible, is the fact that they then decided to hide those implications, in the interests of prolonging the life of the fossil fuel industry (Franta 2018). Further, the industry deliberately set in train a process of deceit and misinformation, designed to throw doubt over the scientific consensus on climate change reached by both their own, and external, experts, and to manufacture controversy aimed at preventing any move toward carbon-emission reductions.

US oil companies were arguably more blatant in this process than their European counterparts, but both were implicated at the highest board and management level (Cook et al. 2019; van Beek et al. 2020). While initially united in their denial of the climate science, over time the absurdity of this position split the industry. European members and more enlightened US corporations began leaving the Global Climate Coalition lobby group which had been formed to provide an industry front in 1989: BP in 1996, Royal Dutch Shell and Dupont in 1997, the Shell Oil US subsidiary in 1998 and the Ford Motor Company in 1999. But the Coalition still played a significant role in persuading President George W. Bush not to sign the Kyoto Protocol.

There was little altruism in these withdrawals. Opposition to climate action just moved on to more subtle expression, articulated by Alex Steffen as the concept of *predatory delay*:

> Those in positions of power understand only too well the need to change, but they simply argue for delay, on the basis that to change too quickly would be unfair to them. This allows them to be seen as responsible and caring. They want change, they claim; they just think we need prudent, appropriately paced change, mindful of economic trade-offs and judiciously studied – by which they mean cosmetic change for the foreseeable future. In the meantime, they fight like hell to delay change of any real magnitude, attacking not only the prospects

> of our kids and kin in the future, but increasingly of our society in the present. Their delay has real, serious human consequences, across generations. They're taking, not creating; the harm they cause is immeasurable (Steffen 2016).

And so in 2014, Royal Dutch Shell could still put out a 'socially responsible investor' briefing which, despite acknowledging that "climate change will continue to rise up the public and political agenda", suggested that global energy demand, from 2014 to 2050, would result in more fossil fuels consumption than had occurred up to that point since the Industrial Revolution (Shell 2014).

Likewise ExxonMobil, previously the most strident climate denier in the industry, finally acknowledged that: "The risk of climate change is clear and the risk warrants action", but with the convenient caveat that "all energy resources, including carbon-based fuels, are necessary to meet future energy demand growth as society manages the risks of climate change. … It is equally essential that society manages the risks of climate change by increasing energy efficiency and by investing in research into technologies to reduce greenhouse gas emissions" (ExxonMobil 2014).

As public awareness of climate risk has improved, with climate impact accelerating, and mounting regulator and investor pressure for action, the industry is gradually being forced to confront reality. It now puts out comprehensive scenario analyses suggesting ways to address climate change more urgently. Unfortunately most of these proposed solutions still advocate a substantial increase in fossil fuel use, offset by complex negative-emission technologies to absorb the increased carbon emissions – technologies which are either non-existent today, or have never been proven at scale. This is the ultimate moral hazard.

The most egregious issue around climate action has been the refusal of the industry to internalise the external costs of its operation. In short, to price carbon emissions to fully reflect the damage they cause, for example from physical climate impact itself, health effects, biodiversity loss, etc. The lack of such pricing provides a massive subsidy to the global fossil fuel industry which the International Monetary Fund recently estimated at around US$5.2 trillion annually, the Australian component being roughly A$42 billion (IMF 2019).

Hence the free market for energy with its supposedly level playing field, so beloved of neoliberal afficionados, is in fact totally corrupted in favour of fossil fuels. As Nicholas Stern put it: "Climate change is the result of the greatest market failure the world has seen" (Stern 2007).

Initially the industry fought vehemently against the very idea of carbon pricing. But as climate impact escalated, this gradually changed as the contradiction with their free market, supposedly polluter-pay, ideology became blindingly obvious. It was only possible to maintain that stance if climate denialism remained absolute, and that was no longer realistic globally, at least in terms of public corporate positioning whatever views executives held behind the scenes.

Widespread carbon pricing is growing closer, but resistance amongst many politicians and corporate hard-heads remains strong, on the grounds it would be added taxation, rather than the true reflection of the cost of fossil fuel use it actually represents. Which puts the industry in the convenient, yet hypocritical, position that it can urge implementation, confident in the knowledge that political inertia will ensure it does not occur for years ahead. In the meantime, many companies claim to use internal carbon pricing when evaluating new projects, although prior to the Covid-19 pandemic this appeared to have little effect as expansion proceeded apace. Yet more predatory delay.

9.7 The Australian Response: Ratcheting Climate Ambition Ever Downwards

In the 1980s, the initial response to climate change concerns in some segments of the Australian coal industry was constructive, as they examined ways in which emission reductions from coal use could be achieved using technology, efficiency improvements and conservation. But that was rapidly overwhelmed by climate denialism across the fossil fuel industry overall, even more pronounced than the extremists in the international oil industry, and masterminded by conservatives such as Western Mining's Hugh Morgan and Ray Evans.

Under their influence, Prime Minister John Howard obligingly set the scene with his 'Safeguarding The Future – Australia's Response to Climate Change' statement in November 1997, just prior to signing the Kyoto Protocol (Howard 1997). It was a defensive document, intent primarily on enhancing Australia's status as a major fossil fuel producer and ensuring Australia did little to address climate change. As former Treasury Secretary Ken Henry put it: "The Australian government saw climate change as a risk, not an opportunity. Our negotiators' approach was not to secure strong emissions reduction, but to ensure the burden on Australia was as small as possible" (Henry 2020).

Pleading special circumstances because of our high-carbon economy, during last-minute negotiations chief negotiator and environment minister Robert Hill insisted on the insertion of the infamous 'Australia Clause' into the Kyoto Protocol, effectively holding the rest of the world to ransom (Hamilton and Vellens 1999). This allowed Australia's 1990 base to include emissions from an elevated level of land clearing, which the government well knew had subsequently dropped off dramatically. Add to that the negotiated allowance to increase emissions by 8% in the first Kyoto commitment period (2008–12), when almost all other countries' emission allowances decreased, meant that from 1990 to 2012, Australia's emission from all sources other than land-clearing increased by 28%, yet we 'met' our Kyoto targets. Our emissions subsequently went down during the short carbon-pricing period to 2014, but thereafter increased: by the end of 2017 they were 31% above 1990 levels. Some reduction (Fig. 9.13).

The Australia Clause was hailed as a negotiating triumph, but it had disastrous consequences. By delaying the need for Australian companies to take serious climate action, it created the complacency which remains today, gave strength to the

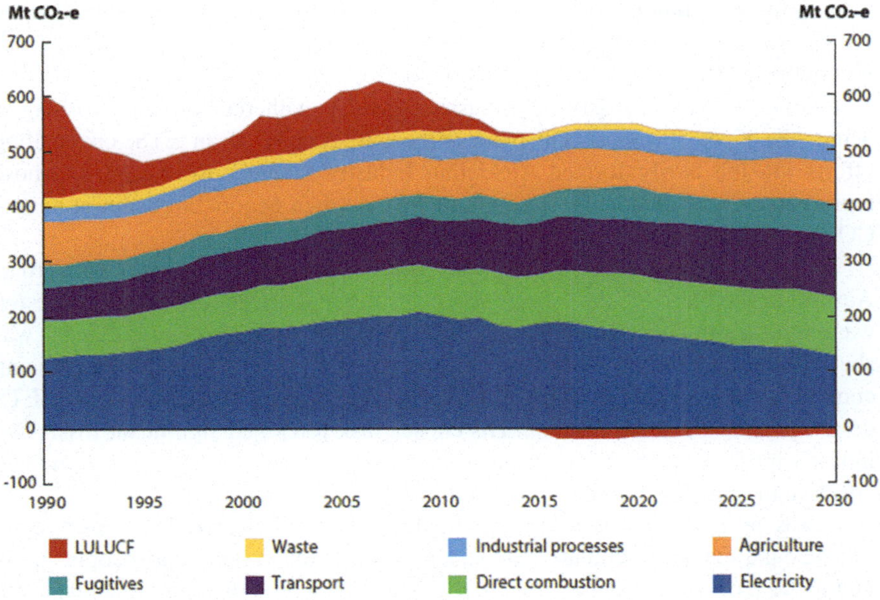

Fig. 9.13 Australia's emissions 1990–2030. (DEE 2019)

denialist lobby and threw away Australia's opportunity to become a world leader in renewable technologies. With the passage of time, it will be seen as the greatest strategic mistake Australia ever made.

In early 2021, the Australian government has still not completely rejected the idea of carrying over unused carbon credits from the Kyoto period which were only gained by this unprincipled subterfuge in the first place, to meet Australia's wholly inadequate 2030 emission-reduction targets under the 2015 Paris Climate Agreement. Not only is this morally and ethically indefensible in a world which must see emergency emission reduction, but it is economic lunacy, yet again undermining our ability to capture the opportunities offered in transforming to a low-carbon world.

Perhaps the worst outcome of the Australia Clause was that it confirmed Australia as a 'free rider', intent upon muddying the waters to prevent any meaningful global climate policy developing, and encouraged others with similar mindset to do likewise. This has been Australia's enduring legacy.

One of the obligations of signing the Kyoto Protocol was the need to develop economic instruments to reduce emissions, the preferred choice of economists being an Emissions Trading System (ETS), which effectively priced carbon emissions to encourage the use of alternatives. Accordingly, the first Australian ETS was designed in 1998, under the auspices of the Australian Greenhouse Office, by an independent expert group (Senate Committee 2000). This received support, ostensibly, from all major business groups and a proposal was submitted for cabinet approval in 2000. Then Prime Minister John Howard sat on the proposal for months, waiting for incoming US president George W. Bush to declare his intentions on the Kyoto Protocol. When Bush decided not to sign, under the influence of US fossil fuel interests, Howard shelved this first ETS proposal.

Behind the scenes, fossil fuel industry denialists gathered strength, working to neuter the Kyoto Protocol and prevent the introduction of an ETS, channelling efforts via the newly formed Australian Industry Greenhouse Network, whose representatives became embedded with the Australian negotiating team at annual COP meetings.

The result of these industry efforts has been a continual ratcheting downward of Australia's ambition to take serious action on climate change. The first ETS design was a robust, simple economic framework with minimal escape clauses. After it was shelved, carbon pricing and any related ETS remained off limits until increasing community concern over climate change forced Howard to initiate a new ETS design prior to the 2006 election. His chosen task force had significant fossil fuel industry representation and produced a design less onerous for the industry, and less effective in reducing emissions. Howard then lost the election.

It was no surprise when Howard finally confirmed that he had never really accepted the science of climate change, although not to the Australian people, but in giving the 2013 annual lecture 'One Religion is Enough' to the climate denialists of the London Global Warming Policy Foundation (Howard 2013).

New Prime Minister Kevin Rudd promptly ratified the Kyoto Protocol, then developed a further ETS variant in the form of the Carbon Pollution Reduction Scheme which did not gain parliamentary support, largely due to the even greater inadequacy of its emission reduction provisions. Finally, having toppled Rudd, Julia Gillard's Clean Energy Act, with a $23/ton carbon price, was introduced in 2012. This reduced emissions, but it was promptly abandoned in 2014 after a conservative government led by Tony Abbott came to power. Thereafter, further political chaos saw a sequence of ineffective schemes including the Clean Energy Target, Emission Reduction Fund and the National Energy Guarantee, ratcheting climate ambition ever downwards and doing little to cut Australia's emissions. The only saving grace in this sorry tale is that, to avoid becoming a complete global pariah, Australia signed and ratified the 2015 Paris Climate Agreement, albeit with wholly inadequate initial emission-reduction commitments.

Another demonstration of fossil fuel industry power was the highly successful Minerals Council of Australia campaign which prevented the introduction of the 'resource super profits tax' proposed in the 2010 Henry Tax Review, ensuring that the Australia community continued to receive inadequate returns from its resource developments (Kehoe 2020).

In 2021 the Morrison government continues in climate denial – if effective action is any guide – notwithstanding the disastrous 2019–20 bushfires on the Eastern seaboard of Australia and the climate-related Covid-19 pandemic. Groups charged with planning recovery from the pandemic, such as the National Covid Commission, are heavily fossil-fuel-biased, emphasising a gas-led recovery, notwithstanding the fact that cheaper, stable and low-emission renewable alternatives are now available.

9.8 From Hard to Soft Denial

In common with the global industry, in the light of the accelerating damage from climate impacts globally, the Australian fossil fuel industry found it increasingly difficult to maintain the hard denial of earlier years, so it morphed into a softer variant. Publicly, the industry now accepts, grudgingly, that climate change is real and caused by anthropogenic carbon emissions; every corporate website has its commitment to 'sustainability' and in many cases a climate change plan.

This has occurred in part because of increasing concerns from global regulators, and the main corporate and financial regulators in Australia, that climate change now represents a real threat to the stability of the global financial system. In turn, there is increasing divestment from fossil fuel operations, with investors becoming reluctant to support new projects.

However, what has yet to be accepted is the urgency for action and any commitment to early emission reductions. It is now fashionable for companies to commit to achieving net-zero-emissions by 2050, which is far too late if the Paris climate objectives are to be met. Likewise, politicians and industry leaders divert attention from the need to mitigate climate change by reducing emissions, and instead encourage adaptation to climate impact and the development of more resilient communities, when they know full well that it is not possible to adapt to the 3–4 °C temperature increases which current policies imply. Prevention via rapid emission reduction is the only solution.

In short, soft denial is yet more predatory delay to extend the life of the fossil fuel industry, irrespective of the damage that might do to society.

9.9 Conclusion: Facing the Climate Emergency

Since the Industrial Revolution, the world has benefited greatly from access to cheap fossil fuels. However, it has been clear for decades that their benefits were rapidly being eroded by the damage caused by the climate change they generate.

The brutal reality is that the world now faces a climate emergency, in large part due to the intransigence of the fossil fuel industry in facing up to this reality, and its refusal to accept the need to constructively manage its own decline.

Because of this inaction, the thinking which still dominates political rhetoric, of a graduated transition away from fossil fuels toward low-carbon alternatives, is no longer realistic. Emergency action to reduce emissions is now essential.

In a recent interview, Shell managing director, Ben van Beurden, when asked for his response to accusations that the company failed to act on climate change despite knowing the consequences, replied: "Yeah, we knew. Everybody knew, and somehow we all ignored it" (Worland 2020).

The industry can ignore it no longer. There is no further place for the duplicitous predatory delay which even the more progressive fossil fuel companies remain guilty of today. The industry has no social licence to destroy our civilisation, for that is what they are currently doing. It should either proactively co-operate in shaping the emergency transition to a genuinely low-carbon world, or it will be rapidly phased out.

The words of Lucius Anneaus Seneca come to mind:

> It would be some consolation for the feebleness of ourselves and our works if all things should perish as slowly as they come into being; but as it is, increases are of sluggish growth, but the way to ruin is rapid (Bardi 2017).

References

Abreu C (2019) COP25 becomes first to use "f" words to call out fossil fuels. The Energy Mix. https://theenergymix.com/2019/12/18/cop-25-becomes-first-to-use-f-words-to-call-out-fossil-fuels/. Accessed 27 Feb 2021

Anderson K (2009) The Scotsman. https://sites.google.com/site/punishclimatecriminals/anderson. Accessed 27 Feb 2021

Australian Dept of Environment & Energy (2019) Australian energy update 2019. https://www.energy.gov.au/sites/default/files/australian_energy_statistics_2019_energy_update_report_september.pdf. Accessed 27 Feb 2021

Australian Dept of Foreign Affairs & Trade (2019) Trade & Investment at a glance 2019. https://www.dfat.gov.au/sites/default/files/trade-and-investment-at-a-glance-2019.pdf. Accessed 27 Feb 2021

Bardi U (2020) The Great EROEI scam: are renewables a good idea? https://cassandralegacy.blogspot.com/2020/05/the-great-eroei-scam-what-is-real.html. Accessed 27 Feb 2021

Bardi U (2017) The Seneca effect: why growth is slow, but collapse is rapid. Springer, Heidelberg

Black BC (2017) How WW1 ushered in the century of oil. The Conversation. https://theconversation.com/how-world-war-i-ushered-in-the-century-of-oil-74585. Accessed 27 Feb 2021

Black JF (1978) The greenhouse effect. Exxon Research & Engineering. https://insideclimatenews.org/news/16092015/exxons-own-research-confirmed-fossil-fuels-role-in-global-warming/. Accessed 27 Feb 2021

BP Statistical Review of World Energy (2019) Energy economics – oil price. https://www.bp.com/content/dam/bp/business-sites/en/global/corporate/pdfs/energy-economics/statistical-review/bp-stats-review-2019-full-report.pdf. Accessed 27 Feb 2021

Brett J (2020) The coal curse: resources, climate and Australia's future. Quarterly Essay. https://www.quarterlyessay.com.au/essay/2020/06/the-coal-curse. Accessed 27 Feb 2021

Cook J, Supran G, Lewandowsky S et al (2019) America misled: how the fossil fuel industry deliberately misled Americans about climate change. https://www.climatechangecommunication.org/wp-content/uploads/2019/10/America_Misled.pdf. Accessed 27 Feb 2021

DEE (2019) Australia's emission projections 2019. Department of the Environment & Energy. https://www.industry.gov.au/data-and-publications/australias-emissions-projections-2019. Accessed 27 Feb 2021

Desmog (2019) Global climate coalition. https://www.desmogblog.com/global-climate-coalition. Accessed 27 Feb 2021

ExxonMobil (2014) Report to shareholders on managing climate risk. https://ir.exxonmobil.com/news-releases/news-release-details/exxonmobil-releases-reports-shareholders-managing-climate-risk. Accessed 27 Feb 2021

Farand C (2020) For all its green talk, the IEA still gives comfort to oil and gas producers. Climate Change News. https://www.climatechangenews.com/2020/07/27/green-talk-iea-still-gives-comfort-oil-gas-producers/. Accessed 27 Feb 2021

Franta B (2018) Shell & Exxon secret 1980s climate change warnings. The Guardian. https://www.theguardian.com/environment/climate-consensus-97-per-cent/2018/sep/19/shell-and-exxons-secret-1980s-climate-change-warnings. Accessed 27 Feb 2021

Freeze B (2003) Coal: a human history. Arrow Books, London

Galey P (2019) Fossil fuel groups destroying climate talks. Phys Org. https://phys.org/news/2019-12-fossil-fuel-groups-climate-ngos.html. Accessed 27 Feb 2021

Gulledge J, McNeill JR, Podesta J et al (2007) The age of consequences: the foreign policy and national security implications of global climate change. CSIS, Washington, DC. https://www.csis.org/analysis/age-consequences. Accessed 27 Feb 2021

Hamilton C (2007) Scorcher: the dirty politics of climate change. Black Inc Books, Melbourne

Hamilton C, Vellens L (1999) Land use change in Australia and the Kyoto Protocol. Elsevier. https://www.sciencedirect.com/science/article/abs/pii/S1462901199000076. Accessed 27 Feb 2021

Henry K (2020) Climate wars. ABC Four Corners, May. https://www.abc.net.au/4corners/climate-wars/12254562. Accessed 27 Feb 2021

Homer-Dixon T (2006) The upside of down: catastrophe, creativity and the renewal of civilisation. Island Press. https://homerdixon.com/writing/books/the-upside-of-down/. Accessed 27 Feb 2021

Howard J (1997) Safeguarding the future – Australia's response to climate change. https://catalogue.nla.gov.au/Record/1699197?. Accessed 27 Feb 2021

Howard J (2013) One religion is enough. Speech delivered to the Global Warming Policy Foundation, November. https://www.thegwpf.org/content/uploads/2013/12/Howard-2013-Annual-GWPF-Lecture.pdf. Accessed 27 Feb 2021

IMF (2019) Global fossil fuel subsidies remain large. https://www.imf.org/en/Publications/WP/Issues/2019/05/02/Global-Fossil-Fuel-Subsidies-Remain-Large-An-Update-Based-on-Country-Level-Estimates-46509. Accessed 27 Feb 2021

International Energy Agency (2008) World energy outlook. https://www.iea.org/reports/world-energy-outlook-2008. Accessed 27 Feb 2021

International Energy Agency (2019) World energy outlook. https://www.iea.org/reports/world-energy-outlook-2019. Accessed 27 Feb 2021

International Energy Agency (2021) History. https://www.iea.org/about/history. Accessed 27 Feb 2021

Jacobs RPWM (1988) The greenhouse effect. Shell Internationale Petroleum Maatschappij. http://www.climatefiles.com/shell/1988-shell-report-greenhouse/. Accessed 27 Feb 2021

Kehoe J (2020) Henry blasts mining tax stupidity. Aust Financial Rev, 3 August. https://www.afr.com/policy/tax-and-super/henry-blasts-mining-tax-stupidity-20200803-p55hzj. Accessed 27 Feb 2021

Kelly S (2020) As Trump leaves Permian oilfield, industry insiders question if 2020 bust marks Texas oil's last big boom. Desmog. https://www.desmogblog.com/2020/07/30/trump-leaves-texas-some-question-if-shale-bust-marks-permian-oilfield-s-final-boom?utm_source=DeSmog%20Weekly%20Newsletter. Accessed 27 Feb 2021

MacLean N (2017) Democracy in chains, scribe publications. https://www.nytimes.com/2017/08/15/books/review/democracy-in-chains-nancy-maclean.html. Accessed 27 Feb 2021

Mason J (2020) The history of climate science. Skeptical Sci. https://skepticalscience.com/history-climate-science.html. Accessed 27 Feb 2021

Meadows DH, Meadows DL, Randers J, Behrens W (1972) The limits to growth. New York, Potomac Associates, Universe Books

Mission 2020 (2019/20) Letters to the IEA. https://www.reuters.com/article/uk-climate-change-energy/investors-scientists-urge-iea-head-to-take-bolder-climate-stance-idUKKBN2351UO?edition-redirect=uk. Accessed 27 Feb 2021

Montgomery CT, Smith MB (2010) Hydraulic fracturing: history of an enduring technology. J Petroleum Technol. https://onepetro.org/JPT/article-abstract/62/12/26/194396/Hydraulic-Fracturing-History-of-an-Enduring?redirectedFrom=fulltext. Accessed 27 Feb 2021

Mushalik M (2019) World crude oil production outside US and Iraq is flat since 2005. https://crudeoilpeak.info/world-crude-production-outside-us-and-iraq-is-flat-since-2005. Accessed 27 Feb 2021

Oil Change International (2018) Off track: how the IEA guides energy decisions toward fossil fuel dependence and climate change. http://priceofoil.org/content/uploads/2018/04/OFF-TRACK-the-IEA-Climate-Change.pdf. Accessed 27 Feb 2021

Paul JA (2002) Great power conflict over Iraq oil: the WW1 era. https://www.globalpolicy.org/component/content/article/185-general/40479-great-power-conflict-over-iraqi-oil-the-world-war-i-era.html. Accessed 27 Feb 2021

Pearse G (2007) High & dry: John Howard, climate change and the selling of Australia's future. Penguin Books, Melbourne

Pearse G, McKnight D, Burton B (2013) Big coal: Australia's dirtiest habit. New South Books, Sydney

PSAC (1965) Restoring the quality of our environment. The White House. http://www.climatefiles.com/climate-change-evidence/presidents-report-atmospher-carbon-dioxide/. Accessed 27 Feb 2021

Ritter D (2018) The coal truth: the fight to stop Adani, defeat the big polluters and reclaim our democracy. UWA Press, Nedlands

Senate Committee (2000) Emissions Trading Ch 9, Senate Committee on Environment. https://www.aph.gov.au/parliamentary_business/committees/senate/environment_and_communications/completed_inquiries/1999-02/gobalwarm/report/c09. Accessed 27 Feb 2021

Shell (2014) Investing for sustainable growth. Ben van Beurden. https://www.shell.com/investors/investor-presentations/2014-investor-presentations/socially-responsible-investors-briefing-london-april-10-2014.html. Accessed 27 Feb 2021

Steffen A (2016) Predatory delay and the rights of future generations. Medium. https://medium.com/@AlexSteffen/predatory-delay-and-the-rights-of-future-generations-69b06094a16. Accessed 27 Feb 2021

Stern N (2007) Climate change is the result of the greatest market failure the world has seen. The Guardian. https://www.theguardian.com/environment/2007/nov/29/climatechange.carbonemissions. Accessed 27 Feb 2021

UNFCCC (1992) United Nations framework convention on climate change 1992. UN, New York. https://unfccc.int/resource/docs/convkp/conveng.pdf. Accessed 27 Feb 2021

UNFCCC (2015) Paris Climate Change Agreement. https://unfccc.int/files/essential_background/convention/application/pdf/english_paris_agreement.pdf. Accessed 27 Feb 2021

van Beek B, Beunder A, Mast J et al (2020) Dutch multinationals funded climate sceptic, follow the money. https://www.ftm.nl/dutch-multinationals-funded-climate-sceptic. Accessed 27 Feb 2021

Watts J (2018) IEA accused of undermining global shift from fossil fuels. The Guardian, 5 April. https://www.theguardian.com/environment/2018/apr/05/iea-accused-of-undermining-global-shift-from-fossil-fuels. Accessed 27 Feb 2021

Wilkinson M (2020) The carbon club: how a network of influential climate sceptics, politicians and business leaders fought to control Australia's climate policy. Allen & Unwin, Sydney

Worland J (2020) The reason fossil fuel companies are finally reckoning with climate change. Time, 16 January. https://time.com/5766188/shell-oil-companies-fossil-fuels-climate-change/. Accessed 27 Feb 2021

Chapter 10
Economic Failures of the IPCC Process

Steve Keen

Abstract Projections of the economic damages from climate change by Neoclassical economists have dominated climate change mitigation policy. These projections have been generated by methods that fail to meet even minimal academic standards. Research by scientists establishing the existence of "tipping points" in climate that will be triggered this century, at levels of temperature increase over pre-industrial levels that have already been exceeded, have been misconstrued to allege that there are no significant tipping points for the next three centuries and for temperature rises below 3 °C. Economists have simply assumed, with no proof, that activities undertaken indoors or underground, which constitute over 85% of GDP, will be unaffected by climate change. They have used data on the weak geographic relationship between temperature and GDP as a proxy for the impact of climate change on the economy. These intellectual failures are endemic to Neoclassical economics, with a methodology that treats critical assumptions as if they are merely simplifying assumptions.

10.1 Introduction

The *Intergovernmental Panel on Climate Change* (*IPCC*) is the premier international body collating the scientific assessment of climate change, and proposals for mitigation. A joint creation of the United Nations agencies the *World Meteorological Organization* (*WMO*) and the *United Nations Environment Programme* (*UNEP*), it brings together scientists from myriad disciplines to assess and summarise the current research on climate change, collating knowledge that is then used to inform governments and politicians. The scientists work on a volunteer basis.

The IPCC relies upon its member governments and "Observers Organizations" to nominate its volunteer authors. This means that, subject to their willingness to volunteer, the most prestigious individuals specialising in climate change in each discipline become the authors of the relevant IPCC chapter for their discipline. They

S. Keen (✉)
Institute for Strategy, Resilience and Security, UCL, London, UK
e-mail: s.keen@isrs.org.uk

S. J. Williams, R. Taylor (eds.), *Sustainability and the New Economics*,
https://doi.org/10.1007/978-3-030-78795-0_10

then undertake a review of the peer-reviewed literature in their field (and some non-peer-reviewed work, such as government reports) to distil the current state of knowledge about climate change in their discipline. A laborious review process is also followed, so the draft reports of the volunteer experts are reviewed by other experts in each field, to ensure conformity of the report with the discipline's current perception of climate change. The emphasis upon producing reports which reflect the consensus within a discipline has resulted in numerous charges that the IPCC's warnings are inherently too conservative (Herrando-Pérez et al. 2019; Brysse et al. 2013).

But the main weaknesses with the IPCC's methodology are firstly that, in economics, it exclusively selects Neoclassical[1] economists (Keen 2011); and secondly, because there is no built-in review of one discipline's findings by another, the conclusions of these Neoclassical economists about the dangers of climate change are reviewed predominantly by other Neoclassical economists. The economic sections of IPCC reports are therefore unchallenged by other schools of thought in economics, and by other disciplines who also contribute to the IPCC's reports.

Given the extent to which economists dominate the formation of most government policies in almost all fields, and not just strictly economic policy (Fourcade et al. 2015; Hirschman and Berman 2014; Christensen 2018; Lazear 2000; Pusey 1991), the otherwise acceptable process by which the IPCC collates human knowledge on climate change has critically weakened, rather than strengthened, human society's response to climate change. This is because, commencing with "Nobel Laureate" William Nordhaus (see Mirowski 2020 for why the "Sveriges Riksbank Prize in Economic Sciences in Memory of Alfred Nobel" is not strictly a Nobel Prize), the economists who specialise on climate change have significantly underestimated the dangers that climate change poses to human civilisation.

10.2 Nobel Oblige

In his 2018 Nobel Prize lecture, William Nordhaus described a trajectory that would lead to global temperatures peaking at 4 °C above pre-industrial levels in 2145 as "optimal" (Nordhaus 2018a, Slide 6) because, according to his calculations, the damages from climate change over time, plus the abatement costs over time (i.e. net costs), are minimised on this trajectory. He estimated the discounted cost of the economic damages from unabated climate change—which would see temperatures approach 6 °C above pre-industrial levels by 2150—at $24 trillion, whereas the 4 °C trajectory had damages of about $15 trillion and abatement costs of about $3 trillion. Trajectories with lower peak temperatures had higher abatement costs that overwhelmed the benefits (Nordhaus 2018a, Slide 7). In a subsequent paper,

[1] Neoclassical economics is what is taught in mainstream economics textbooks, and is believed by about 85% of professional economists. It dominates the public perception of economics as well. This is despite numerous fundamental critiques of it, which I collate in my book *Debunking Economics* (2001).

Nordhaus claimed that even a 6 °C increase would only reduce global income by just 7.9%, compared to what it would be in the complete absence of global warming.

This sanguine assessment of the costs of climate change contrasts starkly with the *non-economic* sections of IPCC reports. The recent *Global Warming of 1.5 °C* Report, for example, predicted that 70% of insects and 40–60% of mammals would lose 50% or more of their range at 4.5 °C (Warren et al. 2018, p. 792). Yet the economic components of IPCC reports concur with Nordhaus that damages from climate change will be slight: the Executive Summary of Chapter 10 of the 2014 Fifth Assessment, "Key Economic Sectors and Services", opens with the declaration that:

> **For most economic sectors, the impact of climate change will be small relative to the impacts of other drivers (*medium evidence, high agreement*).** Changes in population, age, income, technology, relative prices, lifestyle, regulation, governance, and many other aspects of socioeconomic development will have an impact on the supply and demand of economic goods and services that is large relative to the impact of climate change. (Arent et al. 2014a, p. 662)

How can such relatively small estimates of economic damages be reconciled with the large impacts that scientists expect on critical components of the biosphere? The answer is that they can't, because the economic studies are *not* based on the scientific assessments of damage from climate change. Instead, the numerical estimates of the impact of climate change on GDP have been made up by economists themselves. I use the expression "made up" advisedly, because there is no more accurate way to characterise how Neoclassical economists have approached climate change. Before I explain how these spurious estimates were manufactured, it is useful to contrast them with some of the more easily understood dangerous consequences of a higher global average temperature.

10.3 The Scientific Assessment

Using Nordhaus's sanguine prediction of a mere 7.9% reduction in global income from a 6 °C increase (Nordhaus 2018b, p. 345) as a reference point, three of the most obvious threats of a 6 °C warmer world are the impact of these temperatures on human physiology, on the survival of other animals, and on the structure of the Earth's atmospheric circulation systems.

A critical feature of human physiology is our ability to dissipate internal heat by perspiration. To do so, the external air needs to be colder than our ideal body temperature of about 37 °C, and dry enough to absorb our perspiration as well. This becomes impossible when the combination of heat and humidity, known as the "wet bulb temperature", exceeds 35 °C. Above this level, we are unable to dissipate the heat generated by our bodies, and the accumulated heat will kill a healthy individual within a few hours. Scientists have estimated that a 3.8 °C increase in the global average temperature would make Jakarta's temperature and humidity combination permanently fatal for humans, while a 5.5 °C increase would mean that *even*

New York would experience 55 days per year when the combination of temperature and humidity would be deadly (Mora et al. 2017, Figure 4, p. 504).

Temperature also affects the viable range for all biological organisms on the planet. Scientists have estimated that a 4.5 °C increase in global temperatures would reduce the area of the planet in which life could exist by 40% or more, with the decline in the liveable area of the planet ranging from a minimum of 30% for mammals to a maximum of 80% for insects (Warren et al. 2018, Figure 1, p. 792).

The Earth's current climate has three major air circulation systems in each hemisphere: a "Hadley Cell" between the Equator and 30°, a mid-latitude cell between 30° and 60°, and a Polar cell between 60° and 90°. This structure is why there is such large differences in temperature between the tropics, temperate and arctic regions, and relatively small differences within each region.[2] Scientists have modelled the stability of these cells and concluded that they could be tipped, by an average global temperature increase of 4.3 °C or more, into a state with just one cell in the Northern Hemisphere, and an average Arctic temperature of 22.5 °C. This abrupt transition (known as a bifurcation), sometime in the next century, would utterly alter weather patterns everywhere in the Northern Hemisphere, profoundly disrupting all forms of life, and it would occur far too quickly for any meaningful adaptation, by natural and human systems alike (Kypke et al. 2020, Figure 6 p. 399).

These three factors and many others, caused not only by industry's CO_2 emissions but also by the myriad other ways we damage the biosphere, would occur together, and interact with each other, at temperature levels of 4–6 °C above pre-industrial levels. Nordhaus's assertion that such devastating changes to the Earth's climate would reduce global GDP by less than 8%, compared to a world without climate change (Nordhaus 2018b, p. 345), is simply risible. How could he arrive at such an absurd conclusion?

In common with most of my peers in non-Neoclassical economics, I initially assumed that the answer was that he applied far too high a "discount rate" to future damages (Hickel 2018). If you think that 99% of the economy would be destroyed in a century from now by the catastrophic effects of a 6 °C increase in temperatures, but discount that back to today's world at a rate of 7% p.a., you get the result that this collapse in future GDP is worth only 0.1% of today's GDP—which is no big deal.

If this had been how Nordhaus had arrived at such a low damage estimate, then the high discount rate could be challenged, but the rest of his analysis could potentially be sound. But our guess was wrong. Nordhaus explained why he used a high discount rate when he strongly criticised the lower discount rate used by the *Stern Review* (Stern 2007). It was not to reduce future catastrophic damages to trivial levels now, but because, if he used a low discount rate, then:

[2] When one controls for other factors apart from latitude, such as altitude, distance from the ocean, etc.

> *the relatively small damages in the next two centuries* get overwhelmed by the high damages over the centuries and millennia that follow 2200. (Nordhaus 2007, p. 202. Emphasis added)

"*Relatively small damages in the next two centuries*"? How did he reach that conclusion? I found out that he and his colleagues ignored or distorted the work of scientists, and instead used their own trivial estimates of the economic damage from climate change. I have spent fifty years of my life being a critic of Neoclassical economics. Neoclassical work on climate change is by far the lowest grade work that I have read in that half-century.

10.3.1 Reading Catastrophe and Seeing Utopia

I am accustomed to shaking my head in wonder at the capacity of Neoclassical economists to come up with ludicrous assumptions to jump over some logical impasse, like this gem by the developer of the Capital Assets Pricing Model, William Sharpe. After meticulously deriving a model of an individual "rational" investor, Sharpe then proceeded to model the entire market by assuming:

> homogeneity of investor expectations: investors are assumed to agree on the prospects of various investments – the expected values, standard deviations and correlation coefficients described in Part II. (Sharpe 1964, pp. 433–34; Fama and French 2004, p. 26)

This is patently absurd. But at least Sharpe conceded as much, when he attempted to justify it via a clumsy rendition of Milton Friedman's dodgy methodological defence of absurd assumptions (Friedman 1953).[3] Fama, who led the empirical defence of this theory (Fama 1970; Fama and French 1996, 2004), fitted it to actual stock market data, which for a time appeared to support the theory.[4]

But in this work on climate change, Neoclassical economists have fitted their absurd-assumptions models *to their own manifestly absurd made-up data.* When they consulted scientists or referenced scientific literature, they frequently distorted the research or drowned the warnings of scientists with the blasé expectations of

---

[3] "Needless to say, these are highly restrictive and undoubtedly unrealistic assumptions. However, since the proper test of a theory is not the realism of its assumptions but the acceptability of its implications, and since these assumptions imply equilibrium conditions which form a major part of classical financial doctrine, it is far from clear that this formulation should be rejected – especially in view of the dearth of alternative models leading to similar results" (Sharpe 1964, p. 434)

[4] Fama and French, who were strident defenders of this theory for 40 years, admitted in 2004 that the theory also assumes these shared expectations are correct! In other words, their model of a "rational" investor is one who can accurately predict the future—which a sane person would call not rational but "prophetic"—and they assume that everyone is "rational" in that sense: "Sharpe (1964) and Lintner add two key assumptions to the Markowitz model to identify a portfolio that must be mean-variance-efficient. The first assumption is complete agreement … on … asset returns…. *And this distribution is the true one*—that is, it is the distribution from which the returns we use to test the model are drawn" (Fama p. 26).

economists. This was nowhere more evident than in Nordhaus's treatment of a key paper on the likelihood that global warming will trigger "tipping points" that cause runaway climate change, "Tipping elements in the Earth's climate system" (Lenton et al. 2008).

Lenton's paper considered whether there were large-scale elements of Earth's climactic system which could be triggered into a qualitative change that would in turn rapidly alter the climate. He and his co-authors identified nine such systems (all but one of which—the Indian Monsoon—could be tipped by temperature increases alone), after applying the conditions that these systems had to be "subsystems of the Earth system that are at least subcontinental in scale" which could be "switched—under certain circumstances—into a qualitatively different state by small perturbations". They excluded "systems in which any threshold appears inaccessible this century" (Lenton et al. 2008, pp. 1786–87).

If "Tipping elements" exist, and if they can be triggered by a temperature rise that can be expected from global warming this century, and if they would have drastic impacts upon the global climate, then the only prudent thing to do is to avoid such a temperature rise in the first place. If that rise occurs and a climactic element is tipped, it will unleash forces that will be far too large for human action to reverse. A qualitatively different climate would result, whose consequences for human civilisation cannot be predicted by extrapolating from what we know about our current climate.

Lenton and his co-authors identified two definite candidates for a tipping point *this* century—"Arctic sea-ice and the Greenland ice sheet"—and noted that "at least five other elements could surprise us by exhibiting a nearby tipping point" (Lenton et al. 2008, p. 1792) before 2100. They could not at the time specify a critical value for the temperature at which the decline in Arctic summer sea-ice would tip the Arctic from being a reflector of solar energy to an absorber, but noted that "a summer ice-loss threshold, *if not already passed,* may be very close and a transition could occur well within this century" (Lenton et al. 2008, p. 1789. Emphasis added).

Nordhaus "summarised" this research with the sentence:

> Their review finds no critical tipping elements with a time horizon less than 300 years until global temperatures have increased by at least 3°C (Nordhaus 2013a, p. 60).

This summary is patently false. Lenton's careful definition itself contradicts Nordhaus's alleged summary, by explicitly excluding systems which could not be triggered this century. Nordhaus's claim that they found no systems which could be triggered in the next three centuries is false: they identified at least two, and possibly seven (out of eight!), which could be triggered this century—though the process of transiting from their current to final states could take several centuries. Nordhaus claimed that the minimum temperature rise that could trigger a "critical tipping element" was 3 °C: they said that Arctic summer sea-ice could be triggered with a rise of as little as 0.5 °C—a level we have already well and truly exceeded.

There are few—if any—scientific or economic issues that are more important than this. It deserves the closest possible attention. And yet, if an undergraduate student of mine had summarised this paper as Nordhaus did, I would have failed him. However, since I was reading the work of a "Nobel Prize winner", rather than an unsatisfactory undergraduate, I found myself having to do forensic research to work out how Nordhaus could have reached his interpretation of this paper—which explicitly warned against using "smooth projections of global change", and which explicitly warned of a likely tipping point in the Arctic—as supporting him trivialising the significance of losing Arctic summer sea-ice, and using a "damage function" that ignored tipping points, in flagrant contradiction of the text of Lenton's paper (see Table 10.1).

The only feasible explanation for Nordhaus's erroneous summary was a table by Lenton in a related publication (Richardson et al. 2011), which Nordhaus also referenced as a source for his interpretation (Nordhaus 2013b, p. 334). Lenton described this table as "A simple 'straw man' example of tipping element risk assessment". Each "tipping element" was given a point score in terms of likelihood *of occurring this century* (Low = 1, Medium = 1.5, Medium-High = 2.5, and High = 3) and relative impact on the climate *over the next millennium* (Low = 1, Low-Medium = 1.5, Medium = 2, Medium-High = 2.5, and High = 3). A risk score was derived as the product of likelihood times impact. Arctic summer sea-ice had the lowest rating on impact (Low = 1), but the highest in terms of likelihood (High = 3), for an overall score of 3—see Table 10.2, which ranks these tipping elements in the descending order of importance that Nordhaus gave them in his table N.1 (Nordhaus 2013b, p. 333).

It appears that Nordhaus's relative ranking was based primarily on Lenton's "Impact" measure, since he ranked Arctic summer sea-ice as the lowest (one star in his Table N.1, or a low ranking of 3 in Table 10.2), below disrupting the Atlantic Thermohaline Circulation (THC) for example (two stars, or a medium ranking of 2), whereas Lenton ranked Arctic sea-ice above the THC in risk assessment terms (3

Table 10.1 The chasm between Lenton's conclusion and Nordhaus's interpretation of this research

Lenton (Lenton et al. 2008)	Nordhaus 2013
Conclusion: Society may be lulled into a false sense of security by smooth projections of global change.	The current version [of the damage function in Nordhaus's DICE model] assumes that damages are a quadratic function of temperature change and does not include sharp thresholds or tipping points, but this is consistent with the survey by Lenton et al. (2008) (Nordhaus and Sztorc 2013, p. 11)
Our synthesis of present knowledge suggests that a variety of tipping elements could reach their critical point within this century under anthropogenic climate change.	Their review finds no critical tipping elements with a time horizon less than 300 years until global temperatures have increased by at least 3 °C. (Nordhaus 2013a, p. 60)
The greatest threats are tipping the Arctic sea-ice and the Greenland ice sheet, and at least five other elements could surprise us by exhibiting a nearby tipping point. (Lenton et al. 2008, p. 1792)	The most important tipping points, in their view, have a threshold temperature tipping value of 3 °C or higher (such as the destruction of the Amazon rain forest) or have a time scale of at least 300 years (the Greenland Ice Sheet and the West Antarctic Ice Sheet). (Nordhaus 2013a, p. 60)

Table 10.2 *Based on Lenton's Table "A simple 'straw man' example of tipping element risk assessment, by Timothy M. Lenton" in (*Richardson et al. 2011*, p. 186), and Nordhaus's* Table N-1, with his ranking of tipping elements used to sort the table (Nordhaus 2013b, p. 333)

Tipping element	Transition Time scale	Threshold warming °C		Ranking (low number means high ranking)		Lenton Point scoring (high number means high likelihood, impact)		
		Min	Max	Nordhaus	Lenton	Likely by 2100	Impact by 3000	Product
Greenland ice sheet	300	1	2	1	1	2.5	3	7.5
West Antarctic ice sheet	300	3	5	1	2	2	3	6
Amazon rainforest	50	3	4	1	3	2	2	4
West African monsoon	10	3	5	2	4	1	3	3
Atlantic THC	100	3	5	2	6	1	2.5	2.5
ENSO	100	3	6	2	6	1	2.5	2.5
Boreal Forest	50	3	5	3	8	1	1.5	1.5
Arctic summer sea-ice	10	0.5	2	3	4	3	1	3

versus 2.5, where a higher score is worse) because of its much higher likelihood of occurring this century.

This may explain how Nordhaus came to classify losing the Arctic summer sea-ice as an event of low concern in climate change. But this is a false reading of Lenton's table. As Lenton explained, "Impacts are considered in *relative* terms based on an initial subjective judgment (noting that *most tipping-point impacts, if placed on an absolute scale compared to other climate eventualities, would be high*)" (Richardson et al. 2011, p. 186. Emphasis added). In other words, while the impact of the loss of Arctic summer sea-ice was *low compared to, for example, the impact of losing the Atlantic Thermohaline Circulation (THC)*, it was not low in any absolute sense: losing the Arctic's summer sea-ice would have a significant qualitative impact on the climate. Nordhaus's interpretation of Lenton's "low" ranking as meaning that Arctic summer sea-ice was not of absolute importance to the climate—it was not "critical", he alleged—is a fundamentally incorrect reading of Lenton's research, as Lenton confirmed to me in personal correspondence (Keen and Lenton 2020).

Nordhaus also misinterpreted the time ranges given (the "Transition Timescale" column in Table 10.2) as indicating that these tipping elements were not going to be triggered for that many years, when in fact they were an estimate of how long it would take from an initial triggering *this century* until the end of the transition process. The complete melting of the West Antarctic ice sheet might well take 300 years from its initial triggering, whereas the Arctic summer sea-ice could disappear over a period measured in decades rather than centuries. But the fate of the West Antarctic ice sheet would be decided this century, if we let increased CO_2 levels drive up global temperature by 3 °C or more—which we are well on track to do, and which we would reach on Nordhaus's "optimum" trajectory by 2085.

In summary, Nordhaus read Lenton's research as a climate change denier would, cherry picking it to find ways to support his preconception that climate change was insignificant. This was a consistent theme in how Nordhaus treated scientific research on climate change, as evidenced by surveys he undertook of scientists in 1994 (Nordhaus 1994), and scientific literature in 2017 (Nordhaus and Moffat 2017).

10.3.2 Drowning Scientists with Economists

Nordhaus's 1994 survey asked people from various academic backgrounds to give their estimates of the impact on GDP of three global warming scenarios, including a 3 °C rise by 2090. The 2014 IPCC Report used this as one data point in Fig. 10.1 (see Fig. 10.2), claiming that a 3 °C temperature rise would cause a 3.6% fall in GDP.

Nordhaus surveyed 19 people, 18 of whom fully complied, and one partially. Nordhaus described them as including ten economists, four "other social scientists", and five "natural scientists and engineers", noting that eight of the economists come from "other subdisciplines of economics (those whose principal concerns lie outside environmental economics)" (Nordhaus 1994, p. 48). This, *ipso facto*, should rule them out from taking part in this expert survey in the first place.

There was extensive disagreement between the relatively tiny cohort of actual scientists surveyed, and, in particular, the economists "whose principal concerns lie outside environmental economics". As Nordhaus noted, "Natural scientists' estimates [of the damages from climate change] were 20 to 30 times higher than mainstream economists'" (Nordhaus 1994, p. 49). The average estimate by "Non-environmental economists" (Nordhaus 1994, Figure 4, p. 49) of the damages to GDP of a 3 °C rise by 2090 was 0.4% of GDP; the average for natural scientists was 12.3%, and *this was with one of them refusing to answer Nordhaus's key questions*:

> I must tell you that I marvel that economists are willing to make quantitative estimates of economic consequences of climate change where the only measures available are estimates of global surface average increases in temperature. As [one] who has spent his career worrying about the vagaries of the dynamics of the atmosphere, I marvel that they can translate a single global number, an extremely poor surrogate for a description of the climatic conditions, into quantitative estimates of impacts of global economic conditions. (Nordhaus 1994, pp. 50–51)

Comments from economists lay at the other end of the spectrum from this self-absented scientist. Because they had a strong belief in the ability of human economies to adapt, they could not imagine that climate change could do significant damage to the economy, whatever it might do to the biosphere itself:

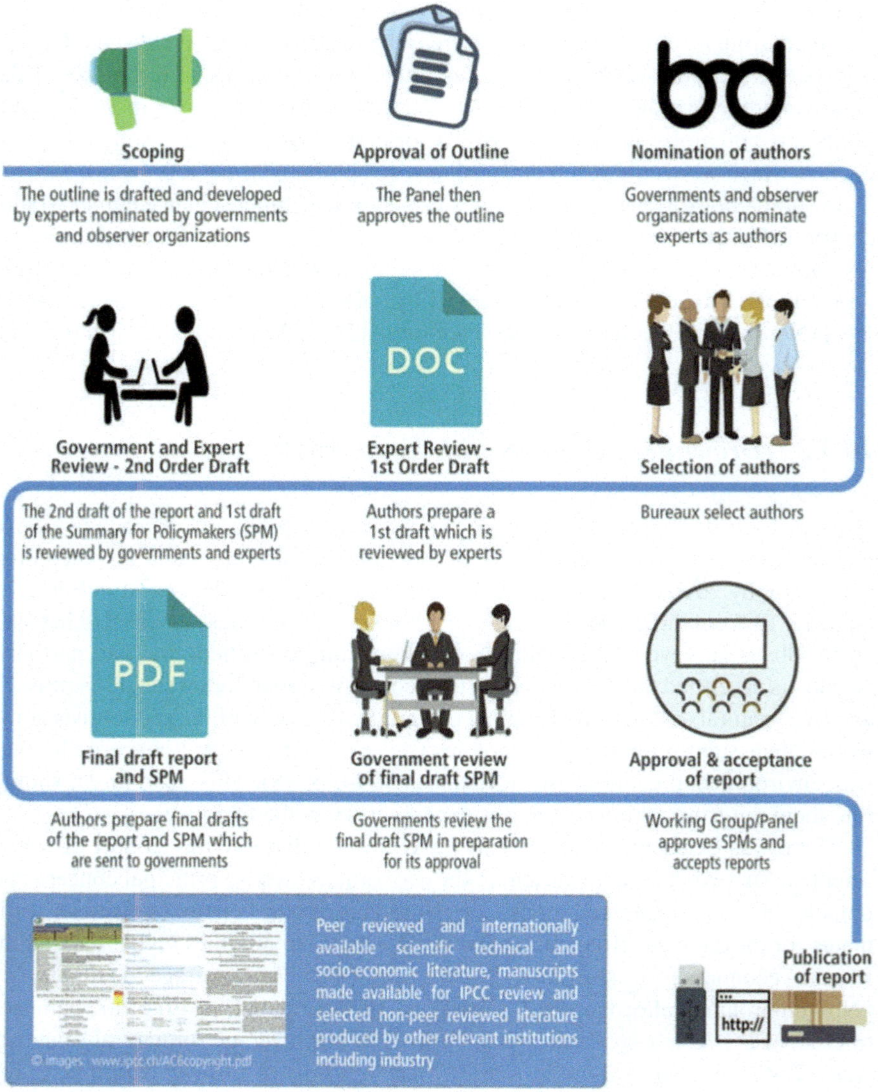

Fig. 10.1 The IPCC's graphic laying out the publication and review process behind its reports. (Source: https://www.ipcc.ch/site/assets/uploads/2018/08/WP_AR6_en_web.jpg)

> There is a clear difference in outlook among the respondents, depending on their assumptions about the ability of society to adapt to climatic changes. One was concerned that society's response to the approaching millennium would be akin to that prevalent during the Dark Ages, whereas *another respondent held that the degree of adaptability of human economies is so high that for most of the scenarios the impact of global warming would be "essentially zero"*. (Nordhaus 1994, pp. 48–49. Emphasis added)

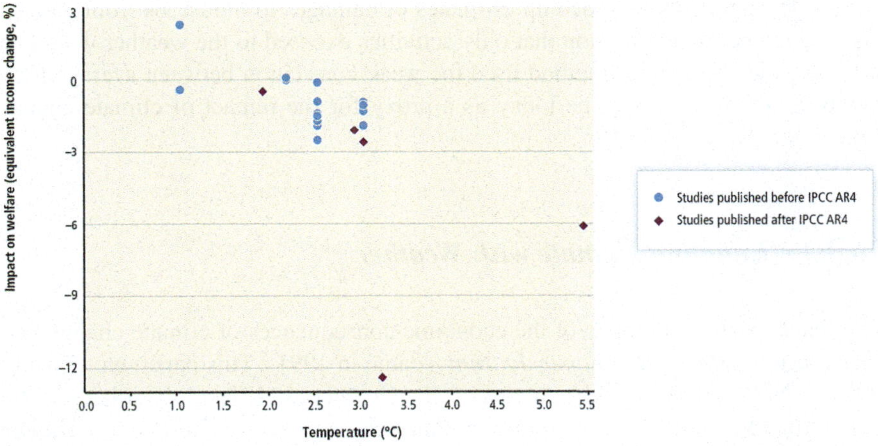

Fig. 10.2 IPCC economic estimates of damages to GDP from global warming. (Arent et al. 2014a, p. 690)

Given this extreme divergence of opinion between economists and scientists, one might expect that Nordhaus's next survey would examine the reasons for it. In fact, the opposite applied: his methodology excluded non-economists entirely.

Rather than a survey of experts, this was a literature survey (Nordhaus and Moffat 2017). He and his co-author searched for relevant articles using the string "(damage OR impact) AND climate AND cost" (Nordhaus and Moffat 2017, p. 7), which is reasonable, if too broad (as they admit in the paper).

The key flaw in this research was where they looked: they executed their search string in *Google*, which returned 64 million results; *Google Scholar*, which returned 2.8 million; and the economics-specific database *Econlit*, which returned just 1700 studies. On the grounds that there were too many results in Google and Google Scholar, they ignored those results, and simply surveyed the 1700 articles in *Econlit* (Nordhaus and Moffat 2017, p. 7). *These are, almost exclusively, articles written by economists*. They did not search a comparable science database like ProQuest Science Journals, where the same too-broad search string (on January 11th 2021) returned 60,315 peer-reviewed full-text articles, and a narrower search string "damage AND climate AND gdp" returned a manageable 2721 hits.

There is therefore no significant science-based content in the papers that generated the "data" on which IPCC economists concluded that "the impact of climate change will be small relative to the impacts of other drivers" (Arent et al. 2014a, p. 662). All of the pairs of numbers in Fig. 10.2 were generated by economists, and all but one predicted an extremely small impact on GDP from global warming, of a less than 3% fall in GDP from temperature rises of up to 3 °C (or a 6% fall for a 5.5 °C rise), compared to what GDP would be in the complete absence of climate change.

These numbers were generated in two main ways, which the IPCC report described as "Enumeration" and "Statistical" (Arent et al. 2014b, Table SM10.2, p.

SM10–4). Enumeration added up estimates of damages to industries from climate change, under the assumption that only activities exposed to the weather would be affected; the "statistical" method used the weak correlation between average temperature and average income today as a proxy for the impact of climate change over time.

10.3.3 Equating Climate with Weather

Nordhaus's first predictions of the economic consequences of climate change in a refereed journal—the *Economic Journal*—came in 1991.[5] This paper, entitled "To Slow or Not to Slow: The Economics of The Greenhouse Effect" (Nordhaus 1991) contained the seeds of all his future work on climate change. He equated climate change over time, due to dramatically increasing the amount of solar radiation retained in the biosphere as heat by increased greenhouse gases, with the geographic variation of *today's* climate across the globe:

> human societies thrive in a wide variety of climatic zones. For the bulk of economic activity, non-climate variables like labour skills, access to markets, or technology swamp climatic considerations in determining economic efficiency. (Nordhaus 1991, p. 930)

He assumed that climate change would only affect economic activities that were directly exposed to the weather:

> The most sensitive sectors are likely to be those, such as agriculture and forestry, in which output depends in a significant way upon climatic variables … Our estimate is that approximately 3% of United States national output is produced in highly sensitive sectors, another 10% in moderately sensitive sectors, and *about 87 % in sectors that are negligibly affected by climate change*. (Nordhaus 1991, Table 5, p. 930. Emphasis added, I reproduce an extract here as Table 10.3)

Table 10.3 Extract from Nordhaus's breakdown of economic activity by vulnerability to climatic change (Nordhaus 1991, p. 931)

Sector	Percentage of total
Negligible effect	
Manufacturing and mining	26.0
Other transportation and communication	5.5
Finance, insurance, and balance real estate	11.4
Trade and other services	27.9
Government services	14.0
Rest of world	2.1
Total "negligible effect"	**86.9**

[5] His first attempt at a quantitative estimate of the economic impact of climate change was in a Cowles Commission working paper in 1980, when he estimated that doubling CO_2—and increasing global average temperatures by 3 °C—would change GDP by between minus 7% to plus 5% by 2100 (Nordhaus 1980, Table 4, p. 49).

Using these beliefs, he derived trivial estimates for the impact of climate change on the economy:

> damage from a 3°C warming is likely to be around ¼% of national income in United States … We might raise the number to around 1% of total global income to allow for these unmeasured and unquantifiable factors, although such an adjustment is purely ad hoc… my hunch is that the overall impact upon human activity is unlikely to be larger than 2% of total output. (Nordhaus 1991, p. 933)

All subsequent papers by Neoclassical climate-change economists replicated the assumption that any activity not directly exposed to the weather would be essentially immune from climate change. The 2014 IPCC Report restated it as a "Frequently Asked Question":

> **FAQ 10.3 | Are other economic sectors vulnerable to climate change too?**
>
> Economic activities such as agriculture, forestry, fisheries, and mining are exposed to the weather and thus vulnerable to climate change. *Other economic activities, such as manufacturing and services, largely take place in controlled environments and are not really exposed to climate change.* (Arent et al. 2014a, p. 688. Emphasis added)

10.3.4 Equating Time with Space

Nordhaus's colleague Robert Mendelsohn (Mendelsohn et al. 1994; Mendelsohn et al. 2000) used Nordhaus's assumption that today's climate and GDP data was relevant to climate change to invent another way to generate numbers, purportedly about the impact of global warming, from today's data:

> An alternative approach … can be called the statistical approach. It is based on direct estimates of the welfare impacts, *using observed variations (across space within a single country) in prices and expenditures to discern the effect of climate.* ***Mendelsohn assumes that the observed variation of economic activity with climate over space holds over time as well;*** *and uses climate models to estimate the future effect of climate change.* (Tol 2009, p. 32. Emphasis added)

This method of generating numbers takes average temperature data and per capita income data, and uses the weak relationship between them to allege that climate change will have a relatively trivial impact upon global GDP. Figure 10.3 illustrates this, using temperature and per capita income for the contiguous United States on a State-by-State basis ("Gross State Product per capita", or GSP_{PC}).

There is no real pattern, but a quadratic can be fitted to the data as shown, with a low correlation coefficient of 0.31. In statistical terms, this means that this function has terrible predictive power. For example, it predicts that States which are 4 °C hotter *or colder* than the USA average will have a GSP_{PC} that is 5% lower. But the states that are between 3.5 °C and 4.5 °C above or below the USA's average temperatures include New York at 30% above the average, and Arkansas at 29% below. If you were trying to win a game of Trivial Pursuit, you wouldn't use this function to supply your answers on US GSP per capita today.

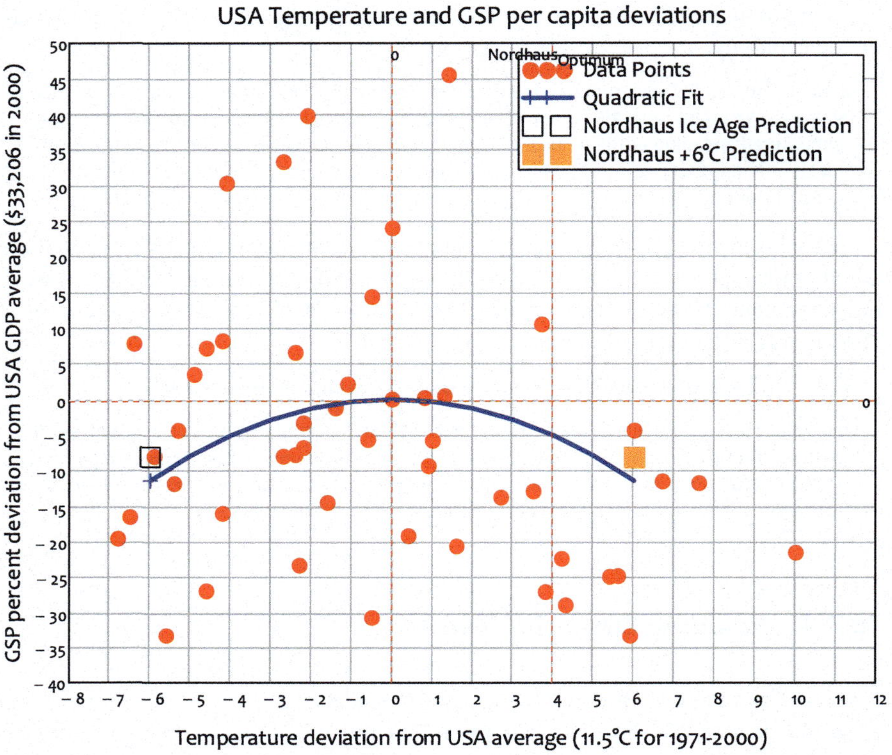

Fig. 10.3 Correlation of temperature and USA Gross State Product per capita. (Sources: https://www.currentresults.com/Weather/US/average-annual-state-temperatures.php, https://apps.bea.gov/iTable/iTable.cfm & http://www2.census.gov/programs-surveys/popest/datasets/2010-2019/national/totals/nst-est2019-alldata.csv)

10.3.5 Trivial Estimates of Serious Damages

Despite this, Nordhaus uses a quadratic, derived from data much like that in Fig. 10.3, but with an *even smaller coefficient*, to "predict" the impact of global warming on "Gross Global Product" (GGP). The equation of the quadratic in Fig. 10.3 is $GSP_{PC} = GDP_{PC} - 0.318\ \% \times Temperature_{Deviation}^{2}$. Nordhaus's "damage function", in the latest version of his global warming model *DICE* (for "Dynamic Integrated Climate and Economics"), is $GGP_{Deviation} = GGP_{NoGlobalWarming} - 0.227\ \% \times Temperature_{Deviation}^{2}$:

> The parameter used in the model was an equation with a parameter of 0.227 percent loss in global income per degrees Celsius squared with no linear term. This leads to a damage of 2.0 percent of income at 3°C, and 7.9 percent of global income at a global temperature rise of 6°C. (Nordhaus 2018b, p. 345)

These predictions are absurd. A 3 °C increase could trigger, and a 6 °C increase would trigger, every "tipping element" shown in Table 10.2. The Earth would have a climate unlike anything our species has experienced in its existence, and the Earth would transition to it hundreds of times faster than it has in any previous naturally-driven global warming event (McNeall et al. 2011). The Tropics and much of the globe's temperate zone would be uninhabitable by humans and most other life forms. And yet Nordhaus thinks it would only reduce the global economy by just 8%?

The absurdity of Nordhaus's damage function can be illustrated by a known experience in humanity's time on Earth, because it is symmetrical—the function predicts the same damages from a fall in temperature as for an equivalent rise. It therefore predicts that a 6 °C *fall* in global temperature would also reduce GGP by just 7.9% (see Fig. 10.3). Unlike global warming, we do know what the world was like when the temperature was 6 °C below twentieth century levels: that was the average temperature of the planet during the last Ice Age (Tierney et al. 2020), which ended 15–20,000 years ago. At the time, all of America north of New York, and of Europe north of Berlin, was beneath a kilometre of ice. The prediction that a transition to such a climate in just over a century would cause global production to fall by less than 8% is ridiculous.

Again, I found myself in the position of a forensic detective, trying to work out how otherwise intelligent people could come to believe that climate change would only affect industries that are directly exposed to the weather, and that the correlation between climate today and economic output today across the globe could be used to predict the impact of global warming on the economy? The only explanation that made sense is that these economists were mistaking the weather for the climate. Ironically, given the calibre of Nordhaus's later contributions, he gave a reasonable, if statistically oriented, explanation of the difference between weather and climate in an early paper:

> When we refer to climate, we usually are thinking of the average of characteristics of the atmosphere at different points of the earth, including the variances such as the diurnal and annual cycle. The important characteristics for man's activities are temperature, precipitation, snow cover, winds and so forth. A more precise representation of the climate would be as a dynamic, stochastic system of equations. The probability distributions of the atmospheric characteristics is what we mean by climate, while a particular realization of that stochastic process is what we call the weather. (Nordhaus 1976, p. 2)

This "probability distribution", as we experience it at any given location on Earth, is affected by the amount of energy in the biosphere, which varies in three primary ways:

1. Variations in the amount of the energy from the Sun that reaches the Earth;
2. Variations in the amount of this energy retained by greenhouse gases; and
3. Variations caused by differences in location on the planet—primarily, differences in distance from the Equator, altitude above sea level, and distance from oceans.

The first factor varies, via cyclical variations in the Earth's orbit, over time measured in thousands of years, and via long-term trends in the Sun's evolution, measured in billions of years. Neither of these are relevant in the timeframe of global warming. Given the Earth's orbit, and how much its surface reflects solar radiation, then in the absence of the second factor—greenhouse gases—the average temperature of the atmosphere would be minus 18 °C (Hay 2014, p. 30; Galimov 2017).

The second is what is at issue with global warming. If the Earth's atmosphere captured and re-radiated all the infra-red energy that the planet's surface reflects back into space (as a relatively dark body, the Earth absorbs high-frequency light and ultra-violet energy from the Sun, and reflects back low-frequency infra-red energy), then the average temperature of the atmosphere would be 29.5 °C.[6] With only naturally-occurring greenhouse gases, the average temperature of the atmosphere at present would be 15 °C. Human industrial activity is adding to this retention of solar energy, primarily by generating additional CO_2 from burning fossil fuels, resulting in more than a 1 °C rise in average surface temperature now (see Chap. 2, this volume).

The third, geographic factor is what is captured by the data shown in Fig. 10.3—*and this has nothing to do with global warming*. And yet this data, plus the belief that only industries which are exposed to the weather will be affected by global warming, is what underpins Nordhaus's "damage function" (and similar constructs by his fellow Neoclassical climate change economists) with its trivial forecasts for economic damage from climate change.

One thing that Fig. 10.3 does establish is that wide variations in temperature within one country today are associated with relatively small differences in income today. The range of average temperatures shown there is 16.8 °C, from 4.7 °C for North Dakota to 21.5 °C for Florida. However, the two States had very similar Gross State Products per capita in 2000 ($26,700 versus $26,000). This cannot be used to argue that, therefore, a huge change in global average temperatures due to global warming will have only a small effect on income—*but that is precisely how Neoclassical economists have used this data.*

That is evident in their predictions, but it helps to also have verbal confirmation of the disconnect between what economists think of climate change, and what it really is. The following statements were made on Twitter by the prominent Neoclassical climate change economist Richard Tol. Tol was one of the two lead co-authors of the economic sections of the 2014 IPCC report on climate change (Arent et al. 2014a), the developer of FUND ("Climate Framework for Uncertainty, Negotiation and Distribution"), one of the "Integrated Assessment models" (IAMs) used to supposedly estimate the impact of climate change on the economy, and editor of the academic journal *Energy Economics*. However, some his Twitter postings are those one might associate with a climate change denier, not an influential economist in the theory and policy of climate change.

In the first tweet, he concludes that climate change is not a problem, because US States with vastly different temperatures today have similar incomes today:

[6] https://geologistsofjacksonhole.org/wp-content/uploads/2018/02/Global-Warming-101-Talk_GJH_MikeAdler.pdf provides a very comprehensive survey of the mathematics of climate.

> 10K is less than the temperature distance between Alaska and Maryland (about equally rich), or between Iowa and Florida (about equally rich). Climate is not a primary driver of income. https://twitter.com/RichardTol/status/1140591420144869381?s=20

In the second, he concludes that global warming can't be a problem, because it is expected to increase temperatures by a small amount compared to the daily temperature variation for any one location on Earth—which averages about 14 °C for the continental USA (Qu et al. 2014):

> People thrive in a wide range of climates. The projected climate change is small relative to the diurnal cycle. It is therefore rather peculiar to conclude that climate change will be disastrous. Those who claim so have been unable to explain why. https://twitter.com/RichardTol/status/1313182006310731776?s=20

In contrast, this is how a key scientific paper (Im et al. 2017`) summarised what a world 4 °C warmer—let alone 10 °C–14 °C warmer—would mean for the over 2 billion human inhabitants of South Asia:

> Human exposure to TW [wet bulb temperatures] of around 35°C for even a few hours will result in death even for the fittest of humans under shaded, well-ventilated conditions … TW_{max} is projected to exceed the survivability threshold … under the RCP8.5 scenario by the end of the century over most of South Asia, including the Ganges river valley, northeastern India, Bangladesh, the eastern coast of India, Chota Nagpur Plateau, northern Sri Lanka, and the Indus valley of Pakistan. (Im et al. 2017, p. 4)

10.4 The Failure of Peer Review

This work is of such low calibre that it should never have been published in peer-reviewed academic literature. How did it manage to not only pass peer review, but ultimately be given the accolade of a Nobel Prize "for integrating climate change into long-run macroeconomic analysis"[7]?

There is one generic defence of the failure of the referees of economic journals to identify this work as without academic merit. Academics are not paid to referee, and the time they are supposedly allotted to do refereeing has been largely eliminated by the efficiency drives that politicians and the managerial class of University administrators have forced upon them. So refereeing is not done as professionally as the image of "peer-reviewed research" implies. Refereeing is also a far lower standard than replication. To referee a paper, all a referee has to do is read it and pass judgment. To replicate, you actually have to independently reproduce the results claimed in the paper.[8]

[7] https://www.nobelprize.org/prizes/economic-sciences/2018/summary/

[8] I once approved for publication a paper criticising Goodwin's growth cycle model (Goodwin 1967) because, though I didn't like the empirical results, they appeared to have been reached professionally (Harvie 2000). Later, when I attempted to use those results in modelling, I found that the numbers they generated were absurd. I contacted the author, and he honestly admitted that he had made "a simple schoolboy error" and confused percentages with fractions. Later researchers corrected this error, and empirically validated Goodwin (Grasselli and Maheshwari 2017).

That said, there is no excuse for referees approving for publication papers that, for example, make the critical and absurd assumption that 87% of GDP will be unaffected by climate change, because it happens indoors (Nordhaus 1991, p. 930). Here, the erroneous party is not Nordhaus alone, but the entire edifice of Neoclassical economics. Only Neoclassical economists, who in general have what Paul Romer—the joint recipient of the 2018 Nobel Prize with Nordhaus—described as a "non-committal relationship with the truth" (Romer 2016, p. 5), would recommend publication of papers that make critical assumptions that are so obviously false.

As I detail in *Debunking Economics* (Keen 2011), Neoclassical economics is riddled with false assumptions, because numerous theoretical and empirical requirements of the underlying theory have been *proven* to be false. Rather than accepting that their initial beliefs were wrong, and then abandoning these beliefs to develop a richer, more complex theory, Neoclassical economists have clung to those beliefs by adding patently absurd assumptions to hide the contrary proofs.

This practice is defended by describing such assumptions as "simplifying", but that is a false description. A simplifying assumption is something that, if it is not made, results in a much more complicated analysis, but changes the result only marginally. For example, Galileo's demonstration that dense bodies fall at the same speed, regardless of their weight, effectively assumed no air resistance. Correctly taking air resistance into account would have resulted in a vastly more complicated experiment, but no significant change in the result.

On the other hand, the type of assumption that Neoclassical economists defend as "simplifying" is frequently critical to the conclusions drawn from the model: if the assumption is false, *then so are the conclusions* (Musgrave 1990). Such assumptions abound in Neoclassical economics, so much so that economists have convinced themselves that it is invalid to criticise a theory on the grounds that its assumptions are unrealistic. The chief promulgator of this belief by economists was Milton Friedman, who stated that a theory cannot be rejected on the basis of its assumptions, because they must be unrealistic, so that "in general, the more significant the theory, the more unrealistic the assumptions":

> To put this point less paradoxically, the relevant question to ask about the "assumptions" of a theory is not whether they are descriptively "realistic", for they never are, but whether they are sufficiently good approximations for the purpose in hand. And this question can be answered only by seeing whether the theory works, which means whether it yields sufficiently accurate predictions. (Friedman 1953, p. 153)

This, as Alan Musgrave explained, is nonsense (Musgrave 1981).[9] However, because they accept Friedman's dodgy methodology, Neoclassical referees regularly recommend the publication of papers in which assumptions are made that are

[9] "Friedman's position, often called the 'F-twist', stems from his failure to distinguish three different types of assumption. Negligibility assumptions state that some factor has a negligible effect upon the phenomenon under investigation. Domain assumptions specify the domain of applicability of the theory. Heuristic assumptions are a means of simplifying the logical development of the theory. It is argued that Friedman's dictum is false of all three types of assumption." (Musgrave 1981, p. 386). See Chapter 8 of *Debunking Economics*, "There is madness in their method", for more.

patently false, if those papers otherwise support Neoclassical beliefs. To such referees, Nordhaus's assumption that "87% [of United States national output is produced] in sectors that are negligibly affected by climate change" (Nordhaus 1991, p. 930) was just a "simplifying assumption", which could not be challenged.

This methodological fallacy is dangerous enough with standard economic issues. But with climate change, it is existentially so. *When* the theory is proven wrong by the failure of its predictions, the consequences of this failure will be both catastrophic and irreversible. As DeCanio eloquently put it, waiting until we know that Neoclassical economists are wrong on climate change "amounts to conducting a one-time, irreversible experiment of unknown outcome with the habitability of the entire planet" (DeCanio 2003, p. 3).

10.5 Conclusion

Neoclassical economics has dominated economics for 150 years, which has given it the advantage of incumbency over its rivals—so much so that most people in authority, journalists, University students, and the general public, think that Neoclassical economics *is* economics. When it is criticised, even by other economists like myself (Keen 2011), the public is unlikely to hear the criticisms in the first place, and likely to regard them as coming from cranks if they do. Q-Anon aside, we usually trust those ordained as experts in their own fields.

This trust is a characteristic of human society which is utterly justified in the complex societies in which we live. Deference to expertise is a necessary feature of life in a complex world. However, with economics, this justifiable deference has helped entrench a fundamentally unscientific school of thought, and has made progress in economics virtually impossible.

I believed, before the Global Financial Crisis of 2008, that the only way economics could be shifted would be by its failure to predict a serious economic crisis. But the transient nature of economic crises—especially in the face of governments determined not to let capitalism fail on their watch—meant that economic theory scraped through that crisis relatively unscathed. That may all change in the next decade, because our trust in expertise has meant that, though scientists led the study of climate change, we have let Neoclassical economists determine our policy response.

Most politicians have studied some economics. Few have studied science, and they are therefore unable to read the science-based parts of the IPCC Reports. Most of their advisers—who actually read the reports for the politicians—are also trained in economics, rather than the sciences. Most political debate is about matters of economics, rather than science. The end result of all this is that, though scientists have led the study of climate change itself, economists have dominated public policy towards it. As Stephen De Canio put it in 2003:

> it is undeniably the case that economic arguments, claims, and calculations have been the dominant influence on the public political debate on climate policy in the United States and around the world… It is an open question whether the economic arguments were the cause or only an ex-post justification of the decisions made … but there is no doubt that economists have *claimed* that their calculations should dictate the proper course of action. (DeCanio 2003, p. 4)

Because these economists, starting with William Nordhaus, trivialised the dangers of climate change, the policy response to climate change has also been trivial. Human civilisation may well not survive Neoclassical economics. It's time it was eliminated, before it eliminates us. At the very minimum, any research by economists which uses the numbers generated by Neoclassical economists should be rejected by those formulating policy on climate change.

References

Arent DJ, Tol RSJ, Faust E, Hella JP, Kumar S, Strzepek KM, Tóth FL, Yan D (2014a) Key economic sectors and services. In: Field CB, Barros VR, Dokken DJ et al (eds) Climate Change 2014: impacts, adaptation, and vulnerability. Part A: Global and sectoral aspects. Contribution of working group II to the fifth assessment report of the intergovernmental panel on climate change. Cambridge University Press, Cambridge, pp 659–708

Arent DJ, Tol RSJ, Faust E, Hella JP, Kumar S, Strzepek KM, Tóth FL, Yan D (2014b) Key economic sectors and services – supplementary material. In: Field CB, Barros VR, Dokken DJ et al. (eds) Climate change 2014: impacts, adaptation, and vulnerability. Part A: Global and sectoral aspects. Contribution of working group II to the fifth assessment report of the intergovernmental panel on climate change IPCC, Geneva

Brysse K, Oreskes N, O'Reilly J, Oppenheimer M (2013) Climate change prediction: erring on the side of least drama? Glob Environ Chang 23(1):327–337. https://doi.org/10.1016/j.gloenvcha.2012.10.008

Christensen J (2018) Economic knowledge and the scientization of policy advice. Policy Sci 51(3):291–311. https://doi.org/10.1007/s11077-018-9316-6

DeCanio SJ (2003) Economic models of climate change: a critique. Palgrave Macmillan, New York

Fama EF (1970) Efficient capital markets: a review of theory and empirical work. J Financ 25(2):383–417

Fama EF, French KR (1996) The CAPM is wanted, dead or alive. J Financ 51(5):1947–1958

Fama EF, French KR (2004) The capital asset pricing model: theory and evidence. J Econ Perspect 18(3):25–46

Fourcade M, Ollion E, Algan Y (2015) The superiority of economists. J Econ Perspect 29(1):89–114. https://doi.org/10.1257/jep.29.1.89

Friedman M (1953) The methodology of positive economics. In: Essays in positive economics. University of Chicago Press, Chicago, pp 3–43

Galimov EM (2017) Role of low solar luminosity in the history of the biosphere. Geochem Int 55(5):401–417. https://doi.org/10.1134/S0016702917050068

Goodwin RM (1967) A growth cycle. In: Feinstein CH (ed) Socialism, capitalism and economic growth. Cambridge University Press, Cambridge, pp 54–58

Grasselli MR, Maheshwari A (2017) A comment on 'testing Goodwin: growth cycles in ten OECD countries'. Camb J Econ 41(6):1761–1766. https://doi.org/10.1093/cje/bex018

Harvie D (2000) Testing Goodwin: growth cycles in ten OECD countries. Camb J Econ 24(3):349–376. http://cje.oxfordjournals.org/

Hay WW (2014) The accelerating rate of global change. Rendiconti Lincei 25(1):29–48. https://doi.org/10.1007/s12210-014-0287-z

Herrando-Pérez S, Bradshaw CJA, Lewandowsky S, Vieites DR (2019) Statistical language backs conservatism in climate-change assessments. Bioscience 69(3):209–219. https://doi.org/10.1093/biosci/biz004

Hickel J (2018) The Nobel prize for climate catastrophe. Foreign Policy

Hirschman D, Berman EP (2014) Do economists make policies? On the political effects of economics1. Soc Econ Rev 12(4):779–811. https://doi.org/10.1093/ser/mwu017

Im E-S, Pal JS, Eltahir EAB (2017) Deadly heat waves projected in the densely populated agricultural regions of South Asia. Sci Adv 3(8):e1603322–e1603322. https://doi.org/10.1126/sciadv.1603322

Keen S (2011) Debunking economics: the naked emperor dethroned? 2nd edn. Zed Books, London

Keen S, Lenton T (2020) Draft paper on Nordhaus et al which cites you. Keen, Steve, Trang

Kypke KL, Langford WF, Willms AR (2020) Anthropocene climate bifurcation. Nonlinear Process Geophys 27(3):391–409. https://doi.org/10.5194/npg-27-391-2020

Lazear EP (2000) Economic imperialism. Q J Econ 115(1):99–146. http://www.mitpressjournals.org/loi/qjec

Lenton TM, Held H, Kriegler E, Hall JW, Lucht W, Rahmstorf S, Schellnhuber HJ (2008) Tipping elements in the Earth's climate system. Proc Natl Acad Sci 105(6):1786–1793. https://doi.org/10.1073/pnas.0705414105

McNeall D, Halloran PR, Good P, Betts RA (2011) Analyzing abrupt and nonlinear climate changes and their impacts. Wiley Interdiscip Rev Clim Chang 2(5):663–686. https://doi.org/10.1002/wcc.130

Mendelsohn R, Nordhaus WD, Shaw D (1994) The impact of global warming on agriculture: a Ricardian analysis. Am Econ Rev 84(4):753–771. https://doi.org/10.2307/2118029

Mendelsohn R, Schlesinger M, Williams L (2000) Comparing impacts across climate models. Integr Assess 1(1):37–48. https://doi.org/10.1023/a:1019111327619

Mirowski P (2020) The Neoliberal Ersatz Nobel Prize. In: Plehwe D, Slobodian Q, Mirowski P (eds) Nine lives of neoliberalism. Verso, London, pp 219–254

Mora C, Bénédicte D, Iain RC, Farrah EP, Rollan CG, Coral RB, WWC C, Bonnie SD, Emily TJ, Leo VL, Matthew PL, Marie MM, Alessandra GS, Han T, Thomas WG, Lisa RL, Ed H, Clay T (2017) Global risk of deadly heat. Nat Clim Chang 7(7). https://doi.org/10.1038/nclimate3322

Musgrave A (1981) 'Unreal assumptions' in economic theory: the F-twist untwisted. Kyklos (Basel) 34(3):377–387. https://doi.org/10.1111/j.1467-6435.1981.tb01195.x

Musgrave A (1990) 'Unreal Assumptions' in economic theory: the F-twist untwisted. In: Wood JC, Woods RN (eds) Milton friedman: critical assessments, Critical assessments of contemporary economists series, vol 3. Routledge, London/New York, pp 333–342

Nordhaus W (1976) Strategies for the control of carbon dioxide. Cowles Foundation Discussion Paper

Nordhaus W (1994) Expert opinion on climate change. Am Sci 82(1):45–51

Nordhaus W (2007) Critical assumptions in the Stern review on climate change. Science 317(5835):201–202

Nordhaus W (2013a) The climate casino: risk, uncertainty, and economics for a warming world. Yale University Press, New Haven

Nordhaus W (2013b) NOTES. In: The climate casino. Risk, uncertainty, and economics for a warming world. Yale University Press, pp 327–364

Nordhaus W (2018a) Nobel lecture in economic sciences. The Ultimate Challenge for Economics. Nobel Prize in Economics Committee, Climate Change. https://www.nobelprize.org/uploads/2018/10/nordhaus-slides.pdf

Nordhaus W (2018b) Projections and uncertainties about climate change in an era of minimal climate policies. Am Econ J 10(3):333–360. https://doi.org/10.1257/pol.20170046

Nordhaus W, Sztorc P (2013) DICE 2013R: introduction and user's manual

Nordhaus WD (1980) Thinking about carbon dioxide: theoretical and empirical aspects of optimal control strategies.. Cowles Foundation Discussion Paper
Nordhaus WD (1991) To slow or not to slow: the economics of the greenhouse effect. Econ J 101(407):920–937. https://doi.org/10.2307/2233864
Nordhaus WD, Moffat A (2017) A survey of global impacts of climate change: replication, survey methods, and a statistical analysis. vol Discussion Paper No. 2096. Cowles Foundation, New Haven, Connecticut
Pusey M (1991) Economic rationalism in Canberra: a nation-building state changes its mind. Cambridge University Press, Cambridge
Qu M, Wan J, Hao X (2014) Analysis of diurnal air temperature range change in the continental United States. Weather Clim Extremes 4:86–95. https://doi.org/10.1016/j.wace.2014.05.002
Richardson K, Steffen W, Liverman D (2011) Climate change: global risks, challenges and decisions. Cambridge University Press, Cambridge
Romer P (2016) The trouble with macroeconomics
Sharpe WF (1964) Capital asset prices: a theory of market equilibrium under conditions of risk. J Financ 19(3):425–442
Stern N (2007) The economics of climate change: the Stern review/Nicholas Stern. Cambridge University Press, Cambridge
Tierney JE, Zhu J, King J, Malevich SB, Hakim GJ, Poulsen CJ (2020) Glacial cooling and climate sensitivity revisited. Nature (London) 584(7822):569–573. https://doi.org/10.1038/s41586-020-2617-x
Tol RSJ (2009) The economic effects of climate change. J Econ Perspect 23(2):29–51
Warren R, Price J, Graham E, Forstenhaeusler N, VanDerWal J (2018) The projected effect on insects, vertebrates, and plants of limiting global warming to 1.5°C rather than 2°C. Science 360(6390):791–795. https://doi.org/10.1126/science.aar3646

Part III
Designing a Safe and Prosperous Future

Chapter 11
An Introduction to Ecological Economics: Principles, Indicators, and Policy

Philip Lawn and Stephen J. Williams

Abstract This chapter begins with a fictional story about a small group of people who become shipwrecked on a bountiful exotic island. The problems encountered as the group transforms itself into an ever-growing society consuming increasing quantities of the island's limited resources provides a simple analogy of what is now occurring at the global level and an ideal basis for an introduction to ecological economics. Following a summary of how ecological economics began and has since evolved, a series of fundamental ecological economics principles are outlined. The principles are grounded on a real-world view of the economy, which starts with the recognition that the economy is embedded in the living world and, as such, humankind's ability to provide for itself on a sustainable basis is constrained by limited stocks of natural resources, biophysical processes, and the physical laws governing production and consumption. Whilst the chapter debunks many myths and reveals a number of harsh realities, it offers a viable and realisable path to a sustainable, just, and prosperous future, albeit one dependent upon growth eventually being abandoned in favour of a qualitatively-improving steady-state economy. Some indicators and basic policy measures to help guide this awkward transition are briefly elucidated.

11.1 Introduction

Imagine you and some friends are shipwrecked on a beautiful island. After the initial shock, you quickly focus on securing basic needs, like fresh water, some primitive shelter, and wood for a fire. During the first few days you discover the island seems to have an abundance of coconuts, wild fruits, yams, fowl, and other game.

P. Lawn (✉)
Measuring Genuine Progress, Adelaide, Australia

Torrens University, Adelaide, Australia

S. J. Williams
Silverwater, NSW, Australia

S. J. Williams, R. Taylor (eds.), *Sustainability and the New Economics*,
https://doi.org/10.1007/978-3-030-78795-0_11

There is also abundant seafood of all kinds, including oysters, clams, fish, and turtles in the lagoon. A nearby waterfall promises any amount of crystal-clean water.

In this early period, you salvage what you can from the abandoned ship, including tools and clothing. You gradually explore every part of the island and marvel at its pristine beauty. As the months turn into years, your group slowly becomes a small society: rules are agreed on, babies are born, and more elaborate dwellings are constructed from the plentiful supply of trees. While the island is only small, its fecundity is great and you start to realise what true wealth is. As fast as your little society consumes resources, they seem to be replaced by nature, although the trees are slow to regrow. But felling trees to make houses and canoes makes space for the village, so that is a win-win situation.

But in time, as the population doubles and then doubles again, it eventually becomes evident that the island is not as abundant as it first seemed, and the elders talk wistfully of the good old days of plenty. It becomes ever more difficult to find the fruits, yams, coconuts, and oysters that were once there for the taking. A species of flightless bird, a culinary staple that even a child could catch, is no longer seen after years of dwindling sightings. It takes longer to catch fish, and they are no longer the prized specimens of yesteryear. One has to walk further to find firewood. The primitive latrines become increasingly unpleasant. Eventually, the graveyard has to be expanded until sea burials are mandated due to space restrictions. Squabbles become common as people accuse each other of being greedy, and eat more than is strictly necessary. Some complain they are being victimised by girth-ism. One faction say there are now too many people, and family size should be curtailed for the good of everyone. Others argue that it is consumption that is the problem, not population. Some of the younger adults talk of sailing their canoes out into the open ocean in search of a new and bigger island, as they see little future where they presently are.

This story, while artificial, clearly describes an unsustainable society because there are, eventually, too many people consuming more than what nature can sustainably provide. It shows that sustainability can be a problem regardless of the size and material affluence of a society. Any school student can understand the problem outlined above, yet the overwhelming majority of people, and virtually all mainstream policymakers, do not recognise that exactly the same thing is taking place on spaceship Earth today – humanity's only home (Harvey 2020). On our fantasy island, nature ultimately limits the number of humans and their consumption levels if humans don't impose limits on themselves. Moreover, once degraded, the island takes an inordinate amount of time to regenerate, if ever. But even before ecological limits are reached, the quality of human life deteriorates. The same thing is happening on planet Earth, with hundreds of millions of people already affected, not to mention the countless other life forms that humans have a duty of care towards. In both cases, the natural environment continually degrades, society fractures, and human wellbeing diminishes long before ecological constraints mercilessly bite.

While mainstream economists call for ever more 'growth' to solve environmental and other problems caused by growth, ecological economists stress that the goal of exponential growth, and our relentless attempts to achieve it, is the problem – that expanding the size of an economy is inseparable from rising rates of resource throughput, growing natural resource scarcities, and, if pursued for too long, inevitable ecological and social collapse.

Ecological economics has deep antecedents, but in the modern period it largely grew out of the work of Nicholas Georgescu-Roegen (1906–1994), his US student Herman Daly (1938–), and other economists, like Kenneth Boulding (1910–1993) and Robert Ayres (1932–). In essence, ecological economics emerged as a reaction against what many academics, including non-economists, saw as the fantasy of mainstream economics – the placing of nature and society outside the economic process and its systematic failure to account for the laws governing physical transformation processes (the first and second laws of thermodynamics).

The groundswell of opposition to the mainstream view of the economic process eventually led to the formation in 1989 of the International Society for Ecological Economics (ISEE) and the publication in the same year of ISEE's official journal, *Ecological Economics* (Elsevier). ISEE's inaugural meeting took place the following year in Washington DC and has since been held every two years.[1] Following ISEE's formation, a number of relatively independent regional chapters/societies have emerged. Most of them convene a biannual conference, usually in the odd-numbered years between the ISEE conferences.

Being a self-described pluralistic school of economic thought, detailed principles and policy prescriptions vary between ecological economics practitioners, often reflecting the transdisciplinary nature of ecological economics and those who have helped develop it. Despite this, the following set of principles are largely representative of this still emerging field of study.

11.2 Principles of Ecological Economics

1. The economy is embedded in society, and society is embedded in the biophysical world, such that healthy economies, societies, and humans are ultimately dependent on a healthy and resilient living planet. Such a planet has high levels of biodiversity.
2. The purpose of the economy is to maximise collective wellbeing, ensuring everybody has a chance to thrive. Wellbeing should be measured using a range of indicators that need continual refining. Gross Domestic Product (GDP), while a useful measure of the volume of economic activity, is not a measure of wellbeing.
3. A nation's wellbeing is a function of the state of its natural capital (natural wealth); the available resources relative to its population size; the health and education of its population; its stock of human-made capital and the technology embodied within it (human-made wealth); the trophic diversity of its economy;

[1] The 2020 conference was cancelled due to the coronavirus pandemic.

and the health of its major institutions, especially those relating to law, governance, and human rights protection.

4. Besides ensuring a sustainable rate of resource throughput, ecological economics is also concerned with the efficient allocation of resources (resources put to their best use) and distributional equity (resources fairly distributed in the form of useful goods and services), ensuring, as a minimum, that everyone's basic needs are met.
5. Given the interdependence between the primary, secondary, and tertiary trophic levels, nations should strive for maximum self-sufficiency, especially in necessities like food and energy. This means avoiding, if possible, dependence on the importation of natural resources and essential goods and services. All countries should aim for exports to equal imports (balanced trade).
6. The past 50 years have seen the rise of globalisation – the integration of national economies into one single global economy through so-called free trade and free capital mobility. Whilst globalisation has helped boost the volume of international trade, it has come at great expense. There is a need to restore global economic arrangements around the principles of internationalisation to protect national sovereignty and to halt the shift in global capital to locations with the cheapest labour and weakest environmental standards.
7. Economists should not be wilfully blind to the basic laws of the natural sciences, especially the laws of thermodynamics, which govern physical transformation processes (production and consumption). They must also familiarise themselves with many ecological indicators calculated by natural scientists, such as biocapacity, which, together with regeneration and waste assimilation rates, determine the maximum rate of resource throughput available for economic purposes on a sustainable basis.
8. Global resource throughput should be significantly less than the planet's biophysical capacity. Biophysical limits should be respected by individual nations as well as by the international community. Exceptions at the national level should be confined to the importing and exporting of natural resources and other permissible but limited actions under an international arrangement that ensures global rates of resource use are eventually reconciled with what the planet can ecologically sustain. Resource-importing countries should remain mindful that reliance on natural resource imports renders them vulnerable to changing international trading circumstances (see point 5).
9. While, up to a point, the resource intensity of production can be reduced via technological progress, the continuous absolute decoupling of throughput from GDP growth, as conventionally perceived, is impossible. Consequently, it is also impossible to prevent GDP growth from degrading the ecosphere.
10. Increasing the size of an economy generates costs as well as benefits. As economies grow, marginal benefits (increasing benefits) diminish while marginal costs (increasing costs) rise. As a consequence, a growing economy eventually reaches a point where the net benefits of economic activity are maximised. This point constitutes the economy's optimal scale. The optimum will be smaller than the maximum sustainable scale, which is where the ecological footprint

required to maintain the economy is equal to biocapacity (i.e., the capacity of the ecosphere to support it). Having reached its optimal scale, the economy should cease growing and enter a steady (non-growing) state. Increases in economic welfare remain possible via qualitative improvements.

11. If an economy exceeds its optimal scale, and certainly if it exceeds its maximum sustainable scale, it should employ degrowth policies combined with policies to increase the efficiency of resource use, improve the quality of goods and services, and ensure everyone's basic needs are met.
12. Human population numbers always affect resource throughput because every human must consume some minimum quantity of resources to survive, and even more to thrive. Whenever growth of the economy occurs, average individual consumption tends to increase rather than decrease over time. It is usually easier to increase human numbers than to compassionately decrease them, although ways to achieve the latter will be required if the number of humans must be reduced to achieve an ecologically sustainable population.

13. Efforts to continuously increase the physical scale of individual economies and the global economy as a whole will significantly hasten the arrival of ecologically determined limits. Exceeding such limits is dystopian and will ultimately result in some form of undesirable collapse. Indeed, simply approaching ecological limits will almost certainly be socially and economically undesirable. Unfortunately, highlighting humankind's impending predicament is failing to incite appropriate human action. It may be more fruitful to appeal to self-interest by pointing out that excessive growth is already suppressing the economic welfare of currently living people. If so, the informational content of indicators, such as the Genuine Progress Indicator, requires extensive publicity and attention.

11.3 Principles in More Detail

11.3.1 The Economy is Embedded in Society, and Society is Embedded in the Biophysical World

System embeddedness may seem obvious to the non-economist, but mainstream economists see things differently. The latter see society and nature as largely irrelevant, or at best, second-order issues in the exchanges that go on between firms, workers, and households, with government playing a mere supporting role (see Fig. 11.1). In this view, humans have an entirely separate character from nature, which is deemed external and something we can bend to our will via technology, rather than a nurturing mother that sets the biophysical rules of the game, especially what constitutes the maximum rate of resource throughput that can sustainably support the economy.

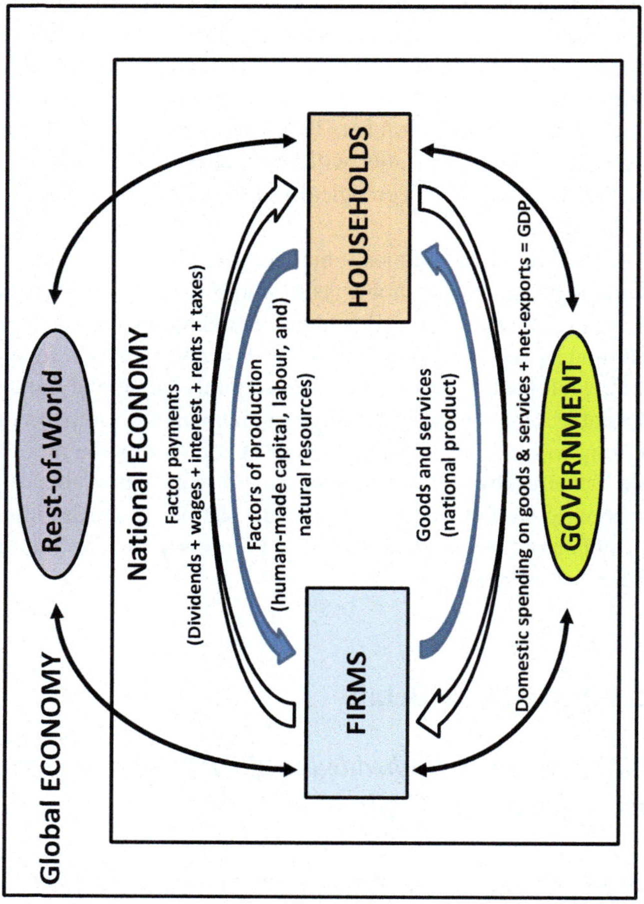

Fig. 11.1 *Mainstream view of the economy. (Source: all figures for this chapter are the lead author's)*

Humans are exceptional animals in many ways, but we are still animals, having evolved over billions of years from our single-celled ancestors. We are totally dependent on the web of life around us, as well as the microscopic web of life inside us, called our microbiome. We require clean air, water, and food if we are to be healthy. We rely on a safe and stable climate to provide certain temperatures, dependable precipitation, and not too frequent or extreme storms. Agriculture is at the mercy of these factors and needs high levels of biodiversity to provide pollination services and to keep pests and diseases in check. Something similar can be said with respect to forestry, fisheries, and critical ecosystem services. There are also propagation and harvesting limits that need to be respected to ensure the food, fibre, and building materials that humanity depends on can be sustainably provided.

With environmental destruction comes increased risks of epidemics and pandemics, as the world is increasingly experiencing (Tollefson 2020). The web of life is complex and highly interdependent, such that damage to one part can have unpredictable consequences elsewhere, and on an unknowable scale. This is why ecological economists accept nature as the basis of the economy's primary trophic level and the foundation of the economy, as a whole.

In a realistic worldview, the economy is recognised as a construct of society and therefore something entwined with culture and cultural values (see Fig. 11.2). We create economic systems to provide the means to satisfy human needs and wants in the knowledge that there are finite resources. It is up to society to determine how to allocate resources and distribute goods and services in a fair and efficient way, where outcomes are determined by a combination of rules and provisioning mechanisms (political and ethical systems), management principles (private businesses and government instrumentalities), and market forces (price signals). In this view, we see that many outcomes, for good or for ill, are largely human choices rather than aftermaths of forces beyond our influence. In short, the human-constructed economy can and should serve humanity rather than the other way round.

11.3.2 The Purpose of the Economy is to Maximise Collective Wellbeing

This seemingly obvious principle is made explicit in ecological economics, rather than assumed, as it is by the mainstream, as an inevitable outcome of efficient markets and the growth of GDP. To ascertain whether wellbeing is increasing, we must continue to develop a range of measuring tools that, however imperfect initially, collectively show trends and a reasonable idea of genuine progress. This must include environmental trends, as we monitor the health of our only home.

Mainstream schools of thought preach that we should continue to increase the physical size of economies by about 2–3% per year to increase wellbeing. This means the size of economies would need to double every 35 years (2% growth rate) or 23 years (3% growth rate), seemingly forever. Under this assumption, human

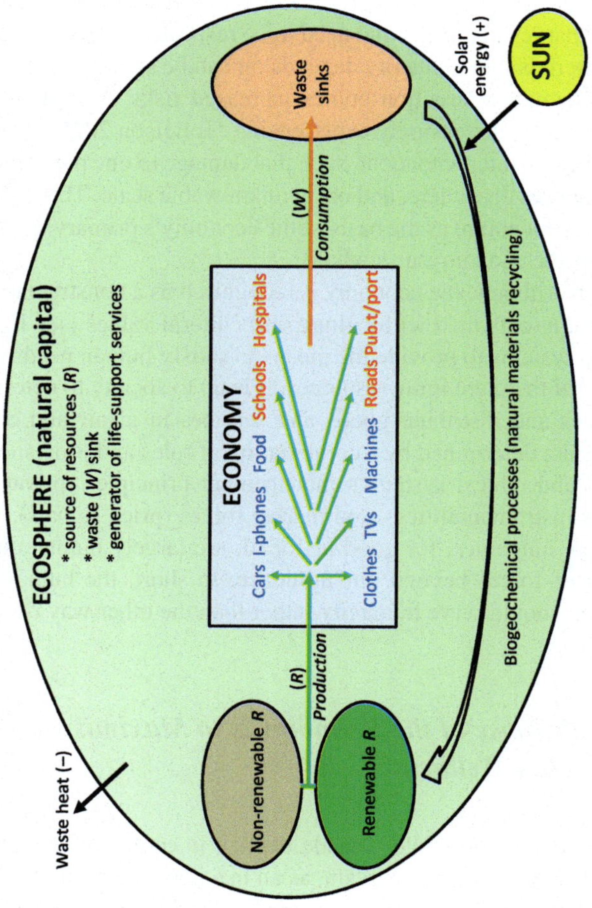

Fig. 11.2 *Real world view of the economy*

populations should also increase in size, assisting the expansion of the economy.[2] Simple maths shows the impossibility of doubling any physical thing indefinitely, or even for a few centuries, especially when such doubling relies on a similar increase in physical resource throughput.

Given this impossibility and the dubious connection between human wellbeing and various macroeconomic indicators, a range of alternative indexes have been developed. These can be classified into: (i) quality-of-life dashboards; (ii) aggregate non-monetary indexes (e.g., the United Nations Human Development Index); (iii) aggregate monetary indexes; and (iv) subjective wellbeing indexes (e.g., the World Health Organization Well-Being Index (WHO-5)) (Jackson 2020). These indexes focus on measuring things we actually desire, like better healthcare, education, law and order, reasonable income and wealth equality, a healthy natural environment, and even subjective feelings of happiness. The size of the economy may or may not grow as we increase wellbeing, since the economy is a means to satisfying our collective ends. That said, once the economy reaches a particular scale, increases in wellbeing will depend, not on growth, but on qualitative improvements in the goods we produce, the manner and the efficiency with which we produce them, and how equitably goods are distributed amongst the population.

While we don't have space to discuss all these indexes, it is worth outlining a relatively new aggregate monetary index – the Genuine Progress Indicator (GPI) – since it not only highlights some obvious problems with using GDP as a metric of progress, it serves as a viable alternative. For instance, in calculating GDP, both the costs and benefits of economic activity – at least those which are measured – are added together. Conversely, the GPI captures more costs and benefits than GDP and subtracts the former from the latter to indicate a net outcome.

The basis of the GPI is, firstly, that all economic activity begins with the extraction of natural resources from the ecosphere (natural capital), which are transformed by human-made capital and labour into new goods and services, including new capital goods. Notwithstanding some materials recycling, the economic process terminates with the insertion of wastes into the ecosphere following the consumption and/or wearing out of human-made goods. It is the initial input of natural resources and the eventual output of wastes which constitutes the *throughput* of matter-energy that sustains the economic process.

Second, the GPI recognises that there is a purpose to economic activity. This purpose is not to produce as many goods and services as possible, but to engage in meaningful productive endeavours and generate more use value for human enjoyment than the ecosphere can generate alone.

Third, the GPI acknowledges that the transformation of natural resources to human-made goods and services comes at various environmental and social costs. Environmental costs arise because making available the throughput of matter-energy to fuel the economic process comes at the expense of some of the ecosphere's source, sink, and life-support services. For example, while transforming

[2] Countries like Japan, with a population decrease, are seen as aberrations.

timber into wooden furniture increases use value, the loss of the trees to provide the timber decreases the use value (ecosystem services) provided by the timber-yielding forest.

Social costs arise because there are inevitable flaws in the design of economic systems and their supporting institutions. These flaws lead to unemployment, under-employment, poverty-induced crime, family breakdown, and overwork. Further costs are incurred when defensive and rehabilitative measures are undertaken to minimise the present and impending costs of environmental degradation and social malaise. These measures are often referred to as 'adaptation' responses. They are best considered welfare-maintaining rather than welfare-increasing and are treated as such when calculating the GPI (i.e., their value is subtracted from some of the GPI benefit items).

Fourth, unlike GDP, the GPI assigns the benefits provided by durable goods and infrastructure to the years in which the benefits are enjoyed. For instance, the GPI for 2020 includes the benefits provided during 2020 by a government-funded school built in 2010. It also includes the benefits yielded in 2020 by a television produced and purchased in 2015. However, the GPI for 2020 does not include the benefits that a road built in 2020 is expected to provide over the period of its useful service. Nor does it include the services that a refrigerator purchased in 2020 will provide in the years following its acquisition. GDP calculations do the very opposite because GDP is a monetary measure of domestic production and thus fails to account for the true magnitude and timing of consumption benefits.

Fifth, the GPI recognises that the welfare contribution of privately purchased consumer goods depends, in part, on the distribution of income. It is well known that the marginal benefit of consumption is less for the rich than it is for the poor (Robinson 1962). Hence, a more equal distribution of income – which marginally reduces the consumption benefits of the rich, but significantly boosts the consumption benefits of the poor – increases the welfare contribution of aggregate consumption. While the welfare impact of a change in the distribution of income is factored into the GPI, it is not so in the case of GDP.

Finally, not all benefits arising from the transformation of natural resources to human-made goods are generated within the formal economy. Many goods and services are produced by unpaid labour. The enormous economic benefits provided by unpaid labour are included in the GPI but completely overlooked in the calculation of GDP.

Table 11.1 reveals the various benefit and cost items that are typically used to calculate the GPI and highlights the difference in the way the GPI and GDP treat each benefit and cost factor. Of note is the fact that, firstly, the only commonly shared items are the welfare benefits of private and government spending on non-durable consumption goods, and secondly, the calculation of GDP involves additions only, irrespective of whether the associated activity is desirable or undesirable.

An indication of the policy-guiding value of the GPI is revealed in Fig. 11.3. The figure illustrates the difference between Australia's per capita GDP and per capita GPI for the period 1986–2016. It shows that the per capita GPI was less than the per capita GDP throughout the 30-year period. Furthermore, Australia's per capita GPI increased only marginally over the 30-year period (23.0% overall and at an average

Table 11.1 *Items used to calculate the Genuine Progress Indicator (comparison with GDP)*

Benefit/Cost Items ($)	GPI ($)		GDP ($)	
Private spending on *non-durable* consumption goods (current benefit)	Counted in full as a current benefit	(+)	Counted in full as a current benefit	(+)
Private spending on new *durable* consumption goods (future benefit)	Omitted	(0)	Counted in full as a current benefit	(+)
Services from existing *durable* consumption goods (current benefit)	Counted in full as a current benefit	(+)	Omitted	(0)
Welfare contribution of consumption weighted as per distribution of income	More equal – Consumption weighted upwards Less equal – Consumption weighted downwards	(+/–)	No adjustment	(0)
Government spending on consumption goods (current benefit)	Counted in full as a current benefit	(+)	Counted in full as a current benefit	(+)
Private spending on *durable* capital goods (future consumption benefits)	Omitted	(0)	Counted in full as a current benefit	(+)
Government spending on *durable* capital goods (future benefits)	Omitted	(0)	Counted in full as a current benefit	(+)
Services from existing *durable* public goods (current benefit)	Counted in full as a current benefit	(+)	Omitted	(0)
Services from unpaid labour	Counted in full as a current benefit	(+)	Ignored	(0)
Defensive and rehabilitative expenditures	Subtracted (assumed to be welfare-maintaining and not welfare-increasing)	(–)	Not subtracted (assumed to be welfare-increasing)	(0)
Cost of unemployment and underemployment	Psychological cost counted as current cost (cost of reduced consumption reflected in other GPI items)	(–)	Ignored	(0)
Cost of crime	Psychological cost counted as current cost	(–)	Ignored	(0)
Cost of family breakdown	Psychological cost counted as current cost	(–)	Ignored	(0)
Cost of overwork	Psychological cost counted as current cost	(–)	Ignored	(0)

(continued)

Table 11.1 (continued)

Benefit/Cost Items ($)	GPI ($)		GDP ($)	
Cost of lost natural capital services	Lost source function – Non-renewable resource depletion; land degradation; forest losses; excessive water use Lost sink function – Air pollution, waste-water pollution, and solid-waste pollution Lost life-support function – Excessive GHG emissions; ozone depletion; biodiversity losses	(−)	Ignored	(0)
GPI = sum of benefits minus sum of costs; per capita GPI = GPI ÷ population				

Note: (+) denotes addition; (−) denotes subtraction; (0) denotes no change

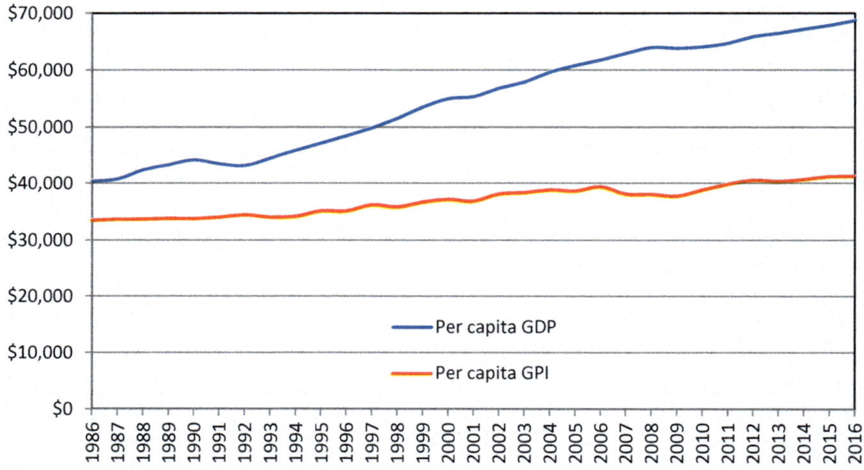

Fig. 11.3 *Per capita GPI versus per capita GDP, Australia, 1986–2016 (2015–16 prices). (Source: Lawn* 2017)

growth rate of 0.69% p.a.), whereas Australia's per capita GDP increased more significantly (70.0% overall and at an average growth rate of 1.79% p.a.). By 2016, Australia's per capita GPI was $41,341 per person, which was much less than the per capita GDP value of $68,808 per person. Strikingly, over the last decade of the period, Australia's per capita GPI rose by a meagre 4.7%, which amounted to a paltry average growth rate of 0.46% p.a.

The GPI results indicate a number of important things. First, Australia's per capita GDP greatly overestimated the magnitude of both the economic welfare delivered by the Australian economy and its annual rate of increase.

Second, towards the end of the 1986–2016 period, the additional benefits to Australia from growing its per capita GDP were almost matched by the additional costs. Close scrutiny of the benefit and costs items comprising the GPI reveals that the additional economic benefits of growth were substantial, albeit tempered by mounting income inequality, but were considerably offset by rising social and environmental costs – the latter due to Australia's very high per capita greenhouse gas emissions, excessive water use on what is the Earth's driest inhabited continent, and heavy reliance upon mining revenues (non-renewable resource depletion) as a means of financing its consumption.

Finally, should Australia continue to grow its GDP, there is good reason to believe that Australia's per capita GPI will eventually decline as the increase in social and environmental costs (marginal costs) eventually outweigh the rising economic benefits (marginal benefits). At this point, the growth in Australia's GDP will cease to be 'economic' and instead become 'uneconomic'. Australia would still be able to increase its per capita GPI, but this would depend upon stabilising GDP, better distributing income, and focusing on qualitative improvements, such as producing better goods and discovering more resource-efficient and less carbon-intensive ways to produce them.

11.3.3 Other Aspects of Wellbeing

From the above brief discussion, we have learned that there are many factors to consider when appraising wellbeing trends, and different societies will prioritise different factors at different times. Regardless, mere economic activity is not an end in itself – it is a means to achieving final ends which we collectively call 'wellbeing' and, hopefully, a life well lived. With this in mind, policymakers should be conscious of the following.

11.3.3.1 Natural Capital (Natural Wealth)

Natural capital refers to a country's stock of natural resources and waste-absorbing sinks, and includes forests, rangelands, soils, rivers, groundwater aquifers, mineral ore deposits, fisheries, wetlands, and so on. The health of natural capital is critical to human health and survival, and therefore must be factored into ecological sustainability assessments. Furthermore, since the depletion of natural capital imposes costs by undermining the economy's primary trophic level, declines in the quantity and quality of natural capital must also be factored into any reasonable economic assessment, as is the case when calculating the GPI.

11.3.3.2 Population Size

As we will soon see, the size of the human population is one determinant of the level of aggregate consumption, where the more people there are for any given circumstances, the greater is the overall level of consumption. What's more, once the maximum sustainable level of consumption is reached, any increase in population reduces the level of per capita sustainable consumption. Unchecked population growth increases the likelihood of unsustainable levels of consumption since, inevitably, society consumes beyond its available biocapacity.

11.3.3.3 Health and Education

Unlike physical consumption, improvements in health and education are less constrained by limited physical resources. While additional resources are consumed to build, staff, and supply more schools and hospitals, improvements in preventative health (diet, exercise, clean air) and education techniques can be achieved via advances in knowledge and science which make better use of, and increase the benefits yielded by, available resources.

11.3.3.4 Human-Made Capital (Human-Made Wealth)

Human-made capital comes in two forms. The first category includes *capital goods*, such as commercial and public buildings, tools, machines, and software, which are used to produce tomorrow's goods and services. Capital goods form part of a nation's productive capacity. The second category includes durable *consumer goods*, such as cars, refrigerators, furniture, and clothes. Durable consumer goods provide useful services over the lifetime of their use. Both categories of human-made wealth can always be improved in terms of design, manufacture, performance, longevity, and recyclability, which can increase the human wellbeing that a given quantity of human-made wealth can deliver.

11.3.3.5 Trophic Diversity

An economy has a trophic structure just like an ecosystem, where energy and matter that is not immediately transformed into waste is transported from lower to higher trophic levels. The economy has three trophic levels – the primary level (agriculture and resource-extraction industries); the secondary level (resource-processing, manufacturing, and materials recycling industries); and a tertiary level (service industries). The non-waste outputs of the primary trophic level become inputs to the secondary level, and the majority of non-waste outputs of the secondary trophic level become inputs to the tertiary level.[3]

Strength and resilience of the domestic economy requires trophic diversity from domestic sources. In other words, a genuinely strong domestic economy will

[3] Some non-waste outputs of the secondary trophic level, such as fuels, farm machinery, and mining equipment, become inputs to the primary trophic level, while others, such as fuels, commercial buildings, and manufacturing equipment, become inputs to the secondary trophic level.

possess a healthy balance of primary, manufacturing, and service industries – the former drawing on abundant high-quality stocks of natural capital capable of providing adequate and ongoing material, energy, and food supplies. An economy with trophic diversity is not overly reliant upon imports to overcome a weak domestic trophic level. It is therefore less vulnerable to external shocks and interference. Such an economy, while hopefully still an active participant in global society, has a level of independence that reduces its exposure to whims and coercion from trading partners. It also wastes less fuel from engaging in international trade, offers a larger range of occupations (i.e., increases lifestyle choices), and reduces its biosecurity risks.

11.3.3.6 Law and Governance

High levels of wellbeing are associated with high levels of transparency and low levels of corruption, crime, addiction, suicide, exploitation, and human-rights abuses. Nations can continually increase their performance in this area without significantly increasing resource demands.

11.3.3.7 Shuffling Financial Assets (Virtual Wealth) Does Not Equate to Generating Real Wealth

A nation generates additional real wealth when it transforms natural resources into useful goods and services (human-made wealth) whilst leaving natural capital (natural wealth) intact. Should a nation deplete its natural capital by extracting natural resources beyond what the ecosphere can sustainably yield, it will simply create human-made wealth at the expense of natural wealth. This may or may not increase human wellbeing at the time, but it will certainly reduce the nation's capacity to sustain a particular level of economic welfare. This is because, from a sustainability perspective, human-made capital is not a productive substitute for natural capital. The two forms of capital are complements, where the latter provides the structured matter-energy transformed in production; the former is used by labour to perform the transformation process.

A financial asset is a claim on real wealth in the sense that it can be exchanged for something real, such as natural resources, human-made capital, or durable and non-durable consumer goods. The ability of a financial asset to be converted to a real asset is what makes it a form of *virtual wealth*, with money being the most liquid of all financial assets.

Seemingly unproductive activities within the tertiary trophic level do deliver something of value, such as insurance, accounting, and useful financial services.[4]

[4]That is not say that all such services are welfare-increasing. Insurance services, for example, are welfare-maintaining and therefore of a defensive nature. They are included in measures of GDP but excluded from alternative welfare indexes, such as the Genuine Progress Indicator (GPI).

Providers of these services should be appropriately remunerated. However, because of financial deregulation, the shuffling of financial assets has proliferated. Much of it is speculative and highly parasitic in the sense that it generates no real wealth. It also tends to redistribute financial claims on existing real wealth towards the rich. It and other speculative or rent-capturing[5] activities (e.g., the buying and selling of real estate for pure capital gain) should be taxed and, in some cases, legally prohibited to encourage wealth-creating ventures and to narrow the gap between rich and poor.

11.3.3.8 International Relations

All societies are members of the international community, where acceptable levels of behaviour should be set by international institutions – in particular, the United Nations. No society should threaten or take advantage of its neighbours, but all should work cooperatively for mutual benefit. This is best achieved by societies pursuing sustainable development – and actively helping others achieve it – rather than engaging in any form of neo-colonialism. Wellbeing cannot be achieved without national security. Nor can it be achieved without the right of countries to introduce welfare-enhancing laws free from subversion by international trading arrangements.

11.3.4 Sustainability, Distribution, and Allocation

Sustainability involves the human pursuit of goals conducted in a way that can continue indefinitely.[6] To achieve sustainability, the resource throughput required to fuel the economic process must remain within the biocapacity of the planet. This requires adherence to three sustainability precepts: (1) renewable resources should be extracted at a rate no greater than they can regenerate; (2) non-renewable resources should be extracted at a rate no greater than renewable resources can be cultivated to replace them (thus keeping the quantity of resource-providing natural capital intact); and (3) the rate of waste generation should not exceed the waste assimilative capacity of the ecosphere (thus preserving the natural biogeochemical processes required to regenerate renewable resources and maintain a planet fit for current life forms).

The pursuit of sustainability requires a shift, in a timely manner, to using renewable resources, but there always remains a limit to the scale of resource use and the

---

Should insurance services mount as the negative impacts of climate change escalate, they would add to GDP but not to economic welfare.

[5] In economics parlance, an economic rent is an amount paid above the minimum necessary to bring a factor of production into operation. An economic rent constitutes unearned income.

[6] This is an anthropocentric view of sustainability.

technical capacity to transform them into goods and services. A wise society would operate within ecological limits at all times (i.e., ensure its ecological footprint does not exceed biocapacity), rather than venture into unsustainable territory in the hope that technological breakthroughs and the importation of natural resources might remedy the situation. Although degrowth offers a solution to nations with economies larger than their maximum sustainable scale, degrowth is a second-best outcome best avoided by remaining within long-run sustainable limits in the first instance.

Distribution refers to the relative share of total income amongst the citizens of a particular jurisdiction. In ecological economics, distribution is equitable if, firstly, each citizen can claim sufficient goods and services to thrive, and secondly, the order-of-magnitude disparity in income of the richest and poorest individuals is deemed socially acceptable. Given that natural resources need to be sustainably managed, overpopulation creates undesirable distributional challenges by accelerating the rate of environmental destruction, intensifying competition for resources, and increasing the likelihood of wars. Because overpopulation affects the yet-to-be-born by altering the distribution of resources across time, it also has an intergenerational impact, invariably an undesirable one.

It is worth stressing that high levels of inequality are usually the result of poor distributive mechanisms – typical of neoliberal ideology – which are the product of an inadequately progressive tax-and-provisioning system. This shortcoming is often compounded by an austerity bias that increases unemployment and reduces the availability of public goods in strong demand by the needy poor. In some instances, high levels of inequality eventually harm the wealthy by creating an unsustainable society where the chances of civil unrest and other wealth-destroying outcomes increase.

Allocation refers to the apportionment of the natural resources extracted from the ecosphere for economic purposes. The allocation question thus posed is: should steel made from extracted iron ore be used to make guns, spacecraft, or schools?[7] If the answer is spacecraft, then a further allocation question arises: what type of spacecraft and by what method? In market economies, allocation within the private sector is largely undertaken within markets and guided by price signals. Prices are invariably determined by price-setting firms, albeit they are influenced by consumer preferences, resource scarcities, society's institutions, and government regulation, including price-altering taxes, subsidies, licence fees, and permit prices. Efficient allocation occurs when resources are put to their best (socially desirable) use.

To achieve an efficient allocation of the incoming resource flow, it is necessary for market prices to reflect, as best as possible, the full benefits and costs of economic activity. This includes benefits and costs of the social and environmental variety. Because of inevitable market failures, accounting for all benefits and costs will often require governments to impose taxes on non-priced spillover costs, such

[7] The question of how much iron ore should be extracted is a throughput concern, not an allocation concern.

as those associated with pollution, or the payment of subsidies on non-priced spill-over benefits, such as those associated with materials recycling. Allocative efficiency will also require governments to obtain a portion of the incoming resource flow (macro-allocation) to provide adequate quantities of non-profitable public goods; to ensure there is sufficient and accurate information for people to make informed decisions; and to enact legislation supported by well-funded regulatory agencies to limit the market power of large corporations and to quell rent-capturing activities (see point 3.7).

Although ecological sustainability, distributional equity, and allocative efficiency require separate policy instruments to be adequately resolved, there is one policy mechanism that can play a significant role in simultaneously achieving all three policy goals. Referred to as a *cap-auction-trade system* it is a policy mechanism that has been fervently promoted by many ecological economists following initial support for the system by Herman Daly (1992).

The system works by placing limits on the use of various resource types as a way of quantitatively controlling a nation's rate of natural resource extraction and pollution generation, especially the generation of greenhouse gases. The government achieves this by setting absolute caps on the rate of resource extraction and pollution to keep them within the ecosphere's regenerative and waste-assimilative capacities – that is, by ensuring they conform to the three previously outlined sustainability precepts. The system then functions to ensure the allocation of extracted resources and the generation of permitted pollution levels are conducted efficiently and in an equitable manner.

Using greenhouse gas emissions as an example of how a cap-auction-trade system would work, a cap would be placed on annual emissions, which would be decreased each year to meet international obligations, such as those pertaining to the Paris Agreement on climate change. The caps would be administered by issuing a limited number of short-lived permits, which would be periodically auctioned by a government agency, thus letting market forces set permit prices. Because the need for polluters to acquire permits increases the cost of generating emissions, polluters would have an incentive to reduce the emissions-intensity of their activities. Consequently, permit prices would serve the same efficiency-inducing function as an emissions tax. Once auctioned, permits would be tradeable – which would allow permit prices to fluctuate – so that permitted emissions are generated by the best-use producers and energy providers.

Properly applied to most natural resources and waste types, cap-auction-trade systems would provide certainty on absolute levels of natural resource use and pollution (ecological sustainability); would assist in allocating resources to their best possible use (allocative efficiency); and, by destroying some spending power of permit buyers, would allow governments to reduce marginal tax rates on low incomes to compensate the poor for consequential increases in goods prices (distributional equity).

11.3.5 Nations Should Strive for Maximum Self-Sufficiency

Recommending national self-sufficiency is not a call for all nations to become totally self-sufficient. Rather, in view of the importance of trophic diversity, the principle says that more self-sufficiency is generally better than less, especially in regard to essential things like food, energy, and technology.[8]

A reliance on international trade – whether importing or exporting – increases risk and decreases resilience. This is because disruptions to, and arguments about, trade are common, for the following reasons. First, an importing country may genuinely change its consumption preferences. Second, a trading partner might disrupt trade for internal political and/or diplomatic reasons. Third, a change in exchange rates might suddenly alter the terms of trade (i.e., increase the cost of imports and reduce export revenue). Fourth, a trading partner might impose tariffs to protect a domestic industry (trade rules notwithstanding). Fifth, trade involves the use of valuable energy for transport, as well as other costs, not always adequately reflected in prices. Sixth, trade in organic materials increases biosecurity risks to the importing country. While these risks can be somewhat ameliorated, they can never be zero, and risk reduction always comes at a cost.

Insofar as nations engage in trade, they should only export surplus goods (often resulting from favourable domestic production conditions) to finance the cost of importing goods in short supply (often resulting from unfavourable domestic production conditions or more favourable conditions elsewhere in the world), so that net exports approximate zero.

Exporting entails a nation sacrificing real wealth to the rest of the world in exchange for foreign currency-denominated financial assets (virtual wealth). Importing entails the rest of the world sacrificing real wealth in exchange for financial assets denominated in the importer's currency. Exports, as goods foregone, are costs, while imports, as goods received, are benefits. Hence, a net-exporter of goods and services incurs greater export costs than import benefits for the sake of accumulating more net financial assets – hardly something to celebrate given that financial assets have zero use value until exchanged for real wealth which, if eventually imported, restores (ironically) a balance in trade.

Rather than net exporting, arrangements should be introduced to balance trade and have the central government – assuming it is a monetary sovereign[9] – purchase the resources that would have been used to produce the 'net-exported' goods and direct them towards building schools, hospitals, and productive infrastructure. This would: (1) increase the quantity of real stuff that is consumed and used domestically

[8]A 2014 study found that 77% of countries are in calorie deficit in terms of domestic food production. See Davis in Clapp (2017).

[9]A monetary sovereign is a central government that issues its own currency which is not pegged to anything real and of intrinsic value (e.g., gold); has no foreign currency-denominated liabilities; and has a floating exchange rate. A monetary sovereign has no financial constraint in that it can always create whatever quantity of its own currency it requires to purchase anything for sale in its own currency.

at no additional resource cost, which would boost domestic productive capacity and economic welfare; (2) maintain the nation's possession of financial assets at pre-net-exporting levels (i.e., the nation accumulates net financial assets denominated in its own currency instead of in foreign currencies); and (3) avoid the perceived need to net-export to maintain employment levels (i.e., job losses from the production of fewer exported goods are offset by job gains to build more schools, hospitals, and productive infrastructure).

The risks associated with low self-sufficiency are exemplified by some countries in the Middle East. With large oil reserves, they have used oil-export revenue to pay for food and other imports and have increased their populations dramatically. In effect, they have exploited their bountiful primary trophic endowment to overcome the lack of a developed secondary trophic level needed to support their larger populations. As the world increasingly changes its energy preferences away from fossil fuels, these oil-exporting countries face a very uncertain future, which may well have negative consequences globally, with cascading civil unrest, terrorism, and refugee flight all made more likely.

There are many examples of countries with economies much weaker than generally perceived due to their low self-sufficiency. Japan, for all its secondary trophic level ascendancy, has a very weak primary trophic level and is heavily dependent on the importation of natural resources to feed its manufacturing industries. International unrest could very easily starve Japan of the natural resources it requires to remain a manufacturing powerhouse. Very differently, Australia has abundant non-renewable resources and therefore a relatively strong primary trophic level (which, in part, helped Australia avoid the worst effects of the 2008 Global Financial Crisis). However, like many high-income countries, Australia has allowed its secondary trophic level to wane. It now relies enormously on the rest of the world to supply it with human-made wealth to support its consumption and tertiary trophic level.

Many high-income nations are not as lucky as Australia. A number of European Union countries, which have largely deindustrialised themselves over the past 40–50 years, have very weak primary and secondary trophic levels and are supported to a large extent by net-exporting countries, such as China, which have foolishly incurred higher export costs than import benefits to accumulate net financial assets denominated in Euros and other 'safe' foreign currencies, such as the US dollar.

Since the Chinese Government is a monetary sovereign, it could have exploited its fiscal capacities to purchase many of the resources that were used to produce the goods it net-exported, which, if well-directed, would have allowed China to build up its secondary trophic level (productive capacity) without having to hand over large quantities of goods for the rest of the world to enjoy. Had China done this, it would have benefited from higher levels of domestic consumption and accumulated Yuan-denominated net financial assets created by the larger spending of the Chinese Government – a small price to pay, if any price at all. The recent push by the Chinese Government to promote the domestic consumption of Chinese-produced goods might just represent an acknowledgement of this past error and a warning to countries lacking trophic diversity that they cannot continue to rely on China's 'generosity' to overcome their economic deficiencies.

11.3.6 Internationalisation and Globalisation

The goal of relative self-sufficiency is in keeping with *internationalisation*, which existed in an imperfect form for around three decades following the creation of the Bretton Woods system of exchange rate management and associated international institutions in 1944.[10] Internationalisation refers to a global economy where national economies exist as separate and autonomous entities tied together in recognition of the importance of international trade, treaties, and alliances.

In an internationalist world, the fundamental unit of concern is the nation state. This explicit emphasis has four important implications for the nation state, its people, and international trade. First, people residing within a nation are viewed as a community of citizens rather than a collection of individual consumers. Second, it is accepted that many national institutions and laws rightfully exist to impinge on economic activities to serve desirable national goals, such as distributional equity, guaranteed minimum wages and conditions of employment, and ecological integrity. Third, international institutions function to serve the interests of their members, which are nation states, not transnational corporations. Fourth, the benefits of international trade are maximised when trade is balanced and there are appropriate restrictions on the mobility of international capital to ensure trade is governed by the desirable economic principle of *comparative advantage* (Daly 1996).[11]

Following the collapse of the Bretton Woods system of exchange rate management in 1971 and the deviation of the former Bretton Woods institutions from their original charter, internationalisation gave way to *globalisation*. Unlike internationalisation, globalisation entails the integration of national economies into one single global economy through so-called free trade and free capital mobility. In a globalised world, the fundamental unit of concern is the corporation and the individual consumer. The consequence of this is a blurring of national boundaries and the bypassing of many national institutions and laws for trade-facilitating purposes. Furthermore, the governing principle of international trade is no longer comparative advantage, but the less desirable economic principle of *absolute advantage*.[12]

Over the half-century that the global economy has been organised around globalisation principles, transnational corporations have offshored a lot of their production activities in search of the cheapest labour and weakest environmental

[10] The Bretton Woods institutions include the World Bank, the International Monetary Fund, and the General Agreement on Tariffs and Trade, now called the World Trade Organization.

[11] It is often overlooked that comparative advantage presupposes balanced trade and capital immobility. In the present world of highly mobile international capital and large trade imbalances, trade is governed by the principle of absolute advantage, which does not guarantee mutual trading benefits at the national level.

[12] With comparative advantage, productive capital moves to where *relative costs* are lowest. With absolute advantage, productive capital moves to where *absolute costs* are lowest. In the latter situation, a 'competitive' advantage can be gained by shifting productive capital to where wages are lowest and environmental regulations are weakest. The threat of this shift – known as 'capital flight' – often induces governments to supress wages growth and dilute environmental standards.

standards. This has resulted in a degenerative 'race to the bottom' to reduce the costs of production, with obvious negative consequences, not the least being the reallocation of global resources to many resource-wasteful and pollution-intensive production locations. It has also led to the hollowing out of manufacturing in many high-income countries, with a corresponding loss of career-pathed employment opportunities and the emergence of insecure, low-wage jobs.

In view of its destructive impact on the primary and secondary trophic levels of some countries, there is little doubt that globalisation has significantly contributed to the world-wide reduction in national self-sufficiency. Clearly, there is an urgent need to restore global economic arrangements around the principles of internationalisation. Provided this is achieved, the domestic economy ought to again be the dog that wags the international trade tail, which would: (1) prevent international trade from dictating terms domestically, both economically and politically; (2) maximise the mutual benefits of international trade; (3) ensure countries remain free from any compulsion to trade; and (4) facilitate the structural adjustments that many countries require to become self-sufficient.

11.3.7 Thermodynamics, Biocapacity and Throughput

From a biophysical perspective, there are three major stages of the economic process. As pointed out earlier, the first stage involves the extraction of natural resources from the ecosphere. The second stage entails the transformation of extracted natural resources by human-made capital and labour into new physical goods. This process is usually referred to as 'production'. Its purpose is to add more use value to the existing use value of the matter-energy embodied in the pre-transformed natural resources.

The third stage involves the transformation of physical goods to waste through the immediate 'destruction' of non-durable goods and the eventual 'depreciation' of durable goods. This process is usually referred to as 'consumption'. Its purpose is to enable humans to enjoy the use value added to natural resources via production.

As physical transformation processes, both production and consumption are governed by the first and second laws of thermodynamics. The implications of these two laws for production and consumption possibilities and ecologically sustainable levels of real output are non-trivial.

The first law of thermodynamics is the *law of conservation of matter and energy*. It informs us that it is impossible to create or destroy matter and energy. The first law reveals how inaccurate it is to describe the second and third stages of the economic process as production and consumption, since nothing physical is ever produced and nothing physical is ever consumed. Matter and energy are simply rearranged in production and further disarranged in consumption. What is created by production is use value; what is destroyed by consumption is also use value. Hence, the terms 'production' and 'consumption' are entirely valid, but not in the manner usually perceived.

Let us consider Q to be the quantity of matter-energy embodied in real GDP. Let us also consider F to be a nation's ecological footprint – that is, the quantity of matter-energy embodied in the natural resources transformed into Q. If we denote E as the technical efficiency of production, then:

$$E = Q / F \tag{11.1}$$

Because of the first law of thermodynamics, it is impossible for the quantity of matter-energy embodied in real GDP to be greater than the quantity of matter-energy embodied in the natural resources used to generate Q, since this would amount to the creation of matter and energy. Clearly, it is impossible for Q to exceed F and for E to exceed a value of one (note: $E = 1$ represents 100% technical efficiency of production).

The second law of thermodynamics is the *Entropy Law*. It informs us that when matter-energy is transformed in any way, its orderliness or structure is diminished. Consequently, the ability of transformed matter to be rearranged in the same way or for transformed energy to perform work in the same way is unavoidably reduced. It is because of this diminished order that production always results in the generation of some waste matter-energy, which we shall denote as W_P, where $W_P > 0$.

Taken together, the first and second laws of thermodynamics dictate that the quantity of matter-energy embodied in real GDP must be *less* than the quantity of matter-energy embodied in the natural resources used to produce it. In other words, the technical efficiency of production must be less than 100% (i.e., E is always <1). In addition, the 100% recycling of waste matter – a process requiring additional energy – is impossible, as is the recycling of any waste energy.

We can therefore represent the production stage of the economic process as:

$$F = Q + W_P \tag{11.2}$$

Furthermore, because consumption involves the eventual transformation of real GDP to waste matter-energy, which we shall denote as W_C, we can represent the economic process at its conclusion as:

$$F = W_C + W_P \tag{11.3}$$

What are the implications of all of this? First, the throughput of matter-energy fuelling the economic process is linear, not circular. Whilst some waste materials can be reconcentrated via recycling, all energy moves in a linear direction from useful to useless, as does all matter eventually.[13] Recycling is beneficial in the sense that it can reduce the extent to which natural capital must be exploited to obtain virgin materials, but only if the additional energy required to recycle waste matter does not extend energy requirements and consequent waste outputs beyond what can be ecologically sustained.

[13] Organic matter is obviously continually recycled by nature into new (useful) biomass.

Second, a minimum and irreducible quantity of matter-energy is required to produce a given level of GDP. Efficiency-increasing technological progress can reduce the waste immediately generated during the production stage of the economic process (i.e., can increase E by reducing W_P). Nonetheless, because E must be something less than one, it cannot do this continuously. This is illustrated in Fig. 11.4, where it can be seen that, of the matter-energy embodied in the natural resources used in production, increases in E can boost the proportion of matter-energy embodied in real GDP (Q), which reduces the proportion of matter-energy embodied in immediate production waste (W_P). However, an asymptotic limit to E (dashed line) means there is an unquestionable thermodynamic constraint on the capacity of technological progress to produce more output from less input.[14]

Third, stocks of natural capital – both renewable and non-renewable – are limited in quantity, while renewable natural capital has limited regenerative and waste assimilative capacities. There is, therefore, only so much matter and energy that the ecosphere can supply on an ecologically sustainable basis. This fact together with the limits imposed by the first and second laws of thermodynamics mean that real GDP cannot grow forever.[15]

In order to know what the GDP limit is, it is essential to know the maximum sustainable rate of throughput. Knowledge of this, at various micro-ecological

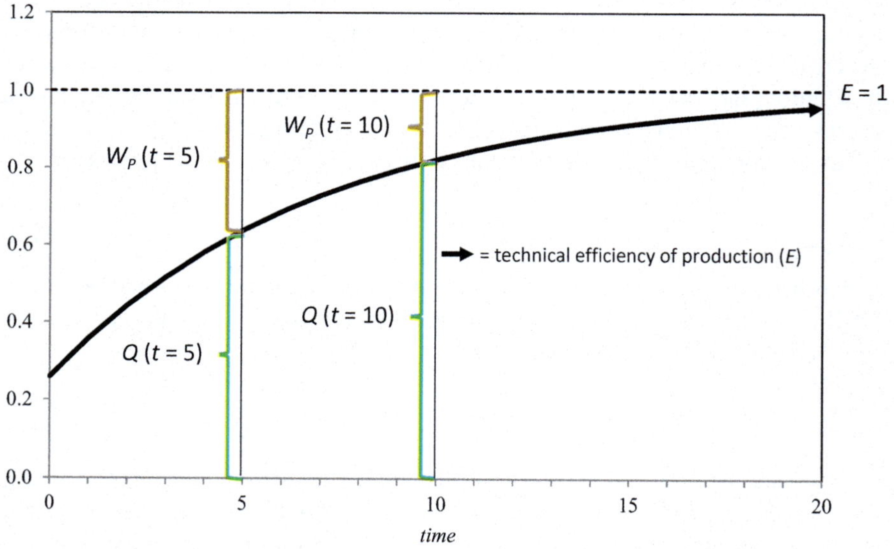

Fig. 11.4 *Thermodynamic limits to increases in the technical efficiency of production (E < 1)*

[14] This fact can also be explained by rearranging Eq. (11.2) to obtain $Q = R - W_P$. Since $W_P > 0$, it follows that Q is always less than R.

[15] Technically, real GDP does not exactly equal the quantity of real output produced. However, the difference is trivial enough that any growth in real GDP, once ecological limits are reached, would be negligible.

levels, requires natural scientists to estimate the regeneration rates of different forms of renewable natural capital (sustainability precept # 1); the speed with which renewable resources can be cultivated to replace depleted non-renewable resource stocks (sustainability precept # 2); the waste assimilative capacities of waste-absorbing sinks (sustainability precept # 3); and the impacts that human activities have on the services provided by natural ecosystems, including the Earth's climate system.

At the macro-ecological level, it is necessary for natural scientists to calculate the biocapacity of the ecosphere, since this provides an estimate of the maximum rate of resource throughput available for economic purposes on a sustainable basis. So, too, is it necessary to calculate a nation's *ecological footprint*. As a measure of a nation's natural resource demands, the ecological footprint should remain below biocapacity unless matched by an offsetting ecological surplus of another country.

11.3.8 Global Resource Throughput Should be Less Than the Planet's Biophysical Capacity

Biophysical capacity or *biocapacity* is the amount of land available to provide the resources, absorb wastes, and still provide essential ecosystem services for a nation (or global civilisation) on an ongoing basis. Measured in global hectares (Gha), it represents ecologically sustainable resource supplies. As alluded to many times, if a nation's ecological footprint exceeds its biocapacity, it is operating beyond what can be ecologically sustained.[16] Although it is difficult to accurately measure biocapacity, techniques are steadily improving with experience. According to some highly respected but conservative calculations (Wackernagel and Beyers 2019), humanity's footprint began to exceed the Earth's biocapacity in 1970 and is now around 1.6 times its sustainable scale.[17] But it may be many times more. Biologist E. O. Wilson (2016) and others (see Chap. 15, this volume) believe that half of the Earth should be set aside to maintain global ecological integrity, which suggests humankind's ecological footprint needs to be significantly less than the planet's biocapacity.

In terms of the mass of raw materials, one study puts global sustainable resource throughput at about 50 billion tonnes per year (Dittrich et al. 2012). In another recent study (https://phys.org/news/2020-01-global-resource-consumption-tops-bn.html), it has been estimated that humanity is using about 100 billion tonnes of resources a year, despite billions of people living in poverty. This again suggests

[16] It is possible for a country to operate for some time with its ecological footprint beyond its biocapacity, but only by depleting natural resource stocks, which is unsustainable, or by importing natural resources (ecological surpluses) from other countries. The latter is perilous given that imports of natural resources and the continual ecological surpluses of other countries can never be guaranteed.

[17] A global ecological footprint 1.6 times global biocapacity implies that sustaining global rates of resource use and waste generation at current levels would require 1.6 Earths.

that humankind's rate of resource use is around twice what the planet can sustain, or more if Wilson's dictum is invoked.

Adopting a very different approach, a group of scientists have identified nine so-called *planetary boundaries* as a way of defining an ecologically 'safe' space for humankind to operate within (Rockström et al. 2009; Chap. 2 of this volume). The nine boundaries have been carefully selected in recognition that their transgression has the potential to trigger catastrophic environmental change at a continental or, worse still, planetary scale. Studies show that four of the planetary boundaries have already been crossed, while the remainder are being approached at a rapid rate (Steffen et al. 2015).

These findings have profound implications for what high-income and overpopulated countries must do to play their part in ensuring there is equitable resource consumption and sustainable rates of resource throughput at the global level (Bringezu 2015). As a starting point, no country should unduly exploit the resources of another nation. Nor should any nation be forced beyond their will to surrender some of their natural resources.

Second, every country should attempt to operate so that its ecological footprint does not exceed its biocapacity, or equivalently, operate within each of the nine planetary boundaries. This means that many high-income countries will need to embrace life-affirming degrowth of their GDP (Hickel 2020). Overpopulated countries, many of which are desperately impoverished, will need to do something to quell their population growth (more on this later).

Of course, it will not be possible for every country to avoid ecological deficits. Because some countries have grown their economies well beyond their maximum sustainable scale, the extent of the degrowth required to operate within biophysical limits will take considerable time to accomplish. For other countries, the little biocapacity they do possess will never support anything but a very small population – in some instances, a population much smaller than they presently have.

So long as the ecological deficits of some countries do not, in the aggregate, exceed the ecological surpluses of others – which is not the case at present – nation-level ecological deficits can be compatible with global sustainability. However, in view of current circumstances, global sustainability is unlikely to be restored through unilateral endeavours. It is probably best achieved within the confines of an international agreement that imposes mutually agreed upon footprint targets and embodies flexibility mechanisms to allow deficit countries (i.e., countries which exceed their national targets) to offset their shortfalls as they undertake the time-consuming but necessary adjustments.

Permissible measures to offset deficits might include: (1) importing resources from willing surplus countries; (2) investing in natural capital at home to increase domestic biocapacity, thereby reducing or perhaps eliminating an existing deficit; (3) investing abroad in natural capital to boost the biocapacities of other countries; (4) investing abroad in efficiency-increasing and pollution-reducing technologies to reduce the ecological footprints of other countries; and/or (5) investing abroad in family planning and fertility-related services to reduce the ecological footprints of overpopulated countries.

Despite its risks, many ecological deficit countries already rely on (1) – for example, Japan, which depends on the importation of natural resources to meet most of its ecological deficit. As for (2), deficit countries would be incentivised to cultivate additional domestic stocks of renewable resources, which would undo a lot of past environmental damage. Most importantly, deficit countries would have an incentive to avoid the depletion of their existing stocks, which is presently common amongst deficit nations. In terms of (3), (4), and (5), individual nations would have a new-found interest in the footprints and biocapacities of all countries, in particular, the impact on ecological footprints of population growth.

Over time, footprint targets would be gradually tightened until the global ecological footprint was no greater than global biocapacity. Assuming widespread compliance, the ecological deficits of some countries would decline until they eventually matched the ecological surpluses of other countries. Because, in the meantime, deficit countries would incur a hefty penalty to offset their target shortfalls, many of them would find it less costly, if not beneficial in a net sense, to undertake a degrowth phase to eliminate their ecological deficits, which would likely include efforts to limit domestic population growth. It is not unreasonable to believe that an arrangement of this nature would significantly reduce the number of ecological deficit countries and do much to limit the growth in global population numbers.

To succeed, the entire arrangement would have to be administered and closely enforced by a new international institution to ensure national commitments, and therefore global ecological footprint targets, are met. The institution would ideally function under the auspices of the United Nations. In many ways, the system would operate along similar principles to the Paris Agreement on climate change, which is a multilateral arrangement with inbuilt flexibility mechanisms purposely designed to equitably reduce global greenhouse gas emissions to 'sustainable' levels.[18]

11.3.9 Continuous Absolute Decoupling is a Dangerous Myth

Much attention has been drawn to the decoupling issue in recent times, sometimes referred to as the dematerialisation of the economic process. We have seen that it is possible for efficiency-increasing technological progress to reduce the immediate waste generated as the matter-energy embodied in the natural resources used in production is transformed into physical goods (see Fig. 11.4). When such progress occurs, the quantity of natural resources required to produce a *given* level of real GDP declines. As we shall see, the key word here is 'given', since decoupling offers

[18] It is highly questionable whether the existing emissions targets embodied in the Paris Agreement are sufficient to stave off catastrophic climate change. Hence, they may not be consistent with ecological sustainability. Some would also question whether national emissions targets – effectively national emissions rights – have been equitably distributed amongst the signatories to the Agreement.

little hope for growthists if an increasing quantity of natural resources is required to produce a *growing* level of real GDP.

It cannot be denied that if the percentage rate of efficiency-increasing technological progress can be kept ahead of the percentage rate of growth of real GDP, it is possible to increase real GDP and reduce the rate of resource throughput and environmental degradation. The problem lies with the fact that progress of this nature cannot continue forever, which means the continued growth of real GDP, at any rate, must eventually be accompanied by rising levels of resource throughput. Many commentators ignore or do not understand this fact and perpetuate the dangerous myth that growth can be ecologically sustained by 'decoupling' real GDP from resource throughput.

To consider the plausibility of such a claim, we first need to understand what is meant by *decoupling*. In conventional parlance, decoupling is the idea that the resource-intensity of GDP can be sufficiently decreased through technological improvements, substitution, and productivity gains to either reduce the rate of resource throughput as GDP rises (referred to as *absolute* decoupling) or slow the rate of growth in resource throughout as real GDP rises (referred to as *relative* decoupling) (Jackson 2009).[19] For the growthists' claim to hold true, it must be possible to maintain absolute decoupling indefinitely.

The difference between the two forms of decoupling is best explained with the aid of the following variation of Ehrlich and Holdren's (1971) famous $I \approx PAT$ identity:[20]

$$F = P \times A \times R \tag{11.4}$$

where F = ecological footprint; P = population; A = material affluence or per capita GDP (GDP/P); and R = resource-intensity of GDP (F/GDP).

[19] It is important to stress that this is a conventional position and does not conform to some views on decoupling. Some hold the view that decoupling literally implies the complete severing of the nexus between real GDP and resource throughput, which is an undeniably impossible circumstance. For the purposes of this chapter, we shall adopt, although not necessarily agree with, the conventional position on decoupling.

[20] In the $I \approx PAT$ identity, P and A also denote population and affluence, but I = ecological impact and T = technology. We believe the $F = PAR$ equation has pedagogical and empirical advantages over the $I \approx PAT$ identity. To begin with, the magnitude of the prevailing rate of resource throughput does not provide an accurate indication of impact (I). A given rate of resource throughput will normally have greater ecological impact the smaller is the prevailing biocapacity. Second, technology (T) can either boost or reduce the rate of resource throughput. *Ceteris paribus*, efficiency-increasing forms of technological progress reduce the rate of throughput, whilst affluence-augmenting forms of technological progress, such as the invention of the car to replace the horse-and-wagon, increase the rate of throughput. For empirical purposes, it is not possible for an independent variable moving in a particular direction to have both positive and negative influences on a dependent variable. The $F = PAR$ equation separates the two forms of technological progress by capturing affluence-augmenting varieties in the affluence variable (A) and efficiency-increasing varieties in the resource-intensity variable (R).

Fully extended, Eq. (11.4) becomes:

$$F = P \times \text{GDP} / P \times F / \text{GDP} \tag{11.5}$$

where, because $(P \times A) = P \times \text{GDP}/P = \text{GDP}$, then:

$$F = \text{GDP} \times F / \text{GDP} \tag{11.6}$$

Now consider Fig. 11.5. The blue line represents real GDP (i.e., $P \times A$) growing at 5% p.a., starting at an index value of 100.0 at time $t = 0$.

Relative decoupling is represented by the red line. It shows the ecological footprint rising, but at a lesser rate than real GDP. As per Eq. (11.4), the rising ecological footprint can be attributed to the fact that the rate of reduction in the resource-intensity of real GDP (2% p.a.) is *less* than the rate of growth in real GDP. Hence, at time $t = 1$, with $(P \times A) = 105.0$ and $R = 98.0$, $F = 102.9$ (all values set to a starting index value of 100.0). That is, F has increased by 2.9%.[21]

With absolute decoupling, which is represented by the green line, the ecological footprint is falling. This is because the rate of reduction in the resource-intensity of real GDP (8% p.a.) is *greater* than the rate of growth in real GDP. Hence, at time $t = 1$, with $(P \times A) = 105.0$ and $R = 92.6$, $F = 97.2$. On this occasion, F has decreased by 2.8%.

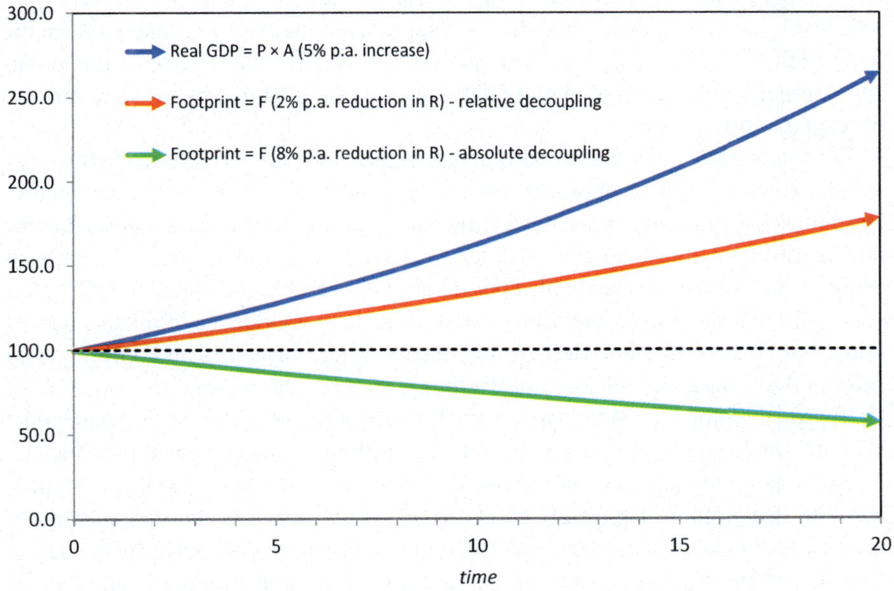

Fig. 11.5 *Conventional view of absolute and relative decoupling*

[21] Some readers might find this confusing. Because the starting values of each variable is 100.0 and not 1, the calculation of $F = 105.0/100 \times 98.0/100 = 102.9/100$.

To show why absolute decoupling cannot be sustained, we need to reconsider the equation representing the technical efficiency of production, which was:

$$E = Q / F \tag{11.1}$$

Equation (11.1) can be rearranged to obtain:

$$F = Q / E \tag{11.7}$$

Because Q represents GDP, we can replace Q in Eq. (11.7) with $(P \times A)$, just as we did to convert Eqs. (11.5) to (11.6). We therefore get:

$$F = (P \times A) / E \tag{11.8}$$

If the product of $(P \times A)$ and F are measured in terms of the matter-energy embodied within them, the only difference between eqs. (11.4) and (11.8) is that R in the former has been replaced by $1/E$ in the latter. That is, $R = 1/E$. This simply means that R is the inverse of E.

To recall, because of the first and second laws of thermodynamics, E must be less than a value of one (i.e., $E < 1$). The flipside of this is that R must be greater than a value of one (i.e., $R > 1$). That is, the matter-energy embodied in the natural resources used in production must be greater than the matter-energy embodied in real GDP. Thus, while technological progress can reduce the value of R, it can only reduce it by so much. More critically, as R approaches one, the percentage rise in the rate of efficiency-increasing technological progress approaches zero. At this point, any non-negligible increase in real GDP must be accompanied by a rise in the rate of resource throughput.

We would hasten to say that continuing to reduce the rate of resource use throughout the time it takes to approach thermodynamic limits is a best-case scenario. Despite plenty of room at present for efficiency increases, evidence suggests that the rate of resource throughput – whether measured in global hectares or units of energy – has risen over an extended period of time in every case where real GDP has grown at a non-negligible rate (Lawn 2013). Yes, there is evidence of relative decoupling, but none of absolute decoupling, except in rare instances and only for very short periods. Virtually all cases of declining resource throughput are the result of GDP growth being stifled by war, internal unrest, a financial crisis, or a pandemic. Overall conclusion? Continuous absolute decoupling is a cornucopian pipedream.

Not to be outdone, many growthists at this point resort to one last leap of faith – absolute decoupling of physical goods from natural resource throughput may be impossible, but what about services? Beginning with the assertion that services have virtually no physical dimension, growthists argue that transitioning from physical goods to non-physical services allows the services share of real GDP to grow as the physical goods share diminishes. This, they preach, allows real GDP to keep expanding as natural resource throughput declines.

The fundamental shortcoming of this proclamation reflects an ignorance of the trophic structure of an economy and its ramifications. Services, which comprise the tertiary trophic level, are never non-physical, as anyone who has had their hair cut by a real hairdresser in a real salon using real appliances running on real electricity will know.

As previously explained, the inputs of the tertiary trophic level are some of the outputs of the secondary trophic level, which, to be produced, require outputs from the primary trophic level. Hence, the natural resource dependence of the tertiary trophic level includes the natural resources expended to make possible the prerequisite activities within the primary and secondary trophic levels. Unfortunately, many studies of the resource dependence of the tertiary trophic level exclude these indirect resource requirements, primarily on the false assumption that the three trophic levels are independent of each other. As one might expect, when indirect resource requirements are taken into account, there is no indication of any natural resource savings to be had from making the shift from physical goods to services, which is an illusionary transition in any case (Lawn 2009).

Evidence of the 'decoupling via services' myth is not difficult to find. Over the past 50 years, services as a percentage of the GDP of high-income countries have increased from around 50% to 70–80% due to the growing reliance on imported manufactured goods (i.e., no transition to services but an increasing reliance upon the secondary trophic level of other countries) and the shifting of many economic activities from the informal (unpaid) economy to the formal economy. Yet, over this same period, the ecological footprints of high-income nations have increased (Lawn 2013; Wackernagel and Beyers 2019). Furthermore, low-income countries, with services constituting a comparatively smaller percentage of their GDP, have much smaller per capita ecological footprints (Lawn 2009). If the growthist claim is valid, should not the per capita ecological footprints of low-income countries be higher than the services-dominated high-income nations? And when will the shift to services start paying ecological dividends? When they reach 90–95% of GDP? Of course, it is the magnitude of real GDP itself, and not its composition, that is the major driving force.

11.3.10 The Optimal Size of an Economy is Where Net Benefits are Maximised

The fact that the composition of real GDP contributes little to the rise in the rate of resource use in no way means the economic process cannot be restructured to increase the benefits yielded by the stock of physical goods at lesser costs. As we have stressed, the net benefits of economic activity can be augmented via redistribution, the production of better goods, increased resource-use efficiency, and reduced carbon emissions. Nevertheless, increases in net benefits will still in large part depend on nations making a well-managed transition to a steady-state (non-growing) economy.

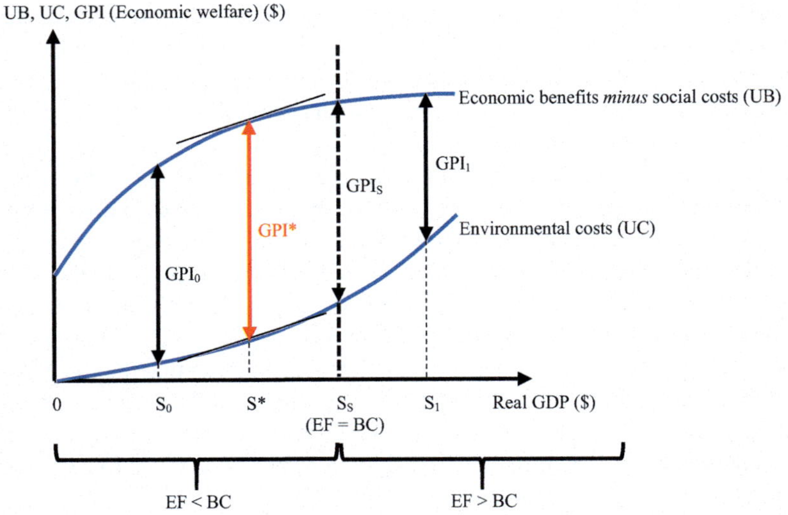

S^* = Optimal scale (OS) – where economic welfare is maximised
S_S = Maximum sustainable scale (MSS) – where ecological footprint = biocapacity (EF = BC)
S_0 < OS < MSS
S_1 > MSS > OS

Fig. 11.6 *The benefits and costs of a growing economy*

To better understand why, consider Fig. 11.6. The UB and UC curves represent the uncancelled benefits and uncancelled costs generated by a growing economy. Collectively, the uncancelled benefits constitute the desirable product of economic activity and are what remain after social costs are netted out against economic benefits (see Table 11.1). The uncancelled costs are the environmental costs incurred from making available the natural resources required to fuel the economic process. As the economy grows, uncancelled benefits increase at a diminishing rate, as per the uncontroversial principle of diminishing marginal benefits. At the same time, environmental costs increase at an increasing rate, which is entirely consistent with the principle of increasing marginal costs.

The economic welfare delivered by the economic process is equal to the vertical distance between the UB and UC curves, and is indicated by the Genuine Progress Indicator (GPI). As real GDP grows from a very low base, such as S_0, economic welfare increases because the marginal benefits of growth are greater than the marginal costs. As growth continues, economic welfare eventually reaches a maximum at S*. This represents the *optimal scale* of economic activity. Growth in real GDP beyond S* reduces economic welfare because the marginal benefits of growth are less than the marginal costs.

In conventional economic parlance, growth beyond S* is 'uneconomic', whereas growth between zero and S* is 'economic'. Note, therefore, that increases in real GDP should never be automatically referred to as economic growth, as is the modern predilection. It can be either economic or uneconomic. Ecological economists strongly oppose uneconomic growth and encourage people to make the distinction to discourage uneconomic growth and to raise awareness of the potential undesirability of GDP growth.

The *maximum sustainable scale* is denoted by S_S. Representing the maximum level of real GDP that can be sustained by an ecologically sustainable rate of resource throughput (i.e., where the ecological footprint equals biocapacity), it serves as a sustainability line-in-the-sand (black dashed line in Fig. 11.6). Any level of real GDP below S_S is ecologically sustainable, since the ecological footprint is less than biocapacity. Conversely, any level of real GDP above S_S generates an ecological deficit and is therefore unsustainable in the long run. Operating beyond S_S should at all times be avoided.

It is important to note that the optimal scale is always smaller than the maximum sustainable scale. This implies that an ecologically sustainable level of GDP, such as any level between S* and S_S, can be uneconomic. Although operating in this zone would not be ecologically hazardous, it would be socially and economically less desirable than operating at the optimum. It is best avoided.

Having reached its optimal scale, the economy should cease growing, whereupon it would remain in a virtual steady state hovering around the optimum.[22] The previously mentioned institutional, policy, and technological advances would thus be entrusted to beneficially shift the UB curve upwards and the UC downwards. Because this would increase the vertical distance between the two curves at the optimum, it would boost economic welfare without the need for quantitative expansion. Sure, real GDP would no longer rise, but the GPI would increase substantially, both immediately and long-term, and eventually leave a stagnant GDP in its wake.

11.3.11 Some Countries Will Need to Implement Degrowth Policies

It should now be abundantly clear that the transition to a qualitatively-improving steady-state economy is a sustainable development imperative. For countries with economies smaller than their optimal scale – especially impoverished nations – growth should be encouraged until the optimum is reached, albeit the growth should be performed in the most equitable and allocatively efficient manner.

For countries with economies larger than the optimum but smaller than the maximum sustainable scale (i.e., between S* and S_S in Fig. 11.6), it will be necessary for the economy to contract to maximise economic welfare. Similarly, countries with economies larger than their maximum sustainable scale (i.e., larger than S_S) will need to shrink the economy to operate sustainably.

[22] A steady-state economy would be dynamic in the sense that it would steadily improve over time with only negligible quantitative changes in the stock of human-made goods.

A strategy to reduce the size of the economy for socially beneficial purposes is referred to as *degrowth*. It differs entirely from an unplanned and undesirable shrinking of the economy, which might be caused by war, internal unrest, a natural disaster, a financial crisis, or ecological collapse stemming from prolonged and excessive ecological deficits.

Exactly how a country's degrowth process should proceed will depend on the prevailing scale of its economy. This is because the maximum sustainable scale is defined by ecological criteria, whereas the optimal scale is defined by economic criteria – namely, where the marginal benefits of economic activity equal the marginal costs.

For a country currently operating beyond its maximum sustainable scale, it will need to regulate the scale of the economic process at its source, which is the rate of natural resource inputs.[23] In other words, it would need to impose explicit quantitative restrictions on the rate of resource throughput. Mind you, even countries with an economy smaller than its maximum sustainable scale should impose the same restriction to prevent the throughput of resources from ever exceeding an ecologically sustainable rate, which, in turn, would prevent the economy from growing beyond its maximum sustainable scale.

Without repeating ourselves too much, the quantitative throughput restrictions should be instituted through the use of a cap-auction-trade system. As previously explained, the caps on the use of various resource types and on the generation of certain types of pollution would serve as a throughput brake on the economy. Of course, it would be infeasible for a country functioning well beyond its maximum sustainable scale to impose caps that immediately bring the rate of throughput into line with the nation's biocapacity. For example, it would mean reducing, overnight, the throughput fuelling the USA economy to around 45% of its current rate. For Japan, it would require an even larger percentage reduction. This would trigger a catastrophic collapse of both economies.

Preferably, but imperfectly, it would be better to gradually tighten the caps on resource throughput until the rate is ecologically sustainable. The process could also be made more flexible and tractable if nation-level cap-auction-trade systems were dovetailed with the earlier described international system of footprint targets. Since the international system would permit the ecological deficits of some countries to be matched by the ecological surpluses of others, it would allow the global economy to achieve sustainability status prior to individual countries achieving similar status. Consequently, global sustainability would not be contingent upon across-the-board national sustainability.

As for instances where the economy needs to further contract to settle at its optimal scale, this process would entail an allocative efficiency adjustment facilitated by institutional and market forces massaged by price-altering taxes, subsidies, and permit prices generated by a cap-auction-trade system. Direct regulation of the scale

---

[23] Because of the first law of thermodynamics, regulating resource inputs also regulates the rate of waste outputs, although not the qualitative nature of wastes.

of the economy to maximise economic welfare would not work, for two reasons. First, unlike the maximum sustainable scale, which can be estimated with a reasonable degree of certainty, the optimal scale cannot be ascertained by bureaucrats. Second, maximising economic welfare at the optimum is not just about the size of the economy. It is also about the state of the economy at the optimum. Regulation beyond what is needed to guarantee ecological sustainability and distributional equity would impede the efficiency of the resource allocation process and reduce the economic welfare delivered at all economic scales, not simply the optimal scale.

Given that degrowth entails reductions in real GDP and very likely a considerable restructuring of the economy, governments must ensure the degrowth process is an equitable one, which means ensuring disadvantaged citizens bear little if any of the inevitable sacrifices. This requires governments – in particular, central governments – to take a number of essential actions, albeit some should be undertaken under customary circumstances. The first is a reduction in the marginal tax rate imposed on low incomes to compensate the poor for price rises caused by instituting a resource throughput (aggregate supply) constraint and from levying charges to internalise the undesirable spillover costs of certain economic activities.

The second is the introduction of an employer-of-last-resort scheme, such as the recently proposed Job Guarantee (see Mitchell and Fazi 2017). A Job Guarantee, which would be sustainable if funded by a monetary sovereign, would make an important contribution towards redistributing a smaller GDP via paid work, rather than via welfare payments. Besides maintaining full employment – especially important given inevitable net job losses – the scheme would guarantee every able person below the retirement age a decent minimum income.[24]

The third and final action is the maintenance and provision of adequate quantities of easily accessible, high-quality, and low carbon-intensive public goods to ensure the equitable delivery of the services that matter most to all people – namely, education, health, transport, energy, and communications services. This would also bolster national productive capacity within the self-imposed ecological sustainability constraint.

11.3.12 Population Size Matters

It would be remiss of us to ignore the attention given by ecological economists to the issue of overpopulation. We recognise that the need for some countries to do something about population growth is an awkward and confronting issue. However, the evidence is stark and the imperatives are compelling. Confronting overpopulation and upholding human rights are not mutually exclusive, but properly

[24] People unable to work and people above the retirement age would receive a welfare payment set at the same minimum income – the latter subject to a means test.

understood, are complementary. For instance, future generations have a right to be born into a world with a healthy natural environment.

Although the population issue is most acute for many low-income countries with ecological deficits, population growth remains an issue for impoverished nations fortunate to have an ecological surplus. This is because they will likely want to grow their per capita footprints should a period of GDP growth be necessary to eliminate widespread poverty. For such countries, population growth will be crucial in that it will not only render the task of poverty eradication more difficult, it will increase the chances of eventually becoming an ecological deficit nation.

The overpopulation problem doesn't just lie with low-income countries. Whilst many high-income nations have large ecological deficits because of bloated per capita rates of resource consumption, for some, the problem is severely compounded by excessive population numbers.[25] If these countries do not wish to substantially reduce their per capita incomes, they will be forced to do something to reduce their population numbers. The choices faced are difficult but, having been brought on by growing the economy beyond its maximum sustainable scale, are unavoidable.

Despite the aforementioned, many claim that population growth is not a major ecological concern on the premise that the growth in per capita material affluence is principally at fault. To refute this, we shall consider the matter at the global level by reusing Eq. (11.4). To recall:

$$F = P \times A \times R \tag{11.4}$$

For our purposes, we shall measure F in terms of global hectares (Gha); P in terms of the number of people on Earth; A in terms of per capita Gross World Product (GWP); and R in terms of the number of global hectares exploited per US dollar of GWP.[26]

Now refer to Fig. 11.7. Panel 11.7a shows the global ecological footprint (F) rising in almost every year between 1961 and 2016.[27] By 2016, F was 2.98 times greater than in 1961. In effect, humankind's demands on the planet increased threefold over the 55-year period.

Panel 11.7b reveals the rise in global population (P), which increased in every year between 1961 and 2016. Overall, it rose by a factor of 2.42, which constituted an average rate of increase of 1.78% p.a.

Panel 11.7c shows per capita GWP (A) rising in all but five years of the 55-year period. By 2016, A was 2.72 times greater than in 1961. In material terms, the average human was close to three times wealthier in 2016 than in 1961.

---

[25] A good example is the USA. With one of the world's highest per capita rates of resource consumption and a population of around 330 million, it has an ecological footprint around 2.25 times its biocapacity.

[26] Gross World Product is the term used to describe GDP at the global level.

[27] The few years that F declined coincided with years in which GWP stagnated, hence indicating no evidence of absolute decoupling.

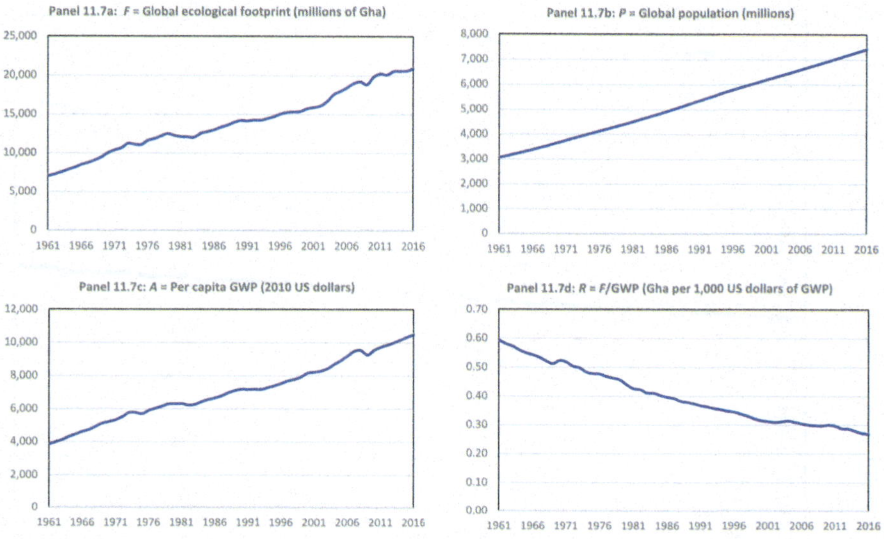

Fig. 11.7 *F = PAR for the global economy, 1961–2016*

Panel 11.7d indicates that the resource intensity of GWP (*R*) was the only variable in Eq. (11.4) to decline over the 1961–2016 period. It fell in all but four of the 55 years, with three of the years occurring towards the end of the period – perhaps indicating that reductions in the resource-intensity of economic activity are becoming, as expected, more difficult to secure. By 2016, *R* was 55% smaller than what it was in 1961, which constituted an average rate of decrease of 1.57% p.a. This means that the area of land required to produce a dollar's worth of real GWP declined by 55% between 1961 and 2016. As significant as this achievement was, it was overwhelmed by the considerable increases in *P* and *A* – indeed, so much so, that the global ecological footprint grew alarmingly, as Panel 11.7a has shown.

Without wanting to downplay the footprint-rising impact of growing material affluence, consider what would have transpired had the number of people throughout the period remained as it was in 1961, and *A* and *R* had changed as they did between 1961 and 2016. Instead of rising by 198%, the global ecological footprint would have only increased by 23%. As alarming as this might seem, the global ecological footprint (F^*) would have remained well below global biocapacity throughout the 1961–2016 period, as the red line in Fig. 11.8 shows. Contrast this with the blue line in Fig. 11.8 representing the actual global footprint (*F*), which cuts the black biocapacity line in 1970 and is 1.7 times greater than global biocapacity by 2016.[28]

To hammer the point home, it is possible to compare the actual population with the maximum sustainable population under the circumstances that prevailed during the 1961–2016 period. If we assume that ecological sustainability is guaranteed so long as the global ecological footprint does not exceed global biocapacity, then the

[28] This assumes that global biocapacity remained as it was throughout the 1961–2016 period.

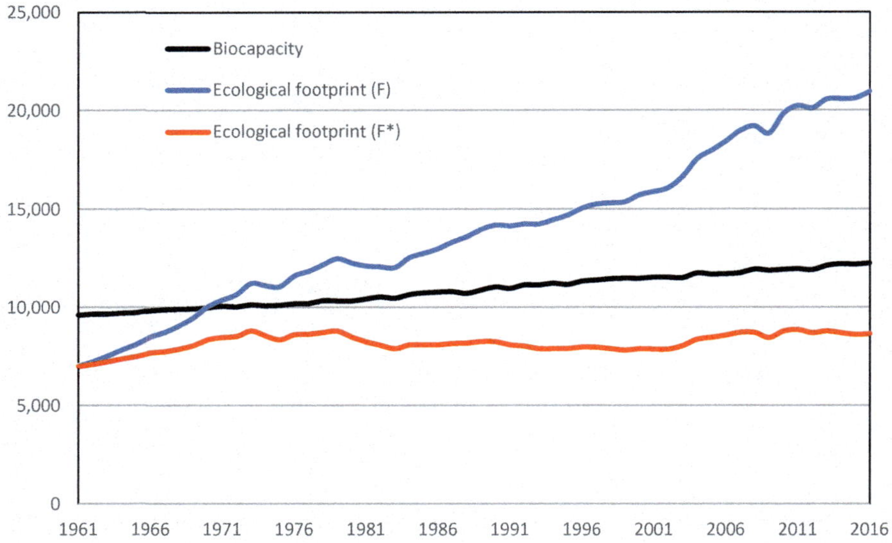

Fig. 11.8 *Global ecological footprint versus global biocapacity at actual and constant 1961 population numbers, 1961–2016 (millions of Gha)*

maximum sustainable population can be estimated by substituting the latter for the former in Eq. (11.4). This yields the following:[29]

$$B = P* \times A \times R \tag{11.9}$$

where A and R remain the same, but B = global biocapacity and $P*$ = maximum sustainable population.

By rearranging Eq. (11.9), one obtains:

$$P* = B / (A \times R) \tag{11.10}$$

By assuming, once again, that A and R changed as they did over the 1961–2016 period, Fig. 11.9 shows that the maximum sustainable population ($P*$), which is represented by the red line, fluctuated between a low of 3.53 billion in 1973 and a high of 4.51 billion in 1999. At 4.34 billion in 2016, $P*$ was only slightly higher than its 1961 value of 4.20 billion. Crucially, the blue line representing the actual population (P) passes $P*$ in 1970, which is the same year the global ecological footprint began to exceed global biocapacity. By 2016, the actual population was 3.08 billion or 71% above its maximum sustainable level. Clearly, population size matters!

[29] If ecological sustainability requires the global ecological footprint to not exceed half that of global biocapacity, one would substitute $0.5B$ for F in Eq. (11.4).

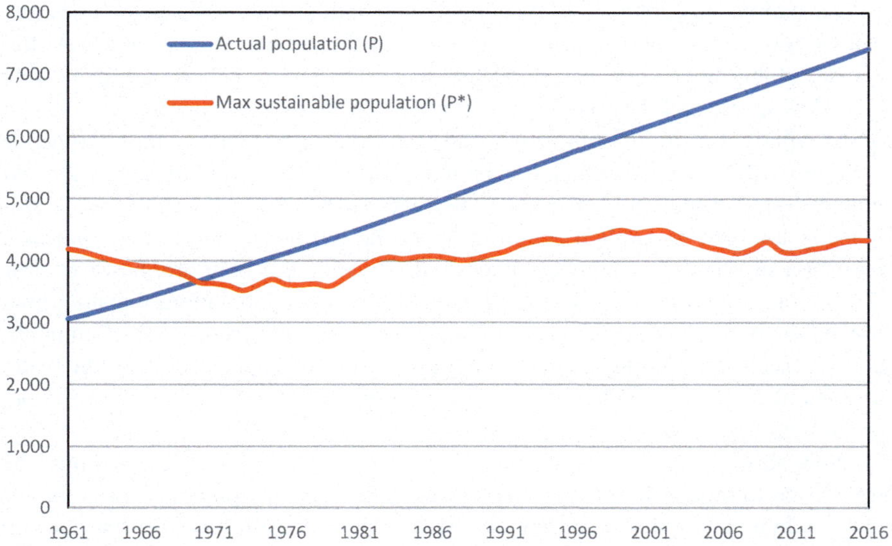

Fig. 11.9 *Actual global population versus maximum sustainable population, 1961–2016 (millions)*

How, then, does one limit population growth or reduce population numbers in countries where it is palpably required? To begin with, the extent to which population numbers need curtailing will depend on what happens to *A* and *R*. Should policies and newly-created institutions significantly increase the economic welfare generated by a given stock of physical goods (i.e., beneficially shift the UB and UC curves shown in Fig. 11.6), then *A* and *R* could be sufficiently reduced to warrant some leniency. That said, for a lot of countries with very large ecological deficits (e.g., Japan), large reductions in *A* and *R* will do little more than temper how much they need to reduce population numbers.[30]

A popular solution to the population problem is the so-called 'demographic transition'. Predicated on evidence showing that wealthy people have much lower fertility rates than the poor, it involves encouraging rapid GDP growth in the belief that, upon making people wealthier, they will have fewer children.

There are many problems with this recommended course of action. First, at the global level, per capita GWP has been rising steadily in recent decades (i.e., people, on average, have been getting wealthier in a material sense), yet global population numbers continue to grow. Second, the demographic transition relies on the rise in one throughput-increasing factor (*A*) to reduce another throughput-increasing factor (*P*). Third, where the demographic transition has succeeded, it has been a very slow process. Furthermore, in almost all such cases, the country in question has been left with an ecological deficit. In other words, the demographic transition has failed to

[30] Japan has an ecological footprint eight times its biocapacity.

deliver nations an ecologically sustainable population. It thus raises the question: can we rely on the demographic transition if, in the decades it takes to run its full course, global population has stabilised above ten billion people and per capita GWP has soared well beyond its current level? The answer is patently no.

To rapidly transition to an ecologically sustainable population, more proactive measures will be needed in high population growth countries and in nations already with excessive populations. For many high-income nations, the fertility rate is sufficiently low that no explicit action to lower the fertility rate is needed, unless a reduction in population numbers is desired or required. There are, however, low fertility rate countries (e.g., the USA and Australia) with populations that are growing mainly because of positive net migration, both past and present, which immediately increases population numbers and boosts the percentage of reproductively able people.[31] Clearly, in these countries, net migration should be cut to zero (more on immigration soon).

What about countries with high fertility rates? Quite simply, measures must be introduced to rapidly lower them. Whatever form these measures take, it is important that humanity joins forces in mutually-beneficial cooperation to ensure they do not violate basic human rights. No doubt, family planning services, easy access to contraception,[32] and gender equality – which gives women greater control over their lives – should be on top of the list. Since most countries with high fertility rates are poor, high-income nations will often be required to provide them with the necessary resources.

Whether the governments of some exceedingly overpopulated countries should impose legal limits on the number of children that people can bear is, of course, extremely controversial and would forever be plagued with all sorts of complications and moral dilemmas.[33] Any such action should require a United Nations mandate to ensure it does not contravene the Universal Declaration on Human Rights. In making its assessment, the United Nations would need to bear in mind that failure to deal with the overpopulation problem means, distressingly, that it is likely to render the transition to ecological sustainability extremely difficult, if not unattainable, and lead to great future suffering. Overall, its goal must be to advance human wellbeing, both present and future; to ensure the sustainability 'window of

---

[31] Although Australia has had a fertility rate at or below the replacement rate of 2.1 since and including 1976 (https://data.worldbank.org/indicator/SP.DYN.TFRT.IN?view=chart), the number of births in Australia each year is around double the number of deaths. This is almost entirely due to continued positive net migration, which extends demographic momentum (where a fertility rate less than the replacement rate is not immediately reflected in more deaths than births, and therefore a shrinking population).

[32] There is an estimated 270 million women with an unmet need for modern contraception (WHO 2020).

[33] It cannot be denied that placing limits on the permissible number of children does limit population growth. China's one-child policy is an example. For all its problems and criticisms, it has been estimated that, without the policy, China's population would have been 360–520 million more than it was in 2015 (Goodkind 2017). If correct, this would have constituted a non-trivial 5–7% of the world's 2015 population.

opportunity' remains open (reconsider Fig. 11.9); and provide impoverished nations with the resources they desperately need and desire (United Nations 2017).

It is not uncommon for recommended action to reduce fertility rates to be labelled elitist and a means of absolving high-income nations of any blame for the ecological impact of their excessive resource consumption. We don't believe we are guilty of these criticisms. With regard to the latter, we were at pains in the previous section to emphasise the need for countries with economies larger than their maximum sustainable scale to undertake a process of degrowth. The majority of these countries are high-income nations.

With regard to the former allegation, what we have urged is not so much an indirect criticism of high fertility rate countries. It is a direct criticism of an exploitative global economy – largely arranged by wealthy nations – that has delayed the onset of the demographic transition in many low-income countries by failing to equitably distribute the world's income. A fairer global economy, such as one based on internationalist principles, would have lowered the fertility rates of low-income countries many decades ago and spared them of any current need to undertake drastic population measures.

Matters don't get any easier when it comes to immigration. It, too, is a topic area loaded with potential prejudices and biases, including desires to increase immigration to fuel GDP growth rather than consider it from a humanitarian perspective.[34] For this reason, any call to massively cut the migrant intake should be prefaced with an upfront qualification that immigration reductions are needed in some countries for ecological and/or quality-of-life reasons, and should never be based on ethnicity or religion.

An unprejudiced policy to reduce net migration is not anti-immigrant. Zero net migration does not mean zero immigration and a smaller number of new migrants could still be drawn from people from a near unrestricted range of backgrounds, histories, and circumstances. Conversely, immigration numbers could be increased while significantly narrowing migration eligibility on ethnic grounds. We'll leave it to readers to determine which of the two is the repugnant immigration policy. It is worth stressing that the former policy would reduce the intake of all people, regardless of their ethnicity. Conclusion? It is numbers that matter.

Second, any high-income country reducing its immigration numbers should reserve as many places as possible for people in genuine need – namely, economic, political, and the growing number of ecological refugees. Besides increasing its humanitarian credentials, this policy would do much to stop the present robbing of the most skilled and productive people from impoverished countries, which is often

[34] Australia is a case in point, where immigration was dramatically increased during the Howard government years (1996–2007) due to misplaced fears of eventual economic calamity arising from a decreasing total fertility rate (Wright 2021). Howard's obsession with high economic immigration was so strong that he imposed punitive measures on boat-arriving asylum seekers in the belief that 'control' of the nation's borders would increase the likelihood of Australians being comfortable with high immigration. Australia can be contrasted with Japan, which has low levels of immigration, takes few refugees, and has a decreasing population.

the result of migrant-accepting countries failing to train sufficient people in areas of need; or demanding that people perform irksome work at wage levels insufficient to attract the interest of domestic labour.

Finally, immigration is not needed to overcome the non-existent problem of an aging population. Central governments, if they are monetary sovereigns, never have a problem financing government-provided pensions or the increasing resource demands of an expanding age-care industry. What does matter is the future availability of real resources and national productive capacity, which will depend on operating within an ecological sustainability constraint and allocating a sufficient portion of the incoming resource flow to suitably boost the productivity of the working population. We believe that an ecological economics-based policy program would guarantee that. We are certain that further population growth to ameliorate the perceived problem of an aging population, as is normally prescribed, would not.

11.3.13 Humankind Needs to Respect the Ecological Limits of the Planet or Face Collapse

The principle of ecological limits to growth is covered in detail in Chap. 8 of this volume, and we have touched on some of the technical aspects in earlier discussions regarding the laws of thermodynamics, production efficiency, and decoupling. Some further general points are worth making here.

The physical world that humans fundamentally rely on for survival, let alone the natural resources to fuel the modern economic process – the soil, the ores, the forests, the rangelands, the fisheries, the climate system, and rivers and aquifers – is finite and largely non-expansionary.[35] Life depends on photosynthesis and crucial biogeochemical cycles driven by a stock-abundant, but flow-constrained, source of solar energy.

Human population numbers have long expanded at the expense of the physical world and our fellow creatures. The history of population and GDP growth, especially since the beginning of the Industrial Revolution, is a history of widespread landscape modification, ecological degradation, and biodiversity loss, now of such magnitude that human-made objects and artifacts exceed all living biomass (Elhacham et al. 2020). At the same time, humankind's natural resource demands have grown to near double what the Earth can sustainably provide. All of this has led many to label the current geological epoch as the Anthropocene (Edwards 2015).

Humankind can continue to disrespect the ecological limits to growth on a finite planet. If it does, it faces the prospect of ecological collapse, if not the prior collapse of civilisation as we know it. An eventual realisation of our existential plight and a desirable altering of our ways cannot be assumed or taken for granted. History

[35] Renewable resource stocks can be expanded to some extent, but not by very much because renewable natural capital has limited regenerative and waste assimilative capacities.

reveals that the eventual collapse of civilisations is the norm, rather than the exception (Tainter 1988; Diamond 2005). Once begun, collapse is a process that is difficult to reverse and entails relatively quick, unplanned, and chaotic changes to society and the supporting ecosphere from complexity to a much simpler and less bountiful super-structure. Being the antithesis of sustainability, it is usually associated with a significant increase in morbidity and mortality.

As the percentage of humanity living in urban areas has increased, more people have become detached from the ecological processes that support them. Notwithstanding the recent coronavirus pandemic, fewer people understand how much their lives, and those of future people, are at the whims of ecological, social, and economic disturbances. Consequently, ecological awareness and 'knowledge' at the societal level, which is part-and-parcel of indigenous culture, has been lost.

Ecological economists don't claim to have a solution to this awareness problem, but they do recognise that a managed and well-designed transition to an alternative pathway is necessary to achieve sustainable prosperity and to avoid an unruly, miserable disaster. Disconcertingly, highlighting humankind's perilous predicament, as many are attempting to do, is failing to incite appropriate human action. One shouldn't have to appeal to self-interest to persuade current generations to abandon the folly of continuous growth in favour of a qualitatively improving steady-state economy. A moral obligation towards posterity ought to be enough. However, recourse to indicators showing that excessive growth is already suppressing the economic welfare of currently living people may be our best hope. If so, it is more crucial than ever to extensively publicise the informational content of emerging new indicators, such as the Genuine Progress Indicator. We would encourage all readers of this chapter to do just that.

11.4 Conclusion

It is lamentable that we need the term 'ecological economics' to distinguish a discipline based on the real world from the dominant one based on irrational, mythical, and often corrupted thinking. The various topics outlined and discussed in this chapter should simply appear under the rubric of 'economics', with the usual subdisciplines of macroeconomics, microeconomics, economic history, and political economy branching off it.

For the foreseeable future, we are stuck with applying 'ecological' as a qualifying term to emphasise the inescapable fact that the economy is embedded in the living world and is circumscribed – certainly in the long run – by what biophysical processes and the laws of physical transformation allow. Whilst human know-how has the potential to deliver long, healthy, and meaningful lives, both now and into the future, nature will always have the last say if not respected, as many civilisations have learned at great cost.

Perhaps the two most important points we have discussed in this chapter are that, firstly, material throughput or humankind's ecological footprint must be equal to or less than the available biocapacity to sustain human wellbeing. This rule, which is currently being breached, can only be transgressed for a relatively short time before disastrous and uncontrollable consequences emerge.

Second, accepting the inevitability of a steady-state economy is no reason for despair. By operating within ecological constraints and focusing on purpose, community, distributional equity, and qualitative improvements, there is no reason why humankind cannot reach its full potential – something it is falling well short of at present.

For all the benefits that growth has delivered, we have witnessed a growing frequency and intensity of its negative side-effects in recent decades, no more so than the recent coronavirus pandemic. We, as a global community, have been fooled by the dogma that continuous growth, which is the root cause of many of our ills, is the solution to *all* our ills. This mindset must change – there is no alternative. As previously suggested, a gestalt change might best be initiated, not by appeals to logic or ethics, as important as they are and will continue to be, but by highlighting that excessive growth is suppressing human wellbeing at this very moment. The indicators to inform us of the full impact of growth now exist. Yes, they require further refinement. And, yes, we need better systems of data collection to improve the reliability of the calculations, especially in relation to social and environmental phenomena, which, as we send off spacecraft to learn more about Mars, is appallingly scant. But their informational content is sufficiently accurate and comprehensive, and their policy implications are so apparent and in need of an appropriate response, that they desperately require extensive public attention and discussion.

References

Bringezu S (2015) Possible target corridor for sustainable use of global material resources. Resources 4(1):25–54

Clapp J (2017) Food self-sufficiency: making sense of it, and when it makes sense. Food Policy 66:88–96

Daly H (1992) Allocation, distribution, and scale: towards an economics that is efficient, just, and sustainable. Ecol Econ 6:185–193

Daly H (1996) Beyond growth. Beacon Press, Boston

Diamond J (2005) Collapse. Penguin Books, London

Dittrich M, Giljum S, Lutter S, Polzin C (2012) Green economies around the world? Implications of resource use for development and the environment. SERI, Vienna

Edwards L (2015) What is the Anthropocene? Eos 96. https://doi.org/10.1029/2015EO040297

Ehrlich P, Holdren J (1971) Impact of population growth. Science 171:1212–1217

Elhacham E, Ben-Uri L, Grozovski J et al (2020) Global human-made mass exceeds all living biomass. Nature. https://doi.org/10.1038/s41586-020-3010-5

Goodkind D (2017) The astonishing population averted by China's birth restrictions: estimate, nightmares, and reprogrammed ambitions. Demography 54:1375–1400

Harvey F (2020) Humanity is waging war on nature, says UN secretary general The Guardian, 3 December 2020

Hickel J (2020) Less is more: how degrowth will save the world. William Heinemann, London

Jackson T (2009) Prosperity without growth: economics for a finite planet. Earthscan, London

Jackson T (2020) Wellbeing matters: tackling growth dependency. Briefing paper no. 3, APPG on limits to growth. https://limits2growth.org.uk/wp-content/uploads/AETW-Policy-Briefing-No-3-digital.pdf. Accessed 10 Nov 2020

Lawn P (2009) The potential conflict between ecological sustainability and full employment. In: Lawn P (ed) Environment and employment: a reconciliation. Routledge, London, pp 24–93

Lawn P (2013) The environmental Kuznets curve: some theoretical and empirical insights. In: Lawn P (ed) Globalisation, economic transition, and the environment: forging a path to sustainable development. Edward Elgar, Cheltenham, pp 227–258

Lawn P (2017) South Australia: progressing in recent years and outperforming the rest-of-Australia: a genuine Progress Indicator study of South Australia, 1986–2016. Prepared on behalf of the Wakefield futures Group for the South Australian Department of environment, water, and natural resources, Adelaide

Mitchell W, Fazi T (2017) Reclaiming the state: a progressive vision of sovereignty for a post-neoliberal world. Pluto Press, London

Robinson J (1962) Economic philosophy. CA Watts & Co., London

Rockström J, Steffen W, Noone K, Persson Å, Chapin FS III, Lambin E, Lenton TM, Scheffer M, Folke C, Schellnhuber H, Nykvist B, De Wit CA, Hughes T, van der Leeuw S, Rodhe H, Sörlin S, Snyder PK, Costanza R, Svedin U, Falkenmark M, Karlberg L, Corell RW, Fabry VJ, Hansen J, Walker B, Liverman D, Richardson K, Crutzen P, Foley J (2009) A safe operating space for humanity. Nature 461(24):472–475

Steffen W, Richardson K, Rockström J, Cornell S, Fetzer I, Bennett E, Biggs R, Carpenter S, de Vries W, de Wit C, Folke C, Gerten D, Heinke J, Mace G, Persson L, Ramanathan V, Reyers B, Sörlin S (2015) Planetary boundaries: guiding human development on a changing planet. Science 347(6223):125985, 1–10

Tainter J (1988) The collapse of complex societies. Cambridge University Press, Cambridge

Tollefson J (2020) Why deforestation and extinctions make pandemics more likely. Nature (7 August) https://www.nature.com/articles/d41586-020-02341-1. Accessed 27 Feb 2021

United Nations (2017) Population facts, https://www.un.org/development/desa/pd/sites/www.un.org.development.desa.pd/files/files/documents/2020/Jan/un_2017_factsheet10.pdf. Accessed 24 Feb 2021

Wackernagel M, Beyers B (2019) Ecological footprint: managing our biocapacity budget. New Society Publishers, Gabriola Island

WHO (2020) Family planning/contraception methods, accessed at https://www.who.int/en/news-room/fact-sheets/detail/family-planning-contraception. Accessed 27 Feb 2021

Wilson EO (2016) Half-earth: our planet's fight for life. Liveright, New York

Wright S (2021) Howard government warned of a baby shortage hitting the budget, Sydney Morning Herald, 1 January. https://www.smh.com.au/politics/federal/howard-government-warned-of-a-baby-shortage-hitting-the-budget-20201228-p56qi3.html. Accessed 27 Feb 2021

Chapter 12
Energy Systems for Sustainable Prosperity

Mark Diesendorf

Abstract Ecologically sustainable energy technologies comprise renewable energy supply together with improved efficiency of energy conversion and use. Together they can mitigate the climate crisis, greatly reduce pollution of air, water and land, create more jobs than are lost in the fossil fuel industries they replace, and contribute to energy independence and social equity. The best technical energy supply strategy is transitioning fossil fuelled electricity to renewables, electrifying most heating and transportation, and producing fuels by using renewable electricity to make hydrogen and ammonia. This technological transition is necessary and urgent, but unlikely to be sufficiently rapid to avoid irreversible climate change. Substantial demand reductions are needed by rich countries, beyond the technological measures of energy efficiency. This would entail an end to growth in energy production, materials extraction, land clearing and population, that is, the creation of a steady-state economy within Earth's biocapacity.

Keywords Renewable energy · Energy efficiency · Australian case study · Barriers to sustainable energy · Steady-state economy · Consumption

12.1 Introduction

Energy from the Sun makes life on Earth possible. Sunlight is captured by the leaves of plants via the process of photosynthesis, absorbing carbon dioxide (CO_2) and storing the solar energy in the form of carbohydrates. That energy comes to us humans by eating plants, or animals that eat plants, or animals that eat animals that eat plants. Thus solar energy drives our metabolic systems. Over millions of years, dead plants have been converted by natural processes into fossil fuels – coal, oil and natural gas – which became the energy source of the Industrial Revolution. Burning

M. Diesendorf (✉)
School of Humanities & Languages, UNSW, Sydney, Australia
e-mail: m.diesendorf@unsw.edu.au

S. J. Williams, R. Taylor (eds.), *Sustainability and the New Economics*,
https://doi.org/10.1007/978-3-030-78795-0_12

these fuels drives the wheels of industry and transportation, giving us the 'goods' and services of modern living.

We have known for more than a century that combusting fossil fuels releases CO_2 which stays in the atmosphere for long periods[1], acting like a selective blanket the allows sunshine to pass through while trapping heat that tries to escape, thus changing Earth's climate. By releasing, over periods of decades, vast quantities of carbon that has been stored underground for millions of years, we have upset the natural balance and now face a human-induced climate crisis (see Chap. 2). Fossil fuels are also the major cause of other types of air pollution, producing many adverse health impacts (Smith et al. 2013; Jacobson 2020), water pollution, over-use of water and land degradation. Fossil fuels are concentrated sources of energy, used to produce electricity from huge power stations, and petrol and diesel from huge oil refineries. They also concentrate political power and create dependence of individuals and the community upon large corporations and external governments.

Clearly, in the interests of our health and wellbeing, and that of the whole biosphere, we must shift to energy sources and energy uses that have very low environmental and health impacts. Renewable sources of energy, together with reductions in energy demand by improvements in the efficiency of energy conversion and use and by behaviour change, can mitigate the climate crisis, greatly reduce pollution of air, water and land, create more jobs than are lost in the fossil fuel industries they replace (Campbell 2018), and contribute to energy independence and social equity. This combination of much cleaner energy supply and demand reduction can be called 'ecologically sustainable energy' or just 'sustainable energy'.

Two renewable energy technologies can have significant adverse environmental and social impacts. One is the construction of large dams for hydroelectricity, which can involve the dislocation of communities and the flooding of ecologically precious valleys. If the valleys are heavily vegetated, the dams can become significant sources of greenhouse gas (GHG) emissions in the form of methane. The other is the production of biofuels from plants, which may involve clearing forests, resulting in GHG emissions and threats to biodiversity, and competition with food production. Biofuels play a minor role in most sustainable energy scenarios. Nevertheless, environmentally sound hydro-electric and biofuel projects can be chosen. With good design and appropriate scale, a renewable energy system can be ecologically sustainable, improve social equity and contribute to sustainable prosperity.

Nuclear technologies are unsuitable for an ecologically sustainable energy future. Nuclear fission power is too dangerous, both in terms of devastating accidents and its contribution to the proliferation of nuclear weapons, too expensive, too slow to build and too inflexible in operation to be a suitable partner for renewable electricity (Sovacool 2011; Diesendorf 2014). Nuclear fusion has not been achieved in a controlled manner on Earth under conditions when energy output is greater than energy input.

[1] If human-induced CO_2 emissions somehow stopped today, a large fraction would remain in the atmosphere for centuries to millennia.

This chapter recognises that changing to 100% renewable energy is necessary and urgent, but not sufficient to achieve sustainability. If a well-designed renewable energy system drives a global socio-economy based on endless growth, adverse impacts of that system on human wellbeing and the planet will continue. Even with 100% renewable energy, which is technologically and economically feasible (Ram et al. 2019; Jacobson 2020), growing industries would continue to clear our forests, destroy biodiversity, degrade our soils, waterways and oceans, pollute our environment with physical and chemical wastes, increase traffic congestion and road casualties[2], and deplete the environment of essential elements such as phosphorus. Therefore, we must reduce the total quantity of energy generation and use as well as greatly improving its quality. Growth in energy demand and the socio-economy more broadly is occurring rapidly in developing countries. To compensate for this growth, part of which is necessary for poverty alleviation, the rich countries must reduce their use of energy, materials and land, while transitioning to renewable energy, and stabilise their populations. In other words, they must transition to steady-state economies with low throughputs via planned de-growth (Daly 1977; Dietz and O'Neill 2013; Jackson 2017; Victor 2019; Hickel 2020).

Although this chapter applies to all countries, Australia is taken as a case study to illustrate some points. It has high fossil fuel use, very high GHG emissions per capita, but high solar and wind resources. Despite resistance from a conservative federal government with strong links to the fossil fuel industries, Australia is transitioning to renewable energy on all three scales: small (residential and small business), medium (community and commercial) and large (utility and industrial).

This chapter first puts forward a broad strategy for transitioning from fossil fuels to sustainable energy (Sect. 12.2). The huge potential for improving energy conversion efficiency in electricity generation and for improving energy efficiency of buildings and appliances, is outlined in Sect. 12.3. Then Sect. 12.4 summarises the current status of renewable energy. Section 12.5 refutes two myths disseminated by sustainable energy critics – the base-load myth and the Energy Return on Energy Invested myth – and then identifies consumption as the main problem that still must be addressed. Section 12.6 is a general discussion and conclusion.

12.2 Global Strategy for Cutting GHG Emissions from Energy

The vast majority of global GHG emissions come from the combustion of fossil fuels. Of these, coal is responsible for the largest share, followed by oil and then natural gas. Even gas is a high emitter (UCS 2014) and must be replaced as soon as possible, along with coal and oil.

[2] It remains to be seen whether driverless vehicles can reduce these impacts.

To develop a strategy for reducing emissions, we must now include all GHGs and address the different sectors that are responsible for the emissions. The share of emissions between sectors varies between countries. Australia is taken as an example of a developed country with high production and use of fossil fuels, so high that it is one of the biggest per capita emitters of GHGs in the developed world. Figure 12.1 shows the breakdown by sector of Australia's emissions from all monitored GHGs in 2019 in CO_2-equivalents. It's not surprising that electricity generation is the largest contributor, because coal still supplied 58% of Australia's electricity in 2018–19, although it has declined from 83% in 1999–2000 (DEE 2019) thanks to the gradual growth of renewable energy and energy efficiency. In second place are transport and heat, each responsible for about 19% of emissions, then fugitive emissions[3]. In total, all energy use is responsible for 82% of Australia's total annual GHG emissions.

To understand the proposed strategy for transitioning the world from fossil fuels to renewable energy, the reader needs three additional items of information:

- Wind and solar photovoltaics (PV) can supply the most electricity at prices below that of new fossil fuelled power and far below that of new nuclear power stations (IRENA 2019; Lazard 2020; Graham et al. 2020).
- Almost all residential, commercial and industrial heating that is currently supplied by burning fossil fuels (mainly gas) could be supplied economically by electric heat pumps and direct electrical heating.

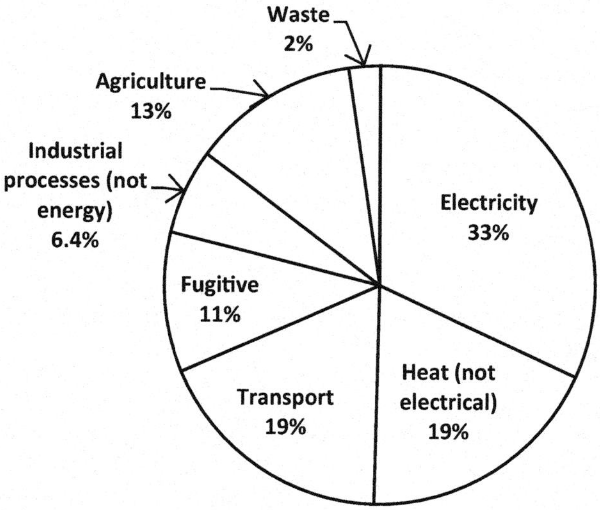

Fig. 12.1 Australia's GHG emissions 2019 by end-use sector. (Source: DISER 2020a, Table 3) Note: Taking account of land use change, which gives an estimated reduction of 3.6% (not shown), total CO_2-equivalent emissions in 2019 was 532.5 Mt

[3] Escapes of GHGs from oil and gas fields, coal mines and gas pipelines.

- Within several years, electric vehicles (EVs) are likely to be economically competitive with internal combustion engine (ICE) vehicles for general transportation purposes. The vehicle transition is being accelerated in several countries by means of emission standards for ICE vehicles, tax incentives and/or a rollout of fast-charge stations.

Hence, the strategy for the global energy transition is:

- Replace fossil fuelled electricity with renewable electricity supplied mainly by wind and solar PV supplemented by storage. This is already occurring in several countries, states and provinces, and many towns, businesses and industries (see Sect. 12.4). Part of this process must be to remove the enormous subsidies to the production and use of fossil fuels, US$317 billion in 2019 according to the estimate[4] of IEA (2020).
- Electrify all heating, apart from solar hot water and passive solar design of buildings.
- Electrify almost all road transportation.
- Greatly increase the energy efficiency of buildings, appliances and industrial processes.
- Upgrade high-voltage transmission lines so that wind and solar farms can supply cities, industries and regions with low-cost renewable energy resources.
- Assist workers in fossil fuel industries and related businesses to retrain, relocate or transition to retirement.

The technologies for this strategy are all commercially available. Some are economically competitive now, while others (e.g. EVs) are likely to become economically competitive in many countries within several years (see Sect. 12.4). The above strategy would eliminate emissions from electricity generation, heating, most transportation and almost all fugitive emissions. In the Australian example, this strategy would eliminate about three-quarters of the country's GHG emissions. The remainder of energy emissions would arise from air and sea transportation and long-distance road transport in remote areas. These energy uses could not be supplied *directly* from renewable electricity. Therefore, the proposed pathway is to use renewable electricity to produce 'green' hydrogen by electrolysis of water and then convert the hydrogen to a more manageable fuel such as ammonia. When hydrogen and ammonia are burned or used to generate electricity in a fuel cell, they do not produce carbon emissions. These 'renewable' fuels (actually a form of energy storage) can also be used to cut emissions in non-energy industrial processes, such as substituting 'green' hydrogen for the use of coal as a chemical to reduce iron ore to iron. The production of 'green' hydrogen and ammonia is at present very inefficient and expensive, although research and development are taking place around the world to bring down costs.

[4] Different researchers give widely differing estimates depending on their assumptions as to what constitutes a subsidy. The IEA estimate is one of the lowest.

The potential rewards for a country like Australia with huge renewable energy resources are great, not least of which is energy independence. An export industry for renewable electricity stored as ammonia in sea tankers that at present carry liquefied natural gas is possible (Finkel 2019), as is export by transmission lines to south-east Asia. The latter project is 'technically viable, with technology that is proven and improving, but economically, socially, politically, and administratively challenging' (Halawa et al. 2018). Australia's current annual export of thermal coal, with energy content about 6×10^{18} joules, can supply about 570 TWh per year of electricity, i.e. roughly double Australia's annual electricity generation of 265 TWh. Australia has ample land to generate and export this quantity of electricity in future, after it has achieved 100% renewable electricity, to replace its thermal coal exports. Thus geographic inequities in renewable energy resources around the world would be reduced by trade, just as geographic inequities in fossil fuels have been handled, except that renewable energy can be traded both by transmission line and sea tanker.

The policies needed within the energy sector to facilitate the transition have been discussed widely (e.g. Diesendorf 2014, 2018; UNECE 2015; Valentine et al. 2019).

12.3 Energy Efficiency and Conservation

12.3.1 Distinguishing Conservation from Efficiency

Energy efficiency is defined as obtaining the same energy service – e.g. warm house in winter; hot shower; cold food; mobility – with a reduced energy use. It is achieved mostly by improved technologies, e.g. home insulation; a water-efficient shower head; an energy-efficient refrigerator; an efficient vehicle or public transport with high occupancy. *Energy conservation* is defined as reducing energy use by accepting a modified or different energy service – e.g. wearing warmer clothes and accepting reduced home heating in winter; taking a shorter shower; buying a smaller refrigerator; walking or cycling to the local shops instead of driving. Thus energy conservation is mostly behavioural. Since it cannot be legislated directly, the challenge is to create a culture of conservation, where the benefits of sufficiency are contrasted with the downsides of consumerism and waste. Both energy efficiency and energy conservation can be implemented quickly and, for many measures, can save money in the long run, as well as energy and emissions. Energy efficiency and conservation are very important because they enable 100% renewable energy to be reached more rapidly than otherwise. Furthermore, energy conservation would be an important part of a transition to a steady-state economy with low throughput.

This section focuses on the efficiency of energy conversion in generating electricity, the efficiency of energy end-use in general and a phenomenon that must be addressed to obtain the maximum benefits from both efficiency and conservation, namely rebound.

12.3.2 Energy Conversion Efficiency

Huge reductions in energy wastage can be achieved by replacing fossil-fuelled electricity with energy efficiency and renewable energy, as illustrated by Fig. 12.2. The left-hand side shows primary energy, the energy sources in nature. When fuels are combusted in a power station, typically about 70% of the primary energy is lost, most as waste heat. The remainder of 'wasted' energy is the electricity used to operate the power station. If one unit of fossil-fuelled electricity is saved by improved efficiency of electricity use, or replaced by renewable energy, then more than three units of fossil fuel primary energy are saved. It is fossil fuel primary energy that's responsible for GHG emissions and air pollution. Thus, even a small reduction in fossil-fuelled electricity received by the customer (the 'end-use energy' or 'final energy consumption') can produce a large reduction in emissions from primary energy.

Huge improvements in efficiency of energy conversion can also be achieved by replacing an ICE vehicle (efficiencies ranging from 15% to 25%) with an EV (efficiencies of 80% to 85%). When the EV is charged with grid electricity in the European Union, with 35% renewables in 2019 (Agora Energiewende 2020), the reduction in CO_2 emissions is significant; when charged with 100% renewable energy, the reduction is huge (Carbon Brief 2020).

Humans actually require energy services rather than the end-use energy *per se*. Primary energy and the pollution it creates can be further reduced by choosing energy services to be a combination of efficiency of energy use and end-use energy supply, for example, for the energy service of a hot shower, using a water-efficient

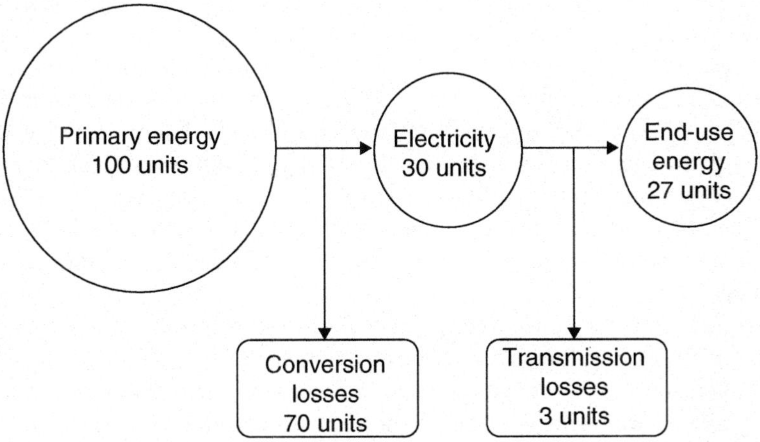

Fig. 12.2 Typical flow diagram for electricity generation from fossil fuels. (Source: the author, based on Australian black coal electricity generators; conversion losses are even greater from brown coal or lignite)

shower-head[5] with solar hot water supply. Many opportunities for doing this are low-cost, with rapid payback, and can be implemented quickly. The International Energy Agency's Sustainable Development Scenario has energy efficiency and renewable energy responsible for almost equal reductions in CO_2 emissions over 2010–2050 (IEA 2019). The American Council for an Energy Efficient Economy ranks several European countries and Japan high in its International Energy Efficiency Scorecard for their policies and programs, but ranks Australia, Brazil, Russia, the UAE and Saudi Arabia low; the USA has a middle ranking (ACEEE 2019). In many countries, energy efficiency initiatives are being weakened, despite the fact that these countries still have large potential for improvement.

12.3.3 Rebound

Within the energy sector, rebound can occur when someone saves money by implementing energy efficiency, e.g. insulating one's home, or conservation measures, e.g. cycling to work instead of driving, and then spends that money partly on using more energy. Thus the net saving of energy is less than expected from the efficiency/conservation gain (Kolstad et al. 2014, pp.249–250). Numerous studies show that rebound exists. For example, if I have solar hot water that's free apart from the sunk capital cost, I may take longer showers than when I used water heated with fossil-fuelled electricity from the grid. The false implication drawn out of this situation by some authors (e.g. Jenkins et al. 2011) is that reducing energy demand is a wasted effort.

In general, the size of the rebound depends on whether the money saved is spent on an energy-hungry or an energy-efficient product. Rebound effects are context-specific, making it difficult to generalize quantitatively on their relative size and importance (Blanco et al. 2014, pp.390–391). Numerous studies have surveyed rebound resulting from specific energy efficiency and conservation measures under specific circumstances. Many of these case studies find rebound to be small compared with the energy saved, e.g. when driving slightly further per year after replacing a fuel-guzzling car with a fuel efficient one, while a few cases find rebound to be greater than energy saved, e.g. from switching to heating the whole house after insulating it, instead of previously just heating a single uninsulated room while it's occupied.

Although data are not currently available to estimate total national or global rebound from energy efficiency and conservation, the methods for reducing rebound are straightforward. Government policies are needed to nudge consumers towards spending on environmentally desirable products and services. In particular, governments can encourage energy efficiency with low rebound by regulations that require energy-using appliances, vehicles and buildings to have energy audits, energy

[5] Saving hot water saves energy as well as water.

labelling and energy performance standards, backed up with information and education programs. Incidentally, a household or community that's transitioning to renewable energy can be given incentives to spend savings received from energy efficiency improvements and conservation on renewable energy, further reducing carbon emissions.

Rebound from energy efficiency and conservation is part of a more general form of rebound, known as Jevons Paradox. This occurs in the macro-economy when technological improvement or government policy increases the quantity of product per unit of currency spent. This is likely to increase production and consumption of these products which results in increased resource use (throughput). Obvious examples in recent decades include cars, computers, smart phones, televisions, and indeed household PV systems. Undesirable consumption could possibly be brought under control by environmental taxes (e.g. carbon tax or emissions trading scheme) or a cap on resource throughput (see Chap. 15).

12.4 Renewable Energy: Current Status

Several scenario studies find that the combination of wind, solar and hydro, together with several new long-distance transmission links, could in future provide 100% renewable electricity and hence all final (end-use energy) to the whole world (Jacobson et al. 2017, 2018; Ram et al. 2019; Jacobson 2020). Several countries and states/provinces with high 'once-through'[6] hydroelectric resources (in some cases together with geothermal energy and small amounts of other renewables) already have 90–100% renewable electricity. These include Norway, Iceland, New Zealand, Costa Rica, Paraguay, Uruguay, Bhutan, Nepal and Tasmania. However, once-through hydroelectric resources are geographically limited, so this chapter focuses on the much more extensive regions where wind and/or solar PV can generate the bulk of electricity. These include Australia, where the Australian Capital Territory now runs on 100% renewable electricity (wind and solar PV), the USA, northern Europe, the Middle East, northern Africa and northwest China. Most other regions can be supplied from these regions by transmission line. Table 12.1 summarises the progress of selected countries and states in replacing fossil-fuelled electricity with wind and/or solar PV.

In practice, it is too difficult and expensive to extend national grids to many remote regions. As the prices of batteries continue to fall, it will become more economically feasible for individual homes and small businesses in these areas to have 24-h off-grid electricity. Indeed, this is already happening in many less developed countries, where a small solar panel and battery powers lighting, a radio and mobile phone charging. Also, some precincts, villages and towns in developed countries are forming their own micro-grids based on solar and batteries, but, in developing

[6] Water flows one-way downhill from a dam on a river through a power station.

Table 12.1 Status of variable renewable electricity (wind and solar PV) in selected countries and states in 2019

Country or state	Renewable electricity as % of total in 2019	Comment
Denmark	RElec 69% of consumption comprising VRElec 50% (mostly wind) + bioenergy from agricultural wastes 19%	100% RElec expected in 2028
Scotland	RElec 56% of generation (VRElec 40%, hydro 10%) and 76% of consumption	VRElec mostly wind, supplemented by solar PV; big electricity exports to England
South Australia	RElec 50% of generation (mostly wind + rooftop PV)	State gov't expects 100% net RElec by 2030
Germany	RElec 46% of public generation (wind 25%; PV 9%; biomass 9%; hydro 4%)	Target: at least 80% RElec by 2050
German states: Schleswig-Holstein & Mecklenburg-Vorpommern	About 100% net generation each – mostly wind	High-capacity transmission links with their neighbours balance variability of VRElec
US states: Iowa, Kansas & Oklahoma	42%, 41% & 35% generation from wind respectively	Texas with 17.5% has largest wind capacity in megawatts
China	RElec 26% (wind 4.8%; solar 3%; hydro 17.8%)	2019 data: thermal (mostly coal) is still 69% & nuclear 4.8%;

Source: the author's compilation
Notes: RElec is renewable electricity. VRElec is variable renewable electricity
Only countries and states where hydro plays a minor role, if any, are listed

countries, these are experiencing difficulties in implementation and operation (Fowlie et al. 2018). Remote mines are increasingly powered by solar backed up with diesel and/or battery.

In 2019 renewable electricity comprised 27.3% of global electricity generation (hydro 15.9%, wind 5.9% and solar 2.8%)—see Fig. 12.3. Renewable generation is growing rapidly, but so is demand for electricity, especially in developing countries. This underlines the need for increased energy efficiency and conservation measures discussed in Sect. 12.3, if renewables are to replace fossil fuels in a timely manner, rather than just supplement them. Since 2014, global annual investment in new renewable electricity capacity has been greater than US$250 billion; in 2019 it was US$302 billion, that is, three times the combined annual investment in fossil-fuelled and nuclear electricity capacity (REN21 2020, p.165). In 2019, 75% of net annual additions to electricity generating capacity were renewables. Therefore, renewable electricity is decreasingly called 'alternative energy'. It is mainstream.

But in 2018 electricity supplied only 20% of total global end-use energy (also known as total final energy consumption or TFEC), which includes transportation and non-electrical heating in addition to electricity. Modern renewables (i.e. excluding traditional biomass) supplied 11% of TFEC and traditional biomass an

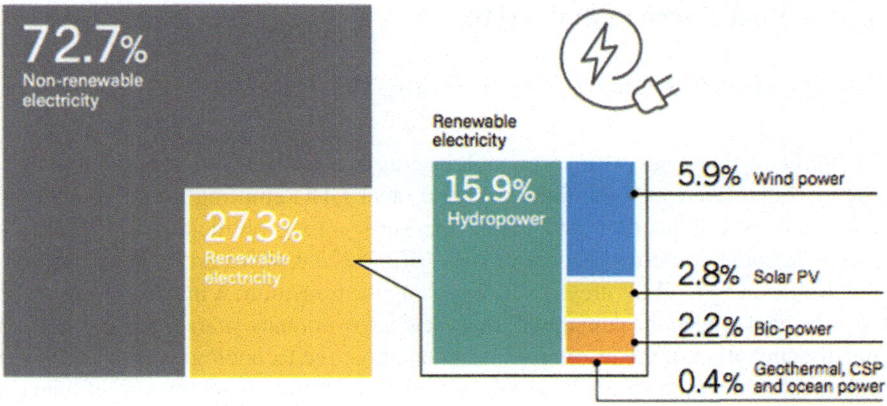

Fig. 12.3 Estimated renewable share of global electricity production, end-2019. (Source: REN21 2020, Fig. 10)

additional 6.9% of TFEC (REN21 2020, Fig. 1). Thus, to transition to a sustainable energy system, the multiple challenges are to electrify transportation and heating while continuing to grow renewable electricity rapidly and simultaneously to strengthen energy efficiency and conservation, i.e. to reduce demand.

Heating is a major part of energy use in high latitudes of the Northern Hemisphere. Changing from natural gas heating to renewable electricity is progressing very slowly, due to the cost of replacing infrastructure, such as hot water boilers in residential and commercial buildings and steam generation in industrial processes. Only a few governments have policies for facilitating the transition of heating to renewable energy. Nevertheless, the high efficiencies of electric heat pumps (e.g. reverse-cycle air conditioning) are likely to drive the replacement of retired gas heating with electric in residential and commercial buildings. (REN21 2020)

EVs are much cheaper to drive and maintain than ICE vehicles. The higher upfront cost of EVs results from the present high cost of batteries. As the market grows, battery prices will decline substantially and EVs will likely become economically competitive for general use by the mid-2020s. They are already becoming competitive for fleets. In Norway, EVs already comprise over half of new car sales assisted by government incentives. Several countries are introducing future bans on ICE sales, Norway commencing as early as 2025.

To sum up, although renewable *electricity* generation is growing steadily on a global scale (and growing rapidly in some regions), the electrification of transport is just commencing and the rate of electrification of fossil-fuelled heating is very low. Furthermore, renewable energy is chasing a retreating target, namely the pre-Covid-19 growth in total global energy demand (see Sect. 12.5.4). Despite having a much higher growth *rate* than other energy sources, renewables were responsible for less than one-third of total global energy demand growth over 2013–2018 (REN21 2020, Fig. 2). Therefore, growth in global energy demand must be curtailed, while growth in renewable energy is accelerated.

12.5 Real Barriers and Myths

12.5.1 Vested Interests Are Resisting the Transition

The growing awareness that sustainable energy can replace fossil fuels is threatening to vested interests – the fossil fuel and other GHG emitting industries and the nuclear energy industry – and their supporters among politicians and the mass media in some countries. Although some of these industries are diversifying into renewable energy, others are actively opposing the transition. A manifestation of the opposition has been the denial of climate science over many years (see Chap. 9) and the dissemination of misleading information about the technologies contributing to solutions to the climate crisis, especially renewable energy and energy efficiency. These myths are propagated by self-styled 'think tanks', such as the Heartland Institute and the Breakthrough Institute in the USA, the Global Warming Policy Foundation and the Institute of Economic Affairs in the United Kingdom, and the Institute of Public Affairs in Australia (Oreskes and Conway 2010). These can reasonably be described as 'renewable energy deniers', in the sense that they deny the evidence that renewable energy can fully replace fossil fuels. Their tactics are very similar to those of climate science deniers – indeed they are often the same groups and individuals. Their misinformation campaigns, while arguably losing credibility as reality blooms, still continue in scholarly literature and are still propagated by the mass media, especially by News Corp, social media, films and videos, and some politicians. Section 12.5 refutes two of their common myths and then discusses a major real barrier to a sustainable future, growth in consumption.

12.5.2 The Base-Load Myth

The most frequently used myth is that large-scale electricity systems need base-load[7] power stations and that renewable energy is too variable or 'intermittent' and so cannot provide base-load. This myth has been refuted time and again by both practical experience and simulation modelling.

Practical experience comes from power systems with high penetrations of wind and/or solar. Denmark's system has operated for more than 300 h (not consecutive) per year on 100% renewable energy of which the vast majority is wind. Nevertheless, it is one of the most reliable systems in Europe. South Australia's system, which no longer has any power stations operating as base-load and has only low-capacity interconnections to a neighbouring state (Victoria), has from time to time operated for several hours to a whole day entirely on wind. It has also operated reliably as an

[7] Base-load power stations can operate 24/7 at rated power (except when they breakdown or undergo planned maintenance). They are mostly coal-fired or nuclear, which are inflexible in terms of varying their power outputs and hence poor partners for wind and solar.

'island' on three occasions when the interconnector failed. There is no problem in supplying base-load demand with renewables, although it can be more challenging to supply the peaks. That's where storage plays its main role.

Scores of research groups around the world have conducted simulation modelling of power systems with 100% renewable electricity in which the majority of annual generation is provided by variable renewables: wind and solar PV. Each hour (or shorter timestep), real historical data (or synthetic data) on wind, sunshine and demand are used to balance renewable energy supply against actual demand over periods of 1–6 years. The results are that well-designed systems are reliable and affordable, even though they have no base-load power stations (Jacobson et al. 2015; Diesendorf and Elliston 2018; Jacobson 2020) and, in some cases, they have quite small amounts of storage (Elliston et al. 2016; Blakers et al. 2017; Diesendorf and Wiedmann 2020).

Generation reliability, which is a property of the whole system and not a single power station, is achieved by some of the following design elements:

- a diversity of types of renewable energy;
- geographic diversity of wind and solar farms linked by transmission lines;
- storage; and
- demand response[8] in a 'smart' grid.

Two different types of storage can be distinguished:

(i) Storage that's integrated with a net generator of electricity, e.g. once-through hydro; concentrated solar thermal power with thermal storage; open-cycle gas turbine[9]. The simulations of the Australian National Electricity Market (NEM) by Elliston et al. (2016) used this form of storage—actually all three of the above.

(ii) 'Passive' storage that's not a net generator, e.g. battery; pumped hydro[10]; compressed air. The simulations of the NEM by Blakers et al. (2017) used this form of storage, specifically new off-river pumped hydro, together with existing once-through hydro.

Big lithium batteries with short-term storage, such as the Hornsdale Power Reserve in South Australia, are already playing valuable roles in frequency control ancillary services and providing very rapid response to disturbances in the system (Parkinson 2019). However, lithium batteries with energy storage capacities of more than a few hours of rated power are at present very expensive. In electricity systems with high penetrations of variable renewables, lithium batteries must be supplemented by technologies with energy storage capacities of days to weeks.

[8] i.e. changing energy use in response to supply and demand conditions in the grid.

[9] i.e. a jet engine; it can burn renewable fuels such as biofuels, 'green' hydrogen and 'green' ammonia.

[10] Water flows in a circuit between two reservoirs at different elevations. Excess generating capacity pumps water from the lower to the higher reservoir. During high demand and/or low supply, water is released from the upper reservoir, generating electricity as it returns to the lower one.

These could be pumped hydro or gas turbines with a renewable fuel store. Hundreds of thousands of technically feasible sites have been identified around the world for small-scale off-river pumped hydro storage schemes (Blakers et al. 2017; Lu et al. 2018).

Because of their high energy density and hence their use in EVs, lithium batteries are at present dominating the battery market. Their principal market is for EVs, but they are increasingly being used in household solar electric systems. Although lithium is abundant in several countries, reuse and recycling businesses will eventually be needed. Meanwhile, batteries with different chemistries are under development and coming onto the market. In particular, flow batteries, which have lower energy density and so are not suitable for cars, are commercially available and may in future become economically competitive with lithium for stationary electricity storage over periods of a day or more.

It turns out that locations that have high wind and/or solar resource and a low demand for winter heating – e.g. Australia, the Middle East, northern Africa, southwest and central USA, northwest South America – need very little if any seasonal storage. They have such large renewable energy resources that they could export to colder regions by transmission line. Colder regions may require seasonal storage in the form of large hydroelectric dams or, in the near future, power-to-gas. In the latter case, excess wind or solar power will be used to produce hydrogen by splitting water (Wulf et al. 2018).

12.5.3 The EROI Myth

Another myth is the claim that building renewable energy technologies requires more life-cycle energy to be invested than they generate over their lifetimes. That claim was never true of wind power and was only true of solar PV panels several decades ago, before they were mass produced. Nowadays a more nuanced version of this claim is still made by some authors of scholarly papers and disseminated by renewable-energy deniers. The modern version is that the energy return on energy invested (EROI) for renewable electricity technologies and systems may be so low that the transition may displace more useful investment in other important economic sectors (Sers and Victor 2018). One study (Capellán-Pérez et al. 2019) even suggests that the transition to renewables may cause complex industrial societies to collapse!

The incorrect EROI claim is based on outdated data on wind and solar PV technologies, on failing to recognise the energy-efficiency advantages of transitioning away from fuel combustion for generating electricity (illustrated in Fig. 12.2), on overestimates of storage requirements, and on inconsistent methods. When these and other errors are corrected, it is found that EROIs of wind and solar PV at appropriate sites are similar to or better than those of fossil-fuelled electricity (Raugei and Leccisi 2016; Leccisi et al. 2016; Brockway et al. 2019; Diesendorf and Wiedmann 2020). Typically, solar PV and wind at good sites generate the life-cycle energy required to build themselves in 1–2 years (PV) and 3–6 months (wind). Their lifetimes are at least 25 years.

Including storage complicates the determination of system EROI. To examine this issue, Diesendorf and Wiedmann (2020) draw upon two published Australian case studies (Elliston et al. 2016; Blakers et al. 2017) to show that the amount of storage required for a reliable power system with 100% renewables is relatively small in Australia and similar climatic regions around the world. Furthermore, Diesendorf and Wiedmann (2020) show that the impact of storage on system EROI depends on the types of storage and the strategies for operating them. Type (i) storage technologies may maintain or, in some cases, increase system EROI. Type (ii) storage technologies play valuable roles in the power system, although they may reduce system EROI to some extent.

So far, this section has discussed EROIs of renewable electricity technologies at a given point in time. Next comes the dynamic problem. Climate science has identified a crisis that demands a logical energy strategy (Sect. 12.2) together with many urgent, rapid, policy responses (Chap. 15). If a system of power stations transitions so rapidly towards zero carbon that the energy invested in building one generation of new clean technologies occurs before the previous generation of clean technologies has 'paid off' (in energy terms) its energy invested, then system EROI will indeed be temporarily lowered. It should be emphasised that, if the transition is sufficiently fast, this dynamic effect on the EROI of the whole electricity system can occur even when the static EROIs of the individual energy technologies within the system are high and that this effect applies to all rapid low-carbon energy transitions (e.g. to nuclear) as well as to renewables. It's the price we should be happy to pay in properly addressing the climate emergency (Diesendorf and Wiedmann 2020).

12.5.4 Consumption—The Real Problem

While renewable energy production, especially renewable electricity generation, has been growing rapidly in some regions of the world, energy demand[11] (and consumption of materials, see Wiedmann et al. 2015) has also been growing quite rapidly, except during the Covid-19 period. From 2013 to 2018, global TFEC[12] grew by 7% or 25 exajoules (EJ). Over the same period, modern renewables (i.e. excluding traditional bioenergy) grew by 21.5%, but this growth from a relatively small base corresponds to only 7 EJ (REN21 2020, Fig. 2). For the Australian case study, the average *annual* growth rate of TFEC for the decade ending 2018–19 was 1.7%,

[11] 'Energy consumption' is the term used by economists, but 'energy demand' is the term preferred by scientists and engineers, who understand that energy cannot be created or destroyed (i.e. 'consumed' in a physical sense).

[12] Total Final Energy Consumption (TFEC) or end-use energy is what remains of the total primary energy demand following all of the losses that occurred during the processes of transformation, transmission and distribution—see Fig. 12.2 for the case of Australia.

while for *all* renewable energy (i.e. including its uses in transport and heating) it was −0.4% (DISER 2020b, Fig. 2.8), despite the fact that for renewable *electricity* it was 10.1% (DISER 2020b, Table 3.2). These figures are explained by the fact that most of the increase in demand for transport and heating energy, in Australia and the whole world, is still supplied by fossil fuels (REN21 2020). Thus, if growth in TFEC continues post-Covid-19, renewable energy (together with its Cinderella sister, energy efficiency) will continue to chase a receding target. And in the current climate crisis, time is of the essence (Lenton et al. 2019; Diesendorf 2020; Steffen (Chap. 2 of this volume)).

Energy demand is correlated with GDP and population size (positive correlations), energy prices (negative correlation), economic structure, population density, energy resources, fuel mix, climate, infrastructure, regulatory standards, culture and more. However, the relationships are complex and causality is difficult to establish. In particular, while global energy demand is strongly correlated with global GDP, countries with essentially the same GDP per person can differ in energy demand per person by a factor of 10, and vice versa (Sorrell 2015).

In terms of the consumption of materials, renewable electricity scenarios generally use less water, but more glass, plastic, concrete, copper and steel than fossil fueled electricity scenarios (Turner et al. 2013). The latter analysis takes account of increased end-use energy efficiency, but not the large reduction in electricity demand resulting from energy conversion efficiency (Fig. 12.2), which reduces the demand for materials. Renewables also require several special materials, such as neodymium for wind turbines and lithium and cobalt for batteries—these are abundant, but supply chains may have to be changed (Chen 2019). For electricity storage, lithium batteries currently dominate the market because of their high energy density that allows use in EVs as well as stationary energy, but alternative batteries based on common materials (e.g. flow batteries which have lower energy densities) are entering the market for stationary energy storage. Reuse and recycling are just beginning to be addressed and require more rapid attention, as does the need to transition to a whole circular economy (European Commission 2020).

There is a large body of evidence that the biggest environmental impacts, and in particular most GHG emissions (Chancel and Piketty 2015), are caused by consumption by the world's rich citizens.[13] They include most of the citizens of the rich countries and the rich elite of the poor countries. Their impacts come from both direct consumption and their savings and investments. Furthermore, they are embedded in an economic system that fosters endless growth in consumption and production. The latest Scientists' Warning on Affluence (Wiedmann et al. 2020) reviews this situation and analyses the different possible responses, some of which involve reforming, but retaining, the capitalist economic system, while others recommend more radical alternatives.

---

[13] This observation supports the point made in Chap. 2 that 'the Great Acceleration was not driven equally by all of humanity but rather by the relatively small number of humans living in OECD countries'.

At the very least, it is necessary for high-income countries to transition to a steady-state economy defined in biophysical rather than monetary terms, that is, 'an economy with constant stocks of people and artifacts [sic], maintained at some desired, sufficient levels by low rates of maintenance "throughput"' (Daly 1977). De-growth of the biophysical global economy should be to a level where its use of energy, materials and land, and its population, are compatible with Earth's biocapacity (Dietz and O'Neill 2013; Jackson 2017; Victor 2019; Hickel 2020). Research by Graham Turner (2011), using a model of the biophysical economy of Australia, and Peter Victor (Victor and Rosenbluth 2007; Victor 2012), using a macroeconomic model[14] of Canada, find that, provided a mix of government policies is implemented, full employment is possible in de-growth scenarios. Further research is needed on the economic, social and environmental impacts of a steady-state economy.

12.6 Discussion and Conclusion

A global ecologically sustainable energy system, based on renewable energy, energy efficiency and energy conservation, is technically and economically feasible and could replace all fossil fuel extraction and use. In this system, the vast majority of energy would be produced and used as electricity from wind, sunshine and water. The remainder awaits further development of technologies to reduce the costs of producing gaseous and/or liquid fuels from renewable electricity for air and sea transportation and non-energy industrial processes.

The current situation is that many countries and smaller regions have achieved annual average electricity generation of 50–100% (or more) from renewable electricity. In a large fraction of these places, *variable* renewables are the major source (Table 12.1). However, the percentages of renewables are much less considered relative to total energy use. Also, many towns, commercial operations and industries have embarked on that pathway: for instance, Australia's capital city, Canberra, now runs on 100% net renewable electricity, mostly wind and solar, and has policies to encourage the electrification of transport. The global market for EVs is expected to expand rapidly from the mid-2020s onwards as battery prices continue to fall and emissions limits or bans are placed on new ICE vehicles. However, progress in expanding the electrification of non-electrical heating has so far been slow almost everywhere in the world.

Continuing cost reductions in wind and solar power, and storage technologies, together with widespread concern about climate change, are making the transition

[14] While I'm sceptical about the validity of conventional macroeconomic models, Victor's research may convince some neoliberal economists that a steady-state economy does not necessarily entail high unemployment. More recent research, from Modern Monetary Theory (see Chap. 14), finds that unemployment and underemployment are policy choices for governments that issue their own currency.

unstoppable. The resulting sustainable energy system could play the major role in mitigating human-induced climate change, provided it is implemented rapidly across the whole world. Because wind and solar technologies are mass produced, they can be rolled out much more rapidly than fossil fuelled or nuclear technologies. Wind and solar farms can be planned and built in 2–3 years. Utility scale batteries can be manufactured and installed in less than a year. Residential rooftop solar systems and energy efficiency technologies are installed in a single day or less. Given the political will, the rate of growth of sustainable energy could be greatly increased. Modern Monetary Theory (see Chap. 14) argues that financial capital is generally no obstacle. The fact that many countries, including most of the larger economies, have committed to net zero emissions by 2050 or so, is encouraging, but substantial reductions on a global scale are needed over the coming decade rather than the promise of big reductions later.

We have collectively wasted critical decades, thanks to a well-orchestrated disinformation campaign. Fossil fuel industries and other big GHG emitters are still exerting strong pressure on governments to delay action, by direct lobbying, political donations, jobs for retiring politicians and influence on the mass media to cast doubt on climate science and renewable energy technologies. Despite these pressures, public opinion generally favours the energy transition. In countries with a nominally democratic system of government where vested interests are politically powerful (e.g. the USA, Australia, Poland, South Africa, Canada, Brazil), stronger community action is needed to push governments to address the above pressures that are potential sources of corruption undermining democratic decision-making (Diesendorf 2021). Global energy demand continues to grow, except during the coronavirus pandemic, driven mainly by growing economies and the growing global population. Energy demand growth, most of which is still fossil fuelled, is slowing the rate at which renewable energy can catch up with demand and energy efficiency can reduce demand. At the current rate of demand growth, irreversible climate change, i.e. where tipping points are reached, could arrive before a global sustainable energy system. Therefore, energy demand and consumption in general must be curtailed while energy is transitioned to renewables.

References

ACEEE (2019) The International Energy Efficiency Scorecard. https://www.aceee.org/portal/national-policy/international-scorecard. Accessed 10 Dec 2020

Agora Energiewende (2020) The European power sector in 2019. https://www.agora-energiewende.de/en/publications/the-european-power-sector-in-2019/. Accessed 22 Dec 2020

Blakers A, Lu B, Stocks M (2017) 100% renewable electricity in Australia. Energy 133:471–482

Blanco G, Gerlagh R, Suh S et al (2014) Chapter 5: Drivers, trends and mitigation. In: Edenhofer O, Pichs-Madruga R, Sokona Y et al (eds) Climate change 2014: mitigation of climate change. Contribution of Working Group III to the fifth assessment report of the Intergovernmental Panel on Climate Change. Cambridge University press, Cambridge. https://www.ipcc.ch/site/assets/uploads/2018/02/ipcc_wg3_ar5_chapter5.pdf. Accessed 19 Feb 2021

Brockway PE, Owen A, Brand-Correa LI et al (2019) Estimation of global final-stage energy-return-on-investment for fossil fuels with comparison to renewable energy sources. Nat Energy 4:612–621

Campbell R (2018) Will-o'-the-ISP: estimating renewable energy employment under the Integrated System Plan. The Australia Institute, Canberra. https://www.tai.org.au/sites/default/files/P648%20Renewable%20job%20estimate%20%5BWeb%5D.pdf/. Accessed 10 Dec 2020

Capellán-Pérez I, de Castro C, González LJM (2019) Dynamic energy return on energy investment (EROI) and material requirements in scenarios of global transition to renewable energies. Energ Strat Rev 26:100399

Carbon Brief (2020). https://www.carbonbrief.org/factcheck-how-electric-vehicles-help-to-tackle-climate-change/. Accessed 10 Dec 2020

Chancel L, Piketty T (2015) Carbon and inequality: from Kyoto to Paris. Paris Sch Econ. https://doi.org/10.13140/RG.2.1.3536.0082. Accessed 21 Jan 2021

Chen A (2019) Where will the materials for our clean energy future come from? The Verge, https://www.theverge.com/2019/2/15/18226210/energy-renewables-materials-mining-environment-neodymium-copper-lithium-cobalt. Accessed 5 Jan 2021

Daly HE (1977) Steady-State Economics. W.H. Freeman, San Francisco

DEE (2019) Australian Energy Statistics, Table O, March. Department of the Environment and Energy, Canberra. https://www.energy.gov.au/sites/default/files/2019_aes_table_o_march_2019.pdf. Accessed 10 Dec 2020

Diesendorf M (2014) Sustainable energy solutions for climate change. Routledge, Abingdon/New York

Diesendorf M (2018) Renewable electricity policy for Australia. The Australia Institute, Canberra

Diesendorf M (2020) COVID-19 and economic recovery in compliance with climate targets. Glob Sustain 3(e36):1–9. https://doi.org/10.1017/sus.2020.32. Accessed 22 Dec 2020

Diesendorf M (2021) Peter Garrett gives Labor a free pass on climate action. https://johnmenadue.com/peter-garrett-gives-labor-a-free-pass-on-climate-action/#more-64077. Accessed 18 Feb 2021

Diesendorf M, Elliston B (2018) The feasibility of 100% renewable electricity systems: a response to critics. Renew Sustain Energy Rev 93:318–330

Diesendorf M, Wiedmann T (2020) Implications of trends in energy return on energy invested (EROI) for transitioning to renewable electricity. Ecol Econ 176:106726

Dietz R, O'Neill D (2013) Enough is Enough. Berrett-Koehler, San Francisco

DISER (2020a) Quarterly update of Australia's National Greenhouse Gas inventory: December 2019. Department of Industry, Science, Energy & Resources, Canberra. https://www.industry.gov.au/sites/default/files/2020-05/nggi-quarterly-update-dec-2019.pdf. Accessed 10 Dec 2020

DISER (2020b) Australian Energy Update 2020, Australian Energy Statistics, September, Department of Industry, Science, Energy and Resources, Canberra. https://www.energy.gov.au/sites/default/files/Australian%20Energy%20Statistics%202020%20Energy%20Update%20Report_0.pdf. Accessed 10 Dec 2020

Elliston B, Riesz J, MacGill I (2016) What cost for more renewables? The incremental cost of renewable generation – an Australian National Electricity Market case study. Renew Energy 95:127–139

European Commission (2020) EU Circular Economy Action Plan. https://ec.europa.eu/environment/circular-economy/index_en.htm. Accessed 5 Jan 2021

Finkel A (2019). https://www.chiefscientist.gov.au/news/hydrogen-australias-future/. Accessed 10 Dec 2020

Fowlie M, Wolfram C, Wolfson D (2018) Are solar microgrids the future in the developing world? Energy Institute at Haas. https://energyathaas.wordpress.com/2018/10/01/are-solar-microgrids-the-future-in-the-developing-world/. Accessed 10 Dec 2020

Graham PW, Hayward J, Foster J et al (2020) GenCost 2019–20. Australia: CSIRO and AEMO, Canberra https://publications.csiro.au/publications/#publication/PIcsiro:EP201952. Accessed 10 Dec 2020

Halawa E, James G, Shi X et al (2018) The prospect for an Australian-Asian power grid: a critical appraisal. Energies 11(1):200. https://doi.org/10.3390/en11010200

Hickel J (2020) Less is more: how degrowth will save the world. Penguin Random House, London

IEA (2019) World Energy Outlook 2019. International Energy Agency. https://www.iea.org/reports/world-energy-outlook-2019/energy-efficiency. Accessed 10 Dec 2020

IEA (2020) Fossil fuel consumption subsidies by country, 2016. https://www.iea.org/data-and-statistics/charts/fossil-fuel-consumption-subsidies-by-country-2016. Accessed 10 Dec 2020

IRENA (2019) Renewable power generation costs in 2018. International Renewable Energy Agency, https://www.irena.org/publications/2019/May/Renewable-power-generation-costs-in-2018. Accessed 10 Dec 2020

Jackson T (2017) Prosperity without growth: economics for a finite planet, 2nd edn. Routledge, London

Jacobson MZ (2020) 100% clean, renewable energy and storage for everything. Cambridge University Press, Cambridge

Jacobson MZ, Delucchi MA, Bazouin G et al (2015) Low-cost solution to the grid reliability problem with 100% penetration of intermittent wind, water, and solar for all purposes. PNAS 112:15060–15065

Jacobson MZ, Delucchi MA, Zack AF et al (2017) 100% clean and renewable wind, water, and sunlight all-sector energy roadmaps for 139 countries of the world. Joule 1:108–121

Jacobson MZ, Delucchi MA, Cameron MA et al (2018) Matching demand with supply at low cost in 139 countries among 20 world regions with 100% intermittent wind, water, and sunlight (WWS) for all purposes. Renew Energy 123:236–248

Jenkins J, Nordhaus T, Schellenberger M (2011) Energy emergence: rebound and backfire as emergent phenomena. Breakthrough Institute. https://s3.us-east-2.amazonaws.com/uploads.thebreakthrough.org/legacy/blog/Energy_Emergence.pdf

Kolstad C, Urama K, Broome J et al (2014) Chapter 3: Social, economic and ethical concepts and methods. In: Edenhofer O, Pichs-Madruga R, Sokona Y et al (eds) Climate change 2014: mitigation of climate change. Contribution of Working Group III to the fifth assessment report of the Intergovernmental Panel on Climate Change. Cambridge University Press, Cambridge. https://www.ipcc.ch/site/assets/uploads/2018/02/ipcc_wg3_ar5_chapter3.pdf. Accessed 19 Feb 2021

Lazard (2020) Lazard's Levelized cost of energy analysis – version 14.0. https://www.lazard.com/media/451419/lazards-levelized-cost-of-energy-version-140.pdf. Accessed 03 Jan 2021

Leccisi E, Raugei M, Fthenakis V (2016) The energy and environmental performance of ground-mounted photovoltaic systems–a timely update. Energies 9:622

Lenton TM, Rockström J, Gaffney O et al (2019) Climate tipping points – too risky to bet against. Nature 575:592–595

Lu B, Stocks M, Blakers A et al (2018) Geographic information system algorithms to locate prospective sites for pumped hydro energy storage. Appl Energy 222:300–312

Oreskes N, Conway EM (2010) Merchants of doubt: how a handful of scientists obscured the truth on issues from tobacco smoke to global warming. Bloomsbury, New York

Parkinson G (2019) Renew economy, 19 February. https://reneweconomy.com.au/tesla-big-battery-earns-another-4m-from-fcas-in-fourth-quarter-65127/. Accessed 10 Dec 2020

Ram M, Bogdanov D, Aghahosseini A et al (2019) Global energy system based on 100% renewable energy – power, heat, transport and desalination sectors. Lappeenranta University of Technology and Energy Watch Group, Lappeenranta/Berlin. http://energywatchgroup.org/new-study-global-energy-system-based-100-renewable-energy. Accessed 10 Dec 2020

Raugei M, Leccisi E (2016) A comprehensive assessment of the energy performance of the full range of electricity generation technologies deployed in the United Kingdom. Energy Policy 90:46–59

REN21 (2020) Renewables 2020 Global Status Report. REN21 Secretariat, Paris

Sers MR, Victor PA (2018) The energy-emissions trap. Ecol Econ 151:10–21

Smith KR et al (2013) Energy and human health. Annu Rev Public Health 34:159–188

Sorrell S (2015) Reducing energy demand: a review of issues, challenges and approaches. Renew Sustain Energy Rev 47:74–82

Sovacool BK (2011) Contesting the future of nuclear power: a critical global assessment of atomic energy. World Scientific, Singapore

Turner G (2011) Consumption and the environment: impacts from a system perspective. In: Newton P (ed) Urban consumption. CSIRO Publishing, Collingwood, chapter 4; reprinted in: Washington H, Twomey P (eds) (2016) *A Future beyond Growth: Towards a steady state economy.* Earthscan from Routledge, London/New York, p 112–128

Turner GM, Elliston B, Diesendorf M (2013) Impacts on the biophysical economy and environment of a transition to 100% renewable electricity in Australia. Energy Policy 54:288–299

UCS (2014) Environmental impacts of natural gas. Union of Concerned Scientists. https://www.ucsusa.org/resources/environmental-impacts-natural-gas. Accessed 5 Jan 2021

UNECE (2015) Best policy practices for promoting energy efficiency. United Nations Economic Commission for Europe, New York/Geneva

Valentine S, Sovacool B, Brown MA (2019) Empowering the great energy transition: policy for a low-carbon future. Columbia University Press, New York

Victor PA (2012) Growth, degrowth and climate change: a scenario analysis. Ecol Econ 84:206–212

Victor PA (2019) Managing without growth: slower by design, not disaster, 2nd edn. Edward Elgar, Cheltenham

Victor PA, Rosenbluth G (2007) Managing without growth. Ecol Econ 61:492–504

Wiedmann TO et al (2015) The material footprint of nations. Proc Natl Acad Sci 112(20):6271–6276

Wiedmann T, Lenzen M, Keyßer LT et al (2020) Scientists' warning on affluence. Nat Commun 11:31017. https://doi.org/10.1038/s41467-020-16941-y. Accessed 10 Dec 2020

Wulf C, Linssen J, Zapp P (2018) Chapter 9: Power-to-gas—concepts, demonstration and prospects. In: Azzaro-Pantel C (ed) Hydrogen supply chains. Academic Press. https://www.sciencedirect.com/topics/engineering/power-to-gas/. Accessed 10 Dec 2020

Chapter 13
Climate Change Litigation and Human Rights

Michael Kirby

Abstract Climate change litigation and jurisprudence, from a slow start in the years before 2000 – when there had been a mere eight specific climate change litigation cases filed globally – rose to 338 cases between 2001 and 2010, reaching 1186 cases between 2011 and 2020. In early 2021, the United States leads in cumulative volume with 1308 cases; Australia ranks second globally with 115 cumulative cases. This chapter outlines the major developments in such litigation, with a particular focus on the role of human rights instruments. It then briefly examines changes in the interpretation of corporate law in respect to climate change. Taken together, there is growing momentum in a search by civil society for legal remedies given the profound seriousness of climate change and the abject disappointment of policy responses.

In October 2021, after this chapter was completed, the Human Rights Council of the United Nations voted to recognise a new human right to a "safe, clean, healthy and sustainable environment".

Parts of this chapter reflect the author's article 'Environmental and Planning Law in the Age of Human Rights and Climate Change' **(2019) 3 *Environmental and Planning Law Journal* 181** but have been adapted and updated.

Justice of the High Court of Australia (1996–2009); President of the NSW Court of Appeal (1984–96); Member of the Group of Experts in International Law, Human Rights and Environmental Law that adopted the ***Oslo Principles on Climate Change Obligations,*** The Hague, Netherlands (2015) ('Expert Group'). The author acknowledges the assistance of Sean Ryan in the preparation of this chapter.

M. Kirby (✉)
Sydney, NSW, Australia
e-mail: mail@michaelkirby.com.au

S. J. Williams, R. Taylor (eds.), *Sustainability and the New Economics*,
https://doi.org/10.1007/978-3-030-78795-0_13

13.1 Introduction

Climate change, including global warming, has been apparent for at least a century.[1] However, the rate of change since the 1970s has been accelerating. Its dangers to humans and the biosphere of planet Earth are frightening, with increasing calls for the recognition of a climate emergency.[2] The nature and extent of the problem is such that, if it is to be solved, a global response is needed.

Truly, our planet and its human populations must think and plan globally; but also act locally in order to alter the trajectory of climate change that will otherwise continue at a dangerous rate. Following the failure of government policy responses to the demonstrated science of climate change, citizens have increasingly turned to the courts for remedies through litigation. This chapter looks at the scope of climate litigation globally, its development within Australia and the future trends drawing on international experience. It will illustrate through key decisions examples of how difficulties of causation and attribution are being overcome. Recently filed cases will be discussed as indicators of future trends and potential directions of further jurisprudence.[3]

13.2 Climate Change Litigation

13.2.1 Obstacles

At the outset of facing the challenge of climate change, it is necessary to acknowledge a number of important obstacles that stand in the way of effective action, including obstacles to the invocation of universal human rights.

The first obstacle is the weakness and divisions of the mechanisms for multilateral cooperation and enforcement when the world faces a danger to the entire human species and to other living things. The institutions of government to secure action are largely divided and selfish in the extent of their cooperation and even hostile to

---

[1] A comprehensive collection of sources, cases and analyses can be found in Donald K. Anton, Jonathan I. Charney, Philippe Sands, Thomas J Schoenbaum and Michael J. Young (eds.) *International Environmental Law: Cases, Materials, Problems*, LexisNexis, Newark, 2007, 1244–5.

[2] Ripple et al. (2019) [https://academic.oup.com/bioscience/article/70/1/8/5610806]; the UN secretary-general has called on all countries to declare a climate emergency: https://www.theguardian.com/environment/2020/dec/12/un-secretary-general-all-countries-declare-climate-emergencies-antonio-guterres-climate-ambition-summit

[3] This chapter is fundamentally about real and potential litigation in courts and tribunals to invoke the deep principles of law and universal human rights so as to contribute to reducing rapidly the perils of climate change (Kirby MD (2019) Multilateralism, Pushback, and Prospects for Global Engagement? Indiana J Legal Studies 27).

the machinery of multilateralism. That machinery is often weak and fragile, particularly at present. Existential dangers are not confined to climate change. They extend to the control and elimination of nuclear weapons; the containment of viral pandemics; the huge growth of human population and resultant shortage of food and potable water; and solutions to endemic poverty. The urgency of the need for action varies in reverse proportion to the strength and effectiveness of the response.

Second, when it comes to responses in the form of litigation, most instances of any relevance to climate change are to be found in sub-national jurisdictions. They cannot easily achieve global outcomes. In these municipal jurisdictions, rules and guidelines of international organisations are mostly not, as such, part of the binding local law. This "dualism" between international and municipal law necessarily limits the effectiveness of global rules. Decisionmakers in municipal litigation are sometimes hostile or dismissive towards international law and its principles.[4]

Third, universal human rights on a global scale constitute a relatively recent development.[5] They arrived substantially with the creation of the United Nations Organisation in 1945. They were partly expressed in the *Universal Declaration of Human Rights* 1948. They were elaborated in the UN treaty law that followed thereafter. While most countries today have constitutional, statutory or other laws to give effect to universal human rights, some, like Australia, have no comprehensive national constitutional charter of human rights. They have only a patchwork of sub-national laws – or sometimes state charters – to which human beings can appeal to uphold their basic rights to life, health, dignity, safety and justice.

These considerations must be acknowledged at the outset of this chapter. Municipal legal systems, including in Australia, have tended to be slow in their response to well-accepted climate change dangers. This is even more true of the overarching planetary boundaries risks, outlined in Chap. 2 of this volume. Human-induced climate change, threats to the biosphere and the need to protect public safety have been well understood since at least 1980 and, in many places, long before. The Intergovernmental Panel on Climate Change (IPCC) published its first assessment report in 1990.[6] A comprehensive and coordinated policy response by national and subnational governments would most likely be the most efficient and effective way of addressing these risks, rather than relying on litigation.

However, international negotiations and national regulations have been timid and inadequate for the urgent problems we face. For example, the aims of the 2016 Paris Agreement include to "keep warming below 2 °C and pursue efforts to 1.5 °C".[7] Yet

---

[4] Kirby MD (2020) Municipal Courts and the international Interpretive Principle: *Al-Kateb v Godwin* (2004) University of New South Wales Law Journal 1, referring to *Al-Kateb v Godwin* (2004 219 CLR 562; [2004] HCA 37.

[5] Charney J (1993) Universal International Law. American J Int'l Law 87:529–543; Higgins R (1994) Problems and Process: International Law and How We Use It. Oxford University Press, Oxford, England, p 205–206 (1994), Anton, above n.1 p 305.

[6] Intergovernmental Panel on Climate Change, 1990, First Assessment Report.

[7] United Nations, *Paris Agreement*, 2015, Article 1(a).

the voluntary (non-binding) Nationally Determined Contributions offered by governments currently equate to a continuance of aggregate global warming of around 3 °C.[8] The 2019 IPCC report 'Global Warming of 1.5 °C' estimates that "Coral reefs, for example, are projected to decline by a further 70–90% at 1.5 °C (high confidence) with larger losses (>99%) at 2 °C (very high confidence)".[9] Thus the current emissions trajectory, under existing international agreements, constitutes an existential threat to the Australian Great Barrier Reef. Aside from its inherent worth, the reef contributes $6.4 billion to the Australian economy and supports for the incomes of approximately 64,000 persons.[10] This indicates that global and national governmental action is currently falling well short of the emissions trajectory presently necessary to avoid some to the dire impacts of climate change, predicted with high probability.

Individuals and groups are therefore beginning to turn, in increasing numbers, to courts and tribunals in subnational jurisdictions to redress current and anticipated harms. The definition of climate change litigation is a matter for widespread academic debate.[11] However, such litigation can be grouped into the following categories:

(a) Mitigation: aimed at reducing humanity's greenhouse gas emissions;
(b) Adaptation: aimed at protecting life, property and nature from the impacts of climate change; and
(c) Compensation and damages: aimed at recovering losses due to climate change from those responsible for failures of mitigation or adaptation.

The cases in municipal courts and tribunals started slowly. The first major climate mitigation litigation happened in the United States in 1986[12] and in Australia in 1994.[13] Adaptation cases soon followed as the impacts of climate change became

[8] Rogelj J, Den Elzen M, Höhne N, Fransen T, Fekete H, et al. (2016) Paris Agreement climate proposals need a boost to keep warming well below 2 °C. Nature 534:631–39; see also Robiou du Pont Y, Meinshausen M (2018) Warming assessment of the bottom-up Paris Agreement emissions pledges. Nat. Commun 9(1):4810. In Chap. 2 of the present volume, Professor Steffen concludes that the Paris target of 1.5 °C is now out of reach.

[9] IPCC (2018) Summary for Policymakers In: Global Warming of 1.5 °C. An IPCC Special Report on the impacts of global warming of 1.5 °C above pre-industrial levels and related global greenhouse gas emission pathways, in the context of strengthening the global response to the threat of climate change, sustainable development, and efforts to eradicate poverty [Masson-Delmotte V, et al. B.4.2, p8.]

[10] Deliotte Access Economics (2017) At what price? The economic, social and icon value of the Great Barrier Reef.

[11] See, for example, Peel J and Osofsky H (2020) Climate Change Litigation. Annu Rev. Law Soc Sci 2020 16:8.1–8.18.

[12] United States Court of Appeals for the District of Columbia Circuit 286 U.S. App. D.C. 78; 912 F.2d 478; 1990 U.S. App. LEXIS 14728; 21 ELR 20170.

[13] *Greenpeace Australia Limited v. Redbank Power Company Pty Limited* [1994] NSWLEC 178.

more apparent.[14] Compensation and damages cases are becoming more prominent in recent years as the financial losses from climate change start to mount.

The Sabin Centre for Climate Change Law, hosted by the Columbia Law School, tracks climate change litigation globally.[15] It lists a mere eight climate litigation cases filed globally before 2000, rising to 338 cases between 2001 and 2010, and reaching 1186 cases between 2011 and 2020. After the United States with 1308 cases since 1986, Australia is home to the largest number of cumulative climate litigation cases at 115.

The impetus for climate change litigation has evolved over time. Initially, the cases relied on appeals in development assessment where statutes provided broad standing rights and facilities for merits assessments measured against specific environmental criteria. Judicial review has also been a common course of action where the failure to consider climate change impacts may arguably be held to constitute a failure of jurisdiction on the part of the decision-maker (that is, it is outside the scope of discretion authorised for the decision). As the climate impacts become more apparent, the scope of causes of action has broadened to include constitutional and human rights grounds.[16] Most recently the clarity of financial impacts of climate change has extended the causes of action into tortious liability (a form of civil wrongdoing), public trust and corporations law.[17]

Each new wave of causes of action, rather than displacing earlier forms of litigation, builds upon them in what may be called a 'harmony' of climate litigation causes of action.[18] Of course defendants would not necessarily conclude that more litigation is harmonious. However, achieving harmony between the climate and with the needs of all living beings is generally the objective of the claimants who initiate such litigation. While motivated by a desire to promote action on, or to seek redress in relation to, climate change, 'climate change litigation' is thus not limited to legal arguments on environmental or climate change points.[19] Some of the key benefits of pursuing relevant litigation include: (i) revealing evidence of pollution or emissions through the court discovery process; (ii) strengthening environmental

---

[14] See, for example, Preston J (2008) The Role of Courts in Relation to Adaptation to Climate Change. A paper presented to: Adapting to Climate Change Law and Policy Conference, http://www.lec.justice.nsw.gov.au/Documents/preston_the%20role%20of%20courts%20in%20relation%20to%20adaptation%20to%20climate%20change.pdf

[15] See http://climatecasechart.com/ (Accessed 11 Jan 2021).

[16] See Sax JL (1971) Defending the Environment: A Handbook for Citizen Action. Vintage Books, New York; Jacqueline Peel (2007) The Role of Climate Change Litigation in Australia's Response to Global Warming. Environment and Planning Law Journal 24:90; Preston P (2009) Climate Change Litigation. Environment and Planning Law Journal 26:169–70; Peel J and Osofsky H (2020) Climate Change Litigation Annu Rev. Law Soc Sci 2020 16:8.1–8.18.

[17] Preston B (2018) Mapping Climate Change Litigation. Australian Law Journal 92:777–81.

[18] Peel J, Osofsky H (2020) Climate Change Litigation. Annu Rev. Law Soc Sci 2020 16:8.1–8.18.

[19] Peel J and Osofsky HM (2015) Climate Change Litigation: Regulatory Pathways to Cleaner Energy. Cambridge University Press, England, p 1; cf. Markell D, Ruhl JB, (2012) An Empirical Assessment of Climate Change in the Courts: A New Jurisprudence or Business as Usual? Florida Law Review 64:27.

conditions imposed on polluters through more accountable government decision-making; (iii) promoting developments in the law; and importantly, (iv) shifting public perceptions concerning climate change.[20] Because litigation generally takes place in courts, and because courts must ordinarily operate in public view, such activity may have an educative function. Educating society is a major necessity in securing action on climate change. In 2019, a special issue of the *Australian Law Journal* devoted to climate change and the law was published.[21] The launch of the special issue was undertaken by Professor Lesley Hughes, professor of biology and Pro Vice-Chancellor of Macquarie University in Sydney. His remarks gained widespread attention at the beginning of 2019. After describing what he called "an existential threat of our own, somewhat inadvertent making", that has led to a huge escalation in climate change, well-known features that appeared to require urgent action were listed. They included the heating of the Earth's oceans that are leading to the "catastrophic bleaching of the world's coral reefs, including our own Great Barrier Reef"; the retreat of glaciers globally; threats to freshwater resources; the rising of sea levels at a rate now triple that of the 1990s; the outbreak of storms of ever-increasing intensity; and the huge increase in forest fires, with some of the worst of them occurring in Australia.

Emphasising the importance of adherence to the *Paris Framework Convention on Climate Change* of 13 December 2015, which Australia ratified on 10 November 2016, promising concerted global action to level the rate of change, Professor Hughes observed:

> Australia has little to be proud of, at least as a nation. We are among the highest per capita emitters in the world, and the highest in the OECD. Our Paris emissions target of 26–28% reduction compared to 2005 by 2030 was not just embarrassingly inadequate but we are failing spectacularly to meet them thus far, having posted increasing emissions in every quarter since March 2015.[22]

Beyond the activities of the legal profession, climate change litigation has contributed to growing community awareness. According to a recent survey, 74% of Australians are concerned about climate change and 80% think we are already experiencing climate change impacts.[23] This has resulted in an increase in political atten-

[20] See Bell-James J, Ryan S (2016) Climate Change Litigation in Queensland: A Case Study in Incrementalism. Environment and Planning Law Journal 33: 515–6 citing Finanzio A SC (2015) Public Participation, Transparency and Accountability – Essential Ingredients of Good Decision Making, Paper presented at The Accountability and the Law Conference, Brisbane, 9 February 2015; Preston B (2010) Climate Change in the Courts Monash University Law Review 36(1):15; Preston BJ (2011) The Influence of Climate Change Litigation on Governments and the Private Sector. Paper presented at The Beyond a Carbon Price: A Framework for Climate Change Regulation in Australia Workshop, Melbourne, 11–12 August 2011 and Peel and Osofsky (n 4) 167, 222.

[21] Special Issue: Climate Change and the Law (2018) Australian Law Journal 92:747. See e.g. Wilder M (2018) Climate Change and the Law – Introduction ALJ 92:747.

[22] Special Issue: Climate Change and the Law (2019) Australian Law Journal 92:8

[23] https://www.theguardian.com/environment/2020/oct/28/overwhelming-majority-believe-australia-is-already-experiencing-climate-change

tion to climate change issues in state and federal elections in Australia, with more than 80% of Australians in the lead-up to the 2019 federal election expecting government to do more on climate change.[24] It is impossible for anyone in Australia with access to the internet, radio or television to be unaware of the serious events that are happening in all parts of the world.

Finally, in the corporate, civil society and judicial contexts, new attention is being paid to climate change in Australia. This has specifically arisen in relation to the production and export of coal from Australia, a resource of which the nation has large deposits and many overseas markets. Corporations have begun to take the risk of litigation relating to climate change seriously and to adjust their business operations, not least because new mining developments have become subject to increasing judicial and other official scrutiny. In early 2019, Glencore Resources Ltd., a Swiss-based company and Australia's biggest coal miner, "announced it would cap coal production at current levels".[25] The decision was not made in a vacuum. It resulted from litigation where "for the first time an Australian judge … refused to allow a new coal mine because it could be contributing to climate change". It is worth paying particular attention to this case.[26]

13.2.2 Climate Change Litigation in Australia: The Rocky Hill Decision

The decision in the Glencore Resources case was delivered by Preston J, Chief Judge of the New South Wales Land and Environment Court. It was announced on 8 February 2019 in the litigation between *Gloucester Resources Limited v Minister for Planning* ('*Rocky Hill Decision*').[27] Preston J concluded that a coal mine proposed by Gloucester Resources Ltd., three kilometres from the picturesque town of Gloucester in New South Wales, should not go ahead. The project involved about 500 hectares of land, which would be mined in three contiguous pits. It was proposed that it would be developed, operated and eventually 'rehabilitated' over 21 years with approximately two million tonnes of coal to be extracted each year for 16 years. It would bring profits to the developer, employment to the district and overseas earnings to the nation. Yet Preston J decided that the mine would be "in the wrong place at the wrong time."[28]

The judge's reasons for upholding the refusal of development consent by the Planning Assessment Commission of New South Wales, relied in part on standard

---

[24] https://www.abc.net.au/news/2019-05-15/federal-election-vote-compass-climate-change/11110912

[25] Coal in Australia. The Dangerous Dark Stuff (2019) The Economist p 57.

[26] Coal in Australia. The Dangerous Dark Stuff (2019) The Economist p 57.

[27] [2019] NSWLEC 7.

[28] *Gloucester Resources* [2019] NSWLEC, 7 at [699] (8 February 2019).

environmental considerations. These included the project's "unacceptable social impacts on the residents and community of Gloucester". However, when Gloucester Resources brought its appeal from the refusal of consent to the specialised Land and Environment Court, a community environmental group sought leave to be joined in the proceedings as a party, to raise the issue of the project's impact *inter alia* on greenhouse gas emissions and climate change. That application was granted by a senior commissioner of the Court in April 2018.[29] It was held that the additional ground of objection to the proposed development would not be adequately addressed if the added party were not joined.[30] The outcome was that the court decided that the development should be refused because of its incompatibility with the residential amenity and other land uses and visual and social impacts of the disturbance.[31] However, the finding of the Court that caught media and international attention was something different. The Chief Judge held:[32]

> The greenhouse gas emissions of the coal mine and its coal product will increase global total concentrations of greenhouse gases at a time when what is now urgently needed in order to meet generally agreed climate targets, is a rapid and deep decrease in greenhouse gas emissions.

Gloucester Resources did not dispute that climate change was real; was happening; and that greenhouse gases had to be reduced to meet internationally agreed targets. Its opposition to the argument based on climate change was that it involved a "side show and a distraction". That submission was rejected by the court. Preston J concluded that:[33]

> The global problem of climate change needs to be addressed by multiple local actions to mitigate emissions by sources and remove greenhouse gases by sinks.

He pointed out that, under the *Environmental Planning and Assessment Act 1979* (NSW), s 4.15(i)(b), he was obliged to consider the broader environmental impacts of a proposed development and to address the issues of "public interest" as the substitute consent authority, standing in the shoes of the commission that had initially refused consent. He rejected arguments relying on the contention that, if Australia did not provide the coal to willing purchasers, other less environmentally protective countries would do so, thereby actually increasing the dangers of greenhouse gases. He held that this argument was neither rational nor persuasive because it relied on theoretical possibilities.[34] He also found that, if Australia were to take the lead on implementing appropriate mitigation measures, developing countries might be encouraged to follow suit.[35] By inference, if Australia did not observe the require-

---

[29] *Gloucester Resources Limited v Minister for Planning* [2018] NSWLEC 1200.

[30] *Gloucester Resources* [2018] NSWLEC 1200 at [31].

[31] *Gloucester Resources* [2018] NSWLEC 1200 [31].

[32] *Gloucester Resources* [2019] NSWLEC 7 [533].

[33] [2019] NSWLEC 7 at [515].

[34] [2019] NSWLEC 7 at [530].

[35] [2019 NSWLEC 7 at [530]–[533].

ments of the *Paris Framework Convention*, it would be unsurprising that developing countries, with fewer resources to absorb such requirements, would decline to do so, citing Australia's inaction.[36]

Though subject to criticism by Gloucester Resources and its supporters, Preston J's decision was widely lauded by professional and academic commentators.[37] The commentaries uniformly pointed to the unique and special features of the legislation in New South Wales and the orthodoxy of Preston J's reasons, viewed in that context. On an earlier occasion, the High Court of Australia had highlighted the special provisions in the New South Wales environmental law and in the powers of the court that administers it.[38] The reference by Preston J to the larger question of global climate change was clearly viewed by specialised practitioners and legal academics as appropriately responding to the specificities of the applicable legislation.

Gloucester Resources did not pursue an appeal against Preston J's decision. Consequently, the reasoning in the case now stands as a strongly persuasive precedent. It was recently followed by the NSW Land and Environment Court upholding the refusal of the Bylong coal mine.[39]

A more recent case of note (ongoing at the time of writing), filed on 22 July 2020, is a claim in the Federal Court of Australia against the Australian Government for not disclosing climate change risks in its issue of Australian Government Bonds.[40]

In light of successful litigation both abroad and in Australia, it can be safely predicted that officials, company directors and judges of the future in Australia will be paying closer attention to international developments in this field than they have done in the past. The use and efficacy of climate change litigation is growing. I therefore propose to discuss recent cases of climate change litigation, exploring how decisions have flowed from renewed attention to the obligations of governments, corporations and civil society.

More specifically, the enactment in three sub-national jurisdictions of Australia of general human rights legislation may give rise to Australian courts following the approach adopted, in that regard, in the United Kingdom. The adoption of international proposals to enliven the attention of municipal judges and lawyers as to the obligations expressed in the *Framework Convention* needs to be examined.

Further, developments in the Australian business sector are now also increasingly suggesting both an obligation on the part of companies to take account of financial risks caused by climate change, and of company directors to exhibit an informed

---

[36] Statement by Gloucester Resources Ltd. noted in Lawyers Weekly, 14 February 2019 (Autumn 2019), 24.

[37] White R, Hallinan J, Raymont B (2018) Climate change takes centre stage in Land and Environment Court. Law Society Journal (NSW) 54:80; Wilder M, Sharona C (2019) The Mainstreaming to Climate Litigation. Asia Jurist 71.

[38] Cf *Oshlack v Richmond River Council* (1988) 193 CLR 72 at 89 [45] per Gaudron and Gummow JJ, 125–6 [139]–[141] per Kirby J.

[39] *KEPCO Bylong Australia Pty Ltd v Independent Planning Commission (No 2)* [2020] NSWLEC 179.

[40] https://www.equitygenerationlawyers.com/odonnell-v-commonwealth/

approach, and due diligence, in their responses to decisions affecting climate change. The emergence of these expanded obligations will also be explored. They suggest some of the future directions of the law in this space.

13.3 International Law, Human Rights and Public Decision-Making

13.3.1 International Law and Climate Change

Over the past 40 years, the international community has struggled to reach agreement about a multilateral response to what has increasingly been accepted as the urgent challenge of climate change.

It was to address this challenge that the 21st Session of the Conference of the States Parties of the *United Nations Framework Convention on Climate Change* (UNFCCC) in Paris was convened in 2015.[41] Just as fundamental human rights, expressed in the *Universal Declaration of Human Rights*[42] and subsequent United Nations treaty law, have come to affect municipal planning and environmental law in many countries, so an increasing number of lawyers globally have begun to consider a human rights-based approach to the dangers of global climate change.[43] Such an approach, they hope, may "facilitate a meaningful response where legal frameworks are otherwise lacking, especially as it broadens the scope of available institutional knowledge, capacity and solutions".

Another international initiative that emerged in 2015 was the *Oslo Principles on Global Climate Change Obligations* (*Oslo Principles*).[44] These Principles were

[41] Paris (France) 13 December 2015, in force 4 November 2016 UNFCC, Report of the Conference of the Parties on its 21st Session, Addendum, undocfccc/cp/2015/10/add1, 29 January 2016. Australia ratified the Paris Agreement on 10 November 2016. In August 2017, the Government of the United States of America submitted a formal notice that the United States intended to withdraw from the Paris Agreement, a process which was completed in November 2020, in accordance with art 28 of the Agreement. See Peel J (2017) Climate Change Litigation: Lessons and Pathways. Judicial Officers' Bulletin (NSW) 29:99. See also McAdam J (2012) Climate Change, Forced Migration and International Law. Oxford University Press, England and Saul B et al. (2012) Climate Change and Australia: Warming to the Global Challenge. Federation, Sydney p 191–232.

[42] *Universal Declaration of Human Rights*, GA Res 217A (III), UN GAOR, UN Doc A/810 (10 December 1948).

[43] Nagra S (2017) The Oslo Principles and Climate Change Displacement: Missed Opportunity or Misplaced Expectations? Carbon and Climate Law Review. Birnie TW, Boyle AE (1992) International Law and the Environment, Oxford University Press, England p 188–195.

[44] *Oslo Principles on Global Climate Change Obligations* (1 March 2015), https://globaljustice.yale.edu/sites/default/files/files/OsloPrinciples.pdf"Oslo Principles". In 2013, the author accepted an invitation from judges of the Hoge Raad der Nederlanden (the Supreme Court of the Netherlands) to join an international expert group to develop legal principles that might meaningfully engage with the problem of climate change, in a way that municipal judges and like decision-makers would understand and find useful. The Oslo Principles were drafted by this group.

expressed by an expert group in which the present writer participated. The expert group addressed the issues of global climate change and emerging human rights law. It included members (judges) of the Hoge Raad der Nederlanden, the highest judicial body of the Netherlands. The original meetings of the expert group took place in The Hague, The Netherlands, and ultimately in Oslo, Norway, where the group's principles were propounded. The *Oslo Principles* were adopted shortly before the Paris meeting summoned to agree upon a binding treaty on the subject. A purpose of the *Oslo Principles* was to support the momentum that was then developing towards an international consensus to support, and increase appreciation of, the urgency of action. This was considered to be in the interests of human beings everywhere and of the biosphere generally. So was a commitment to political and legal action by recruiting civil society organisations and concerned individuals to offer support. It was hoped that judges and lawyers, together with civil society, would have an important role to play concerning the "leeways for choice" presented to them, deciding environmental matters in accordance with their own respective national law but viewing that law in the context of developments in the international law of human rights.

This is not the occasion to explain all of the provisions of the *Oslo Principles*. Suffice it to say that the expert group resolved:

> 6. States and enterprises must take measures based on Principle 1 to ensure that the global average surface temperature increase never exceeds pre-industrial temperature by more than 2 °C.

The ambit and application of the *Oslo Principles* were expressed to be subject to the 'Precautionary Principle' set out in Principle 1. Relevantly, Principle 1 stated:

> 1. There is clear and convincing evidence that the greenhouse gas (GHG) emissions produced by human activity are causing significant changes to the climate and that these changes pose great risks of irreversible harm to humanity, including present and future generations, to the environment, including other living species, and the entire natural habitat, and to the global economy.

(a) The Precautionary Principle requires that:

1. GHG emissions be reduced to the extent, and at a pace, necessary to protect against the threats of climate change that can still be avoided; and
2. The level of reductions of GHG emissions required to achieve this, should be based on any credible and realistic worst-case scenario accepted by a substantial number of eminent climate change experts.

(b) The measures required by the Precautionary Principle should be adopted without regard to the cost, unless that cost is completely disproportionate to the reduction in emissions that will be brought about by expending it.

The *Oslo Principles* concluded that the implementation of the Precautionary Principle required that:

1. GHG emissions be reduced to the extent, and at a pace, necessary to protect against the threats of climate change that can still be avoided; and
2. The level of reductions of GHG emissions required to achieve this should be based on … [the] credible and worst-case scenario accepted by … eminent climate change experts.

In paragraph 25 of the *Oslo Principles,* certain procedural obligations of nation states and corporate enterprises were stated in order to achieve the substantive requirements:

> 25. States must accept the jurisdiction of independent courts or tribunals in which the state's compliance with its obligations set forth in these principles may be challenged and adjudicated.

(a) States must participate in those proceedings in good faith and ensure that such proceedings are fair and efficient.
(b) In such proceedings, the state, whose compliance with its obligations has been challenged, must fully disclose the ways in which it has effected compliance, in order to enable the court or tribunal to determine whether the state has complied with the relevant obligations and, where it is found that the state has not complied, to determine the state and nature of the state's failure to comply.

13.3.2 The Influence of International Law on Domestic Decision-Making

The *Oslo Principles* proved prescient in respect of legal developments that occurred soon afterwards in the Netherlands. In July 2015, a District Court in that country upheld a class action that had been brought by an environmental group, Urgenda, on behalf of 886 named individuals. The claimants alleged that the then announced plans of the Netherlands' Government to cut emissions of all agencies under its control by 14–17% (based on 1990 levels), in time for 2020, were inadequate. It was argued that such a cut would not adequately protect the public of the Netherlands from the adverse effects of climate change.

As a low-lying country, renowned for its protective sea walls and large areas of reclaimed land that lie below sea level, the Netherlands is particularly exposed to the danger of the global rise in sea levels. Especially is this so because of hazardous situations arising from notorious North Sea storms. The argument of Urgenda was that the Government of the Netherlands should be ordered to cut greenhouse emissions by at least 25% by 2020.[45]

On 20 December 2019, the Hoge Raad upheld the trial court's decision to order the Government to cut emissions more rapidly, although for different reasons. While the trial court had based its decision on the government's general duty of care to its citizens, the Hoge Raad applied the *European Convention of Human Rights* (ECHR). Specifically, it held that Articles 2 and 8 of the ECHR, which guaranteed the

[45] See Baxter T (2017) Urgenda-style Litigation has Promise in Australia. Australian Environment Review 32(3):70. In common law countries such proceedings are likely to be brought under administrative review remedies. Cf *Thomson v Minister for Climate Change Issues* [2017] NZHC 733; <www.courtsofnz.govt.nz/cases/thomson-v-the-minister-for-climate-change-issues/atatimages/filedDecision?r=732.86082992, (accessed 27 Nov 2017). The Court found that the results of New Zealand's election, installing a Labour Coalition Government, rendered the decision moot as the new Government had pledged to review and reduce the country's 2050 carbon target.

claimants a right to life and right to respect for private and family life, required the Government of the Netherlands "to take suitable measures if a real and immediate risk to people's lives or welfare existed and the state was aware of that risk."[46] The court held that those preconditions were fulfilled.

This specific obligation in the ECHR allowed the plaintiffs to claim that the Netherlands Government had failed to do 'its part' in combatting climate change. The Court emphasised that, under the UNFCCC and the Paris Agreement, a State Party could not excuse its unambitious target on the basis that other countries would continue to pollute at a larger scale – according to the Court, "no reduction is negligible".[47] Thus, the Supreme Court of the Netherlands accepted that, notwithstanding the fact that the Netherlands' share of emissions was minor by comparison to other countries, it still had an identifiable responsibility to its citizens to take measures to reduce the risk of severe climate change.

To that end, the Urgenda decision mirrors the similar acknowledgement, expressed in the *Rocky Hill Decision* in Australia, that a country's comparatively minor contribution to global emissions does not excuse its individual responsibility to engage in meaningful emissions reduction as part of a multilateral global commitment. Achievement of the global objective would only be attained if all parties played their part. The court rejected the submission that this approach and outcome constituted an impermissible breach by the court of the separation of powers doctrine, dividing the legislative power from the judicial.

The Urgenda Case also exemplifies the convergence in international law of environmental protection and climate change action, and human rights law. It is striking that the Netherlands' Supreme Court relied on human rights obligations under the ECHR, interpreted with the assistance of more specific international law instruments such as the UNFCCC and Paris Agreement, to hold the Government of the Netherlands accountable. Although no fundamental human right to a healthy environment has yet been spelled out by the United Nations, as Professor John Knox, the UN Special Rapporteur on Human Rights and the Environment, pointed out in his 2019 final report:

> In total, at least 155 States are legally obligated, through treaties, constitutions, and legislation, to respect, protect, and fulfil the right to a healthy environment. This provides a compelling basis for the United Nations to move expeditiously to provide global recognition of the right to a healthy and sustainable environment, as recommended by both the previous and current Special Rapporteurs on human rights and the environment.[48]

[46] *State of the Netherlands v Urgenda* (see unofficial translation here: https://www.urgenda.nl/wp-content/uploads/ENG-Dutch-Supreme-Court-Urgenda-v-Netherlands-20-12-2019.pdf)

[47] Ibid [5.7.9].

[48] Knox J (2019), *Issue of human rights obligations relating to the enjoyment of a safe, clean, healthy and sustainable environment – Report of the Special Rapporteur*, Human Rights Committee, 40th sess, Agenda Item 3, UN Doc A/HRC/40/55 (8 January 2019) 3 [16].

13.3.3 International Law in Australian Litigation

Australia is unique in being the only developed nation to lack a comprehensive human rights instrument at the national constitutional level. Save for the requirements of "just terms" imposed on the federal compulsory acquisition of property, and a few other scattered provisions not presently material, there is no general constitutional Bill of Rights in Australia. This was not because of a lack of respect in Australia for human rights generally, or for the right to own property in particular. Such rights were recognised and upheld by the common law and also by detailed statutory provisions.

However, the general theory of Australian law, at the time of federation in 1900 when the *Australian Constitution* was adopted and long thereafter, was that the imposition of restrictions on legislative power, in the name of human rights, was unnecessary. This was because the legislature, answerable to the electors in regular elections, could be relied upon to uphold such shared basic rights, including the right to own and enjoy property. Where, accidently or otherwise, a derogation from such shared rights arose, Australians (so it was believed) could rely on the elected legislatures to rectify the impairment. That was the main reason why provisions for enforceable human rights – constitutional, statutory or otherwise – were generally considered not only unnecessary but also undesirable. Experience since the *Australian Constitution* was adopted has sometimes justified the popular faith of the founders in the legal protection of basic rights by Parliament.[49] However, in many other instances, the gap in Australia's constitutional arrangements has been manifest. It can now be regarded as serious.[50]

As a result, planning and environmental law in Australia is influenced primarily by statutory law; to some extent, common law principles; rules of interpretation; and since the late 1990s, native title law. It is only very occasionally that a planning and environmental case will venture into constitutional law or (in the Australian Capital Territory,[51] Victoria[52] and now Queensland,[53]) human rights law.

Therefore, unlike the Netherlands, Australia's constitutional and legislative framework offers no directly available method to invoke international law instruments in support of climate change litigation.[54] The most useful indirect method that

[49] See for example *Migration Act 1958* (Cth) which adopted a non-racial right to seek immigration to Australia, commencing the repeal of 'White Australia' laws. Contrast *Pacific Island Labourers Act 1901* (Cth). See also *Australian Communist Party v The Commonwealth* (1951) 83 CLR 1 which invalidated the *Communist Party Dissolution Act 1950* (Cth) which purported to dissolve the Communist Party but was held to be unconstitutional.

[50] See e.g. *Al-Kateb v Godwin* (2004) 215 CLR 562 (validity of indefinite detention under federal statutory law and executive orders).

[51] *Human Rights Act 2004* (ACT), 22.

[52] *Charter of Human Rights and Responsibilities Act 2006* (Vic), s 23.

[53] *Human Rights Act 2019* (Qld).

[54] Although noting that under *Human Rights Act 2019* (Qld), s48(3), international law and the judgments of domestic, foreign and international courts and tribunals relevant to a human right *may* be considered in interpreting a statutory provision.

has emerged in the case law is the 'public interest' consideration by which decision-makers are often required to turn their mind to in coming to a planning or environmental development decision.[55] That consideration may direct attention to the general environmental concerns of human beings and of the entire biosphere. Owing to an expanding universe of human rights and environmental law jurisprudence in the international law sphere, decision-makers and courts are increasingly invoking international law under the criterion of the requirements of a statutory 'public interest'.

As has been said, the growing number of Australian jurisdictions that have now adopted, or are considering, human rights charters make it relevant for Australian lawyers and environmental planners – certainly in Victoria, the Capital Territory and, since 2019, Queensland – to increase their awareness about the way in which parallel legal regulation has affected planning decisions in other jurisdictions. Many case decisions;[56] case analyses;[57] and legal texts[58] explore the ways in which appeals to universal principles of human rights in planning and environmental litigation have sometimes controlled, or influenced, the outcomes of such disputes when brought to litigation. In the future, planning lawyers in Australia will increasingly be obliged, especially in the three jurisdictions that have adopted general human rights charters, to familiarise themselves with the decisions made overseas on general human rights grounds. They will probably pay particular attention to the United Kingdom where the legal system is so similar to Australia's, save for the general enforceable municipal and international legal regimes of human rights protection.

The starting point, at least for Australian lawyers new to human rights reasoning, is to understand the way in which provisions in human rights law have come to interact in other countries with specific planning law provisions, in a way that affects the outcome of a dispute submitted to litigation.[59] Thus, in the United Kingdom, which was an original signatory to the ECHR, it was not until 1998 that the substance of that Convention was incorporated into municipal domestic law.[60]

[55] See, e.g., *Environmental Planning and Assessment Act 1979* (NSW) s 4.15(1)(e); *Environmental Protection Act 1994* (Qld) s 175, sch 4 (meaning of 'standard criteria'); *Environment Protection Act 1970* (Vic) s 67 AC; *Development Act 1993* (SA) s 85(7); *Planning and Development Act 2005* (WA) s 247(1).

[56] See e.g. *Tesco Stores v Secretary of State for the Environment* [1995] 1WLR 759 (HL); *R v Secretary of State for the Environment, Transport and the Regions; ex parte Holding and Barnes plc* [2001] JPL 291; [2001] UKHRR 270.

[57] (2011) Case Law Analysis, Planning Law and the Human Rights Act 1998. Journal of Environmental Law. Oxford University Press 13:3.

[58] Hart D (2000) The Impact of the European Convention on Human Rights on Planning and Environmental Law. Journal of Planning & Environmental Law 117; Purdue M (2002) Human Rights and Planning Law, the UK Experience. Asia Pacific Law Review 10:195.

[59] Ferguson A (2000) Human Rights and Planning Law Collide. Scots Law Times 27:211; Enemark S, Hvingel L, Galland G (2014) Land Administration, Planning and Human Rights. Planning Theory 13:331; Purdue M (2004) The Human Rights Act 1998, Planning Law and Proportionality Environmental Law Review 6:161.

[60] *Human Rights Act 1998* (UK).

Immediately this happened, it resulted in applications by which local lawyers sought to invoke the United Kingdom equivalent to a Convention right. These cases have involved the entitlement to the peaceful enjoyment of property;[61] the right to enjoy privacy and family relations in relation to such property;[62] the right not to suffer immaterial discrimination in the enjoyment of legal entitlements expressed in general terms;[63] and the entitlement, in the determination of rights and obligations at law and of any criminal charge, to have a fair and public hearing by an independent and impartial tribunal.[64]

The application of the foregoing human rights principles in the law of the United Kingdom has sometimes been controversial. In *R v Secretary of State for the Environment, Transport and the Regions; ex parte (Alconbury Developments Ltd),*[65] the question arose as to whether procedures under town and country planning law in the United Kingdom, for the determination of planning applications and appeals by the Secretary of State, were compatible with Article 6 in the *Human Rights Act 1998* (UK). In the English Court of Appeal, Tuckey LJ and Harrison J held that decisions on planning applications and appeals by a Minister could not be considered a determination by "an independent and impartial tribunal". This was because a Minister would necessarily appear to be biased in favour of his or her own polices and desired outcomes. Such bias would not be entirely cured by the availability of judicial review. However, that decision was quickly overturned by the House of Lords. Their Lordships held that the administrator's suggested lack of impartiality could be adequately cured by the provision of a right of access to impartial judicial review. The fact that judicial review was not equivalent to a hearing *de novo* on the merits was not regarded as determinative.

Appeals to human rights principles forbidding discrimination have been invoked by Roma (gypsy) applicants in the United Kingdom, and elsewhere, who had been refused planning permission for the residential use of local land so as to continue their traditional nomadic way of life. In recent years such applicants have had many disputes with planning authorities in the United Kingdom and elsewhere in Europe.[66] In the case of Roma, Article 14 of the European Convention is clearly relevant because it prohibits discrimination on grounds such as sex, race, colour, language, religion, political or other opinion, national or social origin, association with a national minority, birth or other status.

In *Chapman v United Kingdom,*[67] the European Court of Human Rights accepted that the Roma minority fell within the ambit of Article 14. However, in respect of

[61] *European Convention*, art 1 (Peaceful enjoyment of property rights).

[62] *Ibid*, art 8 (Privacy and family rights).

[63] *Ibid*, art 14 (Freedom from discrimination in respect of protected rights).

[64] *Ibid*, art 6 of the first Protocol to the *European Convention*; cf. UNHCR, art 19 (Freedom of opinion and expression).

[65] [2001] UKHL 23 at [48]–[53], [60], [70], [139].

[66] Purdue M (2004) The Human Rights Act 1998, Planning Law and Proportionality Environmental Law Review 6:164.

[67] (2001) 10 BHRC 48.

claims for relief, it was pointed out that, under Article 8, a derogation was permissible where it was in accordance with law and was deemed necessary in a democratic society in the interests of national security, public safety or the economic wellbeing of the country, for the prevention of disorder or crime, for the protection of health or morals, or for the protection of the rights and freedoms of others. Judges dealing with such cases have pointed out that the excuse of justification for interference in the "rights of others" has to be "proportionate".[68] But if it was, it was acceptable. In the result, in many cases, the impact of the *Human Rights Act* in the United Kingdom has been less dramatic in terms of the substance of the issues in dispute when compared with the impact on the procedural rights of the parties (such as the right to reasons for adverse decisions); their legal right to have relevant considerations taken into account; and the need for the decision-maker to observe the principle of "proportionality" in reaching a conclusion.

Under English law, challenges by way of judicial review to the rationality of the administrator's decision have, in previous decades, generally had to run the gauntlet of so-called *Wednesbury* unreasonableness.[69] To secure a court's intervention on the judicial review of an adverse administrative decision upon this ground, proof to an extremely high standard was required: namely irrationality, such that 'no rational decision-maker could have reached such a conclusion'.

Derived from German law, and now increasingly entering English law, originally by way of decisions of the European Court of Human Rights and the European Court of Justice, a different test of "proportionality" has increasingly been accepted. Thus, under the European remit, in cases involving Roma, national courts have been obliged "at least [to] evaluate whether the balance struck by the administrator is proportionate".[70] Thus, the reviewing court has been instructed that it should "review how the decision-maker has applied the proportionality principle". Yet the courts have gone on to emphasise that a:

> … considerable measure of deference should be accorded to the judgment of the decision taker as to proportionality. This is because the courts do not have expertise in the planning field and the balancing exercise involves matters of planning policy. [Moreover] … the courts do not, like the Inspector, have the opportunity of looking at the site and hearing the evidence of special circumstances first hand.

---

[68] *R (On the application of Michael and Jenny Boyd and English Nature)* [2004] EWHC 1105 at [19]–[20].

[69] *The Associated Provincial Picture Theatre Houses Ltd v Wednesbury Corporation* [1948] 1 KB 223. See now in Australia *Administrative Decisions (Judicial Review) Act 1977* (Cth), s 5(1)(e), 5(2)(g), 6(1)(e) and 6(2)(g). In England, the test was earlier expressed as "so unreasonable that it was irrational". *Council of Civil Service Unions v Minister for the Civil Service* [1985] AC 374 at 410, 415. Subsequently, a test has been reformulated in terms of proportionality.

[70] See *R (On the Application of Egan) v Department of Transport, Local Government and the Regions* [1991] AC 749. In *R v Secretary of State for the Home Department; ex parte Brind* [1991] AC 749 at [2] Lord Slynn of Hadley observed that the time had come for the principle of proportionality to be recognised as part of the English law.

Even a cursory glance at the development of planning law in the United Kingdom shows that, since the *Human Rights Act 1998* (UK), the principles and modes of reasoning of the European courts have come to influence quite significantly the approach of United Kingdom courts. Indeed, this influence of the European courts was one of the emotional arguments sometimes raised by the supporters of withdrawal from the European Union (Brexit): advocating the necessity to abandon the 'alien' European Union approach and to sever links with such 'foreign' institutions and their values.

The severance of such links will now be successful in the case of the European Court of Justice, because that court is an institution of the European Union. However, it is less certain that it will be successful in the case of the European Court of Human Rights, an institution of the Council of Europe. In their most recent *Political Declaration setting out the framework for the future relationship between the European Union and the United Kingdom*, both parties agreed to the following wording: "The future relationship should incorporate the United Kingdom's *continued commitment to respect the framework of the European Convention on Human Rights*".[71] Thus, the prospect of the United Kingdom withdrawing from the ECHR, which predated the creation of the European Union, seems unlikely, at least in the immediate future. Should this prove so, the ongoing influence of the ECHR on the general law of the United Kingdom (and thus on planning law) appears likely to continue.[72] As such, it will probably provide a regular stimulus, and an example, to Australian lawyers, especially if sub-national human rights statutes are adopted in further jurisdictions of Australia to import some of the same language, concepts and values as the ECHR.

If this prediction proves accurate, it will become useful, as it has been found in the United Kingdom, to have regard to decisions of the European Court of Human Rights in environmental and planning matters coming before at least some of the courts of Australia. From a modest start, with only two such cases decided in 1966, the workload of the European Court of Human Rights has increased enormously. By 1998 some 4000–5000 cases were awaiting decision. That number has continued to expand. Just as the court's jurisdiction has increased, so has the membership of the Council of Europe, which gave birth to the ECHR. From the original ten member countries, the Council of Europe now has 38 members. As well, it now has a developed jurisprudence which is increasingly detailed and sophisticated.[73] Cases in the

[71] United Kingdom Government, *Political Declaration setting out the framework for the future relationship between the European Union and the United Kingdom* (19 October 2019) [7] (emphasis added).

[72] Compare the references in the High Court of Australia in *Minogue v Victoria* (2018) 92 ALJR 668 at [63]; [2018] HCA 27 citing *Vinter v United Kingdom* (2013) 63 EHRR 1 a case concerning prisoners' rights under the *Charter of Human Rights and Responsibilities 2006* (Vic). See also Groves M (2018) A Life Without Hope – The Victorian Charter and Parole. Criminal Law Journal 42:364.

[73] Hart D (2000) The Impact of the European Court of Human Rights on Planning and Environmental Law. Journal of Planning and Environmental Law 117.

European Court of Human Rights that have been directly relevant to planning law include *Guerra v Italy;*[74] *LBC v United Kingdom*;[75] *Osman v United Kingdom*;[76] and *Lopez Ostra v Lane.*[77]

In the last-mentioned case, the state (Spain) had granted a subsidy to several tanneries to build a waste treatment facility only 12 metres from the applicants' home. The European Court of Human Rights found a violation of the ECHR notwithstanding the absence of any specific finding as to the likely cause of the health problems complained of by the applicant. In doing so, the court stated that "severe environmental pollution may affect the individuals' well-being and prevent them from enjoying their homes in such a way as to affect their private and family life adversely, without, however, seriously endangering their health."[78] Having found a violation of the *European Convention*, the court in *Lopez Ostra* awarded the applicants four million Spanish pesetas, assessed on an equitable basis, "to cover both pecuniary and non-pecuniary damage". Inevitably, the approach of the European Court of Human Rights, to which the English courts are still answerable, has had an impact upon subsequent analogous decisions in the United Kingdom.

Without an equivalent foothold in Australia for arguments of this kind, derived from substantive treaty law, it would be rare, at the present time – certainly outside the Australian Capital Territory, Victoria and now Queensland – for analogous submissions based specifically on human rights law to be raised successfully in litigation involving planning decisions in Australia. However, the European reasoning is clearly a development that will need to be reviewed, in the event that human rights statutes continue to spread in Australia. The likelihood of that happening is substantial, at least in the long-term. At present, Australia is the only developed country without a body of binding human rights legislation and case law. While that remains the case, the prospects of climate change litigation relying on human rights law will be limited to the State level. However, much will depend on the knowledge of European developments amongst advocates; and the awareness and attitude to such developments amongst judges and other decision-makers.

A test case in climate change litigation relying on human rights legislation has recently come before the courts in *Youth Verdict Ltd v Waratah Coal Pty Ltd.*[79] Youth Verdict are a group of young Queenslanders who considered themselves to have been impacted by climate change. They objected to the grant of a mining lease and environmental authority for a large thermal coal mine to be built in central Queensland. Their objection was based on the impact on their human rights resulting from the climate change caused by the burning of coal extracted from the mine.

---

[74] (1998) 56 EHRR 357. See Hart above n.43 at 122.

[75] (1998) 4 BHRC 447.

[76] (1999) 5 BHRC 293 at 321 [116].

[77] A/303-C, [1994] ECHR 46. See also *McGinley v United Kingdom* (1998) 27 EHRR 1; and *Pine Valley Developments v Ireland* (1991) 14 EHRR 319.

[78] (1994) A 303-C para [51].

[79] *Waratah Coal Pty Ltd v Youth Verdict Ltd & Ors* [2020] QLC 33

An application to strike out the human rights grounds was dismissed by the Court[80] and not appealed: the full hearing is expected in 2021. The case will be a novel test of the scope of the new Queensland *Human Rights Act 2019* to address climate change impacts.[81]

13.4 Corporations and Climate Change: Holding Emitters Accountable

13.4.1 Tortious Actions

Corporations are often the largest emitters of greenhouse gases, with just 100 active corporate fossil fuels producers, described as the 'carbon majors', accounting for 71% of global industrial greenhouse gas emissions.[82] In consequence, it is likely that meaningful climate change litigation will seek to hold large emitters accountable for their business decisions, particularly where these are likely to exacerbate or contribute to climate change.[83] Such litigation has had mixed success internationally. There are few relevant cases in Australia.

One significant case arose recently in Germany. In 2015 a Peruvian house owner (Mr Luciano) commenced proceedings in Germany against RWE AG, a German corporation responsible for 0.47% of aggregate global CO_2 emissions.[84] Mr. Luciano sought a court order in Germany that RWE bear the costs of protection measures in Peru designed to safeguard his home there against a 'Glacial Lake Outburst Flood' (GLOF). Mr. Luciano feared that a GLOF would be released from Lake Palcacocha in the Peruvian Cordillera Blanca.

At first instance, the trial court in Germany held that "in all likelihood, in the case of a flood wave, the house of the claimant would be flooded". It also held that climate change could be responsible for such a flood if it were to occur. However, it rejected the claim for protection measures on legal grounds.

On appeal the German Higher Regional Court in Hamm rejected RWE's argument that German law, in principle, did not cover the consequences of climate change because it was "too complex" or because the sources of greenhouse gas emissions were "multiple and undifferentiated". The High Court upheld the legal

[80] *Waratah Coal Pty Ltd v Youth Verdict Ltd & Ors* [2020] QLC 33

[81] Bell-James J, Collins B (2020) Queensland's Human Rights Act: A New Frontier for Australian Climate Change Litigation? UNSW Law Journal 43(1). http://www.unswlawjournal.unsw.edu.au/wp-content/uploads/2020/03/01-Bell-James-Collins.pdf

[82] Griffin P (2017) The Carbon Majors Database: CDP Carbon Majors Report 2017, available at https://climateaccountability.org/pdf/CarbonMajorsRpt2017%20Jul17.pdf

[83] See Hormio S (2017) Can Corporations Have (Moral) Responsibility Regarding Climate Change Mitigation? Journal of Ethics, Policy and Environment 2(3): 314.

[84] http://carbonmajors.org

arguability of Mr. Luciano's claim, based on the *German Civil Code.*[85] Specifically, the court relied upon the provision imposing a legal obligation on a wrongdoer that interferes with the property of another. Having found the arguability of such a legal claim, the High Court committed the matter to re-trial on evidence "to determine [whether] the respondent is liable, proportionately to its level of an impairment of 0.47%, to cover that proportion of the expense of appropriate safety precautions" as undertaken by the claimant to protect his property from a GLOF from Lake Palcacocha, in so far as the claimant would be burdened with such costs.

The court did not decide the conflict in the evidence. It merely accepted the legal possibility that a case might be established by Mr. Luciano, on the basis of the German legal standard invoked by him and the testimony offered in support of the contention that he could, in principle, satisfy that legal standard. While RWE was not the sole potential cause of the risks against which the claimant sought compensation, the High Court concluded that the share of RWE's greenhouse gas emissions was not insignificant for the purposes of the *German Civil Code.* It therefore held that the claim could not be terminated summarily. It had to proceed to the reception of evidence so as to answer the remaining question expressed by the Court. The matter has not yet progressed to a trial, as the German Court is awaiting the consent of Peru to allow investigations to enter its territory. RWE has also indicated its rejection of settlement discussions. The final outcome of the matter may therefore be unknown for some time given pandemic-related travel restrictions.

13.4.2 Corporations Law: Financial Disclosure

In addition to holding corporations to account directly through causes of action in torts or equivalent remedies, there have also been significant developments in doing so through corporations law itself. One prominent area concerns financial disclosure. Corporations, in specified publicly listed companies, are generally required by law to inform their shareholders of significant financial risks.[86] The purpose of disclosure is to promote transparency in the management of the corporation and to ensure that directors are acting in the best interests of the shareholders.

There has been growing professional and academic commentary to the effect that the existing disclosure regime now also requires directors to inform shareholders of risks associated with climate change. Specifically, two types of risks have been identified: (i) transition risks as energy and environmental policy changes; and (ii) physical risks, caused by increasingly disruptive weather patterns.[87] The increased success of climate change litigation, such as in the *Rocky Hill Decision* in Australia,

[85] *German Civil Code* (BGV) art. 1004.

[86] See, eg, *Corporations Act 2001* (Cth) s 299A(1)(c).

[87] See Caldwell Z (2020) Corporations and Climate Change: An Investigation of Mandatory Climate Risk Disclosure in Australia. EPLJ 37:5.

would fall under the former, while business downturns caused for instance by increasingly frequent bushfires would fall under the latter.

Internationally, this new avenue for holding company directors to account for their decisions has enjoyed some success. Most notably, in Poland, a new coal-fired power plant was successfully challenged by a minority shareholder on the basis that the plant would become a stranded asset and would therefore pose a financial risk to the company and thus its shareholders.[88]

It is unclear whether the same success would be seen in Australia. One noteworthy case is a shareholder action brought against the Commonwealth Bank of Australia (CBA) in 2017.[89] The litigating shareholders alleged that CBA had failed to disclose climate change-related risks in its 2016 annual report and therefore breached statutory disclosure rules. The case did not proceed to trial, but arguably can be seen as a success for the claimants because it was only withdrawn by them after CBA voluntarily reported climate change risks in its 2017 annual report.[90]

Though the applicable legal provisions differ, another relevant case is an action brought against REST Super for failing to disclose the climate change-associated risks of its superannuation products.[91] The case settled by REST agreeing to take further steps to identify and manage financial risks posed by climate change.[92]

More recently, following the leadership of its former governor Mark Carney, the Bank of England (the United Kingdom's central bank) continues to increase its warnings to other banks and corporations in respect to due diligence on climate change action and reporting.[93]

It is likely that increased scrutiny of disclosure will form part of a cultural shift rather than increase the arsenal of a public interest litigant's toolbox. Nevertheless, it is important to note that the disclosure requirement is enforceable – not just by concerned shareholders or customers, but also by the Australian Securities and Investments Commission (ASIC) which is able to prosecute public companies for failing to disclose financial risks,[94] or for making false or misleading statements in their disclosure.[95] The fact that ASIC's latest guidance suggests climate change risks

---

[88] *ClientEarth v Enea* (2019). See https://www.clientearth.org/world-first-climate-risk-case-launched-over-major-coal-plant-in-poland/

[89] *Abrahams v Commonwealth Bank of Australia* (case withdrawn)

[90] See Hutchens G (2017) Commonwealth Bank shareholders drop suit over nondisclosure of climate risks. The Guardian Australia.

[91] *McVeigh v Retail Employees Superannuation Pty Ltd* (case withdrawn). The plaintiff alleged a breach of *Corporations Act 2001* (Cth) s 1017C which requires a superannuation supplier to provide information to a concerned person if it is requested and relevant to their making an informed judgment about using the supplier.

[92] https://rest.com.au/why-rest/about-rest/news/rest-reaches-settlement-with-mark-mcveigh

[93] See the Bank of England's webpage on climate change, including discussions on financial disclosure and financial risk: https://www.bankofengland.co.uk/climate-change (accessed 14 Feb 2021).

[94] *Corporations Act 2001* (Cth) s 344.

[95] Ibid s 1308.

should form a part of these disclosures bolsters the potential for this area to promote greater corporate consciousness about, and action against, climate change.[96]

13.4.3 Corporations Law: Directors' Duties

A further development in Australian corporations law and practice is the emerging view that officers of corporations themselves should be held accountable for advice and foresight about developments affecting corporate responsibility for, and awareness about, the impact of climate change. While this may be a novel proposition, it grows out of orthodox statutory sources, including in Australia the provisions of the *Corporations Act 2001* (Cth) s 180 (1). In 2019 the former High Court Justice and Royal Commissioner, Hon. K.M. Hayne QC, summarised the emerging duty as follows:

> neither helplessness (whether that is learned or real) nor short termism provides any answer to the director's duty to act in the best interests of the company. Indeed, each points plainly towards the need for boards
>
> - to recognise both the nature and extent of climate related risks and the speed with which change will have to be made;
> - to develop strategic plans in response; and
> - to report to shareholders and the wider market about what they have done, are doing, and will do in response.[97]

This is a relatively new development. However, the sub-section of the *Corporations Act* relied on imposes general duties of care and diligence on the part of company directors. Given the fast-moving developments affecting potential corporate liability for environmental degradation, does this statutory provision (or the general legal duty of care and vigilance on the part of corporate officers) require new initiatives on the part of company officers to which attention needs to be drawn in the circumstances considered here?

In 2016, Centre for Policy Development (CPD) and the Future Business Council, through Minter Ellison lawyers, sought an opinion on the issue from a leading Australian lawyer and former President of the New South Wales Bar Association, Mr. Noel Hutley SC, co-signed by Mr. Sabastian Hartford-Davis. The lawyers' opinion could not be rejected as coming from a "bleeding heart" or "political" source. The question posed for their opinion was whether the statutory duty expressed in s 180(1) of the *Corporations Act 2001* could "be expected to feature in any future related litigation against company directors" so far as it relates to the

[96] ASIC provided more guidance to company directors in February 2021 with its "four high-level recommendations relating to climate-risk management and disclosure": see https://asic.gov.au/about-asic/news-centre/articles/managing-climate-risk-for-directors/. (Accessed 14 Feb 2021).

[97] See Fernyhough J (2019) Hayne rebukes directors on climate risk failure. Australian Financial Review; Sales P (2020) Directors' Duties in a Post-Hayne World: "The Company" as More than the Sum of Its Shareholders. Australian Law Journal 94: 936, 944–946.

responses of Australian corporations to the risks of climate change?" The opinion concluded that the answer to the question should be in the affirmative:[98]

In summary, the legal opinion stated:[99]

> 3.1 Climate change risks … are capable of representing risks of harm to the interests of Australian companies, which would be regarded by a Court as being foreseeable at the present time.
> 3.2 "Climate change risks" may be relevant to a director's duty of care and diligence to the extent [that] they represent a corporate opportunity or foreseeable risks to the company or its business mode.
> 3.3 For the avoidance of doubt, company directors are certainly not legally prohibited from taking into account climate change and related economic, environmental and social sustainability risks, where those risks are, or may be, material to the interests of the company.
> 3.4 To the contrary, company directors certainly *can*, and in some cases *should*, be considering the impact on their business of "climate change risks".
> 3.5 It is conceivable that directors who fail to consider "climate change risks" now could be found liable for breaching their duty of care and diligence in the future.
> 4. Directors who do turn their minds to the impact of "climate change risks" on their business will need to form their own assessment and make their own decisions as to what action, if any, is to be taken. This is likely to include obtaining and relying upon information and advice provided by employees and experts [*Corporations Act 2001* (Cth) s 189]. Directors, who are proactive in this regard, even if they decide on a properly informed and advised basis not to act, may have the protection of a statutory defence such as the "business judgment rule", under s 180(2) of the Act.
> 5. Finally, whether or not they decide to act, directors who perceive that climate change does present risks to their business should also consider the adequacy of the disclosure of those risks within the company's reporting frameworks.

In some ways, the situation now presented in relation to directors' obligations in respect of climate change risks is comparable to that faced in earlier decades in the minerals industry, in relation to the risks of exposure of investors, employees and customers to asbestos risks. Likewise, the risks presented to the manufacturers and suppliers of tobacco products, as the growing scientific consensus began to accumulate concerning the consequences of smoking tobacco or being otherwise exposed to that product. In the earliest days of research on asbestosis, mesothelioma and lung cancer, directors of a relevant corporation could reasonably plead the 'business judgment rule'. However, as increasing scientific research emerged and was publicised, that defence became increasingly unpersuasive. At least in the case of

[98] Hutley N SC and Hartford-Davis S (2016) Memorandum of Opinion – Climate Change and Directors' Duties. Centre for Policy Development and the Future Business Council, Melbourne, Victoria. The authors supplemented and re-affirmed this opinion in a *Supplementary Memoranda of Opinion* published on 26 March 2019 by the Centre for Policy Development. This relied on the increasing attention to climate risk and disclosure by Australian financial and regulatory bodies; the changes in financial reporting frameworks relevant to the disclosure of climate risk in Australia; the increased investor and community pressure on the topic; and notable development in the state of scientific knowledge. The fact that these developments had been widely publicised persuaded the joint authors to reaffirm the "exponential" increasing exposure of individual directors of corporations to climate change litigation.

[99] Ibid, 2 [3]–[4].

exposure to tobacco inhalation, it had to be replaced with a defence based on the provision of prominent warnings and arguments respecting individual informed rights to choose. Exposure to global climate change is not analogous to the perils of exposure of the individual to lung cancer. Individual decisions of consumers to avoid, or accept, the dangers of lung cancer from tobacco inhalation can scarcely be a substitute for the prudent, properly informed decisions of responsible corporate actors. This point was made in the joint legal opinion of Mr. Hutley and Mr. Hartford-Davis.

The authors' supplementary opinion in 2019 pointed to an increase in litigation risk as one of five key changes since their first opinion. Of particular note, according to the authors, was the advancement in "event attribution science" that could identify the link between climate change and individual extreme weather events. Further, both investor and community concerns about climate change has increased, with the Reserve Bank of Australia, the Australian Securities and Investment Commission and the Australia Prudential Regulation Authority all acknowledging the financial and economic significance of climate risks. This has prompted "shareholder actions on corporate climate targets and disclosures and co-ordinated efforts by institutional investors to target improvements by major emitters."[100] Those trends are in line with the successes in recent cases in the Netherlands, Germany and Poland discussed earlier in this chapter.

The addendum ultimately summarises its resulting position as follows:

> In our opinion, these matters elevate the standard of care that will be expected of a reasonable director. Company directors who consider climate change risks actively, disclose them properly and respond appropriately will reduce exposure to liability. But as time passes, *the benchmark is rising*.[101]

13.5 Conclusion

It would be unwise, in Australia and elsewhere, to wait for future litigation to reveal the neglect, indifference and carelessness of those in responsible positions of decision–making who have sometimes been sleepwalking while potentially important problems relating to climate change were presenting themselves. Such problems may be found by later generations to have required prudential attention in the here and now. Business leaders, politicians – but also judges and lawyers and leaders of civil society – need to be alert to these developments. It is inevitable that litigation will be brought in the future to assess the extent of legal liability on the part of relevant decision-makers. This is a trend line that is to be welcomed, not condemned; especially so if one considers the indifference, inaction and apathy on the part of officials engaged in the political process. The growing possibility of decision-maker

[100] Hutley N SC and Hartford-Davis S (2019) Supplementary Memorandum of Opinion – Climate Change and Directors' Duties, p 5.

[101] Ibid 3 (emphasis added).

liability is to be welcomed for the contribution it may make to the taking of individual steps by those with the power and responsibility to act with awareness and prudence so as to safeguard future generations.[102]

In Australia, a number of recent developments in the political, administrative, corporate and judicial spheres show that, like the ice in the polar regions, the law on climate change is moving. True, the movement in the law is slow. But it is generally in the one direction. Litigation and law will continue to do its part to reverse the trajectory and speed of the risks and dangers of climate change. Individual cases and decisions, even before final national courts (*Urgenda*) or involving enormous potential liability (*Luciano v RWE*) cannot alone reverse the perils of global climate change. However, just as the risks of climate change arise out of millions of individual decisions, escape from the dangers depends on a multitude of multilateral initiatives (Framework Convention) and wise and responsible individual and corporate decisions in litigation (*Rocky Hill*). None of us is exempt from the obligation to respond. And that includes judges and lawyers.[103]

[102] D'Amato A (1990) Do We Owe a Duty to Future Generations to Preserve the Global Environment? American J Int'l Law 84:190–198; Shelton D (1991) Human Rights, Environment Rights, and the Right to Environment. Stanford J Int'l Law 28:103–105.

[103] Preston B (2021) "Climate Conscious Lawyering". Australian Law Journal 95:51, 69–5.

Chapter 14
Paying for a Green New Deal: An Introduction to Modern Monetary Theory

Steven Hail

Abstract Modern monetary theory (MMT) is the brand name for a school of thought which has emerged since the year 2000 as the first serious challenge to orthodox macroeconomics in many years. Although the brand name dates back only to the 1990s, the economics has long historical roots, going back more than a century. It is based on a detailed analysis of the mechanics of modern monetary systems, and a model of the economy as a set of inter-locking balance sheets evolving over time, needing to be managed to ensure full employment and price stability are maintained and financial instability is avoided. The distinction between currency issuers and currency users is of central importance, as is the role of taxation in creating a demand for a currency and maintaining the value of that currency over time. MMT emphasises that unemployment is always a political choice and never a necessity, and that the key issue with the implementation of a Green New Deal is not paying for it but planning for and organising the real productive resources necessary to bring it about. What is technically possible can always be paid for by a monetary sovereign government – including a Green New Deal.

14.1 Introduction

The economist Hyman Minsky (1919–96) once wrote that "the game of policy making is rigged" because proposals are devised, and debates are conducted, using a frame for understanding the economy which is misleading and biased[1]. In planning for a Green New Deal (GND), it is crucially important to avoid using a biased frame. We need to unrig the game. A new approach to macroeconomics offers us the means to do so. It clarifies the issues that matter, distinguishing the real constraints

[1] In *Stabilising an Unstable Economy* (1986, page 110).

S. Hail (✉)
Business School, Torrens University Australia, Adelaide, Australia
e-mail: steven.hail@modernmoneylab.org.au

S. J. Williams, R. Taylor (eds.), *Sustainability and the New Economics*,
https://doi.org/10.1007/978-3-030-78795-0_14

with which we are faced from those that are imaginary. It turns out the problem is not how to pay for a GND. Finding the money is very easy. The difficulty lies in resourcing a GND.

Modern monetary theory (MMT) is this new approach. Although it has long historical roots, drawing on the work of Georg Friedrich Knapp (1842–1926), Alfred Mitchell-Innes (1864–1950), John Maynard Keynes (1883–1946), Michal Kalecki (1899–1970), Abba Lerner[2] (1903–82), Hyman Minsky[3], Wynne Godley[4] (1926–2010) and others (stretching back, in the cases of Knapp (1905) and Innes (1913, 1914), more than a century) it was only developed into a coherent school of thought in the 1990s. The first generation of MMT economists included Warren Mosler (1996), Randall Wray (1998), William Mitchell (1998), and Stephanie Kelton (Bell 2000, 2001), among others. These people independently re-discovered insights developed by those earlier economists. They then identified those historical roots, both modernising and adding institutional detail to what had gone before (Wray 2012a) and developing MMT into a coherent and powerful body of work (Kelton 2020). There have always been alternative perspectives to orthodoxy within macroeconomics, including the Post-Keynesian school[5] from which most MMT economists emerged, but none has had the momentum which MMT enjoys in the early 2020s (Mitchell et al. 2019).

The Global Financial Crisis of 2008 and the subsequent severe economic contraction in the euro-zone highlighted serious flaws in the orthodox approach. These were emphasised once again by the economic consequences of the Covid-19 pandemic of 2020. The neoclassical perspective, still dominant in most universities, provided no warning of financial crisis, no guide to the rapid restoration of full employment, no useful advice to those charged with reacting to the near collapse of the eurozone, and no recommended course of action for governments to pursue when reacting to a full-blown pandemic. Neoclassical macroeconomics would be worse than useless as a framework within which to plan for and execute a Green New Deal, in response to the threat of climate disruption and general ecological breakdown, in the years ahead. The state of orthodox macroeconomics, as leading US economist Paul Krugman (2009) admitted during the Great Recession, is not

[2] In a letter to fellow economist James Meade, written in 1943, Keynes said of Abba Lerner, the father of functional finance, and so for some, if not the grandfather of modern monetary theory, at least a favourite uncle, "His argument is impeccable. But heaven help anyone who tries to put it across" (Hail 2018 p143). Modern monetary theorists know exactly what Keynes meant by those words.

[3] Wray's *Why Minsky Matters* (2016) provides a comprehensive and accessible summary of the evolution and the significance of Hyman Minsky's work.

[4] The essentials of Godley's approach to macroeconomics, which is discussed in simplified form later in the chapter, as the sectoral-balances model, and are fundamental to MMT, can be found in Godley and Cripps (1983).

[5] The Post-Keynesian economists were those who retained the essentials of Keynes' approach to economics after his death in 1946. All his closest colleagues were Post-Keynesians, including Joan Robinson, Nicholas Kaldor and Richard Kahn. Lerner, Godley and Minsky were also Post-Keynesians, although Minsky described himself as a financial Keynesian.

good. Neoclassical macroeconomics, to use a phrase borrowed from another Nobel prize-winner, Paul Romer (2016), "is in trouble".

This is not the place to delve in depth into the many logical, empirical, and philosophical difficulties with what has, in one form or another, been the dominant paradigm for economic analysis for well over a century. Instead, our purpose is to describe the fundamentals of MMT, arguing that it is the superior description available of how economies work, and therefore should replace neoclassical macroeconomics. However, we do need to say a few words about that which MMT should replace, to clarify the reasons why a paradigm shift is urgently required.

14.2 Mainstream Macroeconomics in a Nutshell

A useful macroeconomics must get the monetary system right – the neoclassical approach does not. It developed out of what was originally a pure barter theoretical model, first fully described more than a hundred years ago, but with roots going back at least to Adam Smith (1723–90) and his notion of an 'invisible hand' guiding the economy. Consequently, money, banking, and finance, when they have been added to this model, have been peripheral to it. They have never been incorporated into neoclassical macroeconomics in a realistic way because there is no logical need for them there. Difficult though this is to believe, before the Global Financial Crisis it was common for macroeconomic models to ignore money entirely (Woodford 1998). This is less common today, but it remains the case that treasuries and central banks around the world continue to use models where the description of monetary and financial institutions differs in crucially important ways from the way those institutions work in the real world.[6]

According to these models, the kinds of financial crises with which we have become very familiar over the past forty years, and especially since 2008, should be impossible. According to them, there is nothing fundamentally wrong with designing a monetary system the way the eurozone was designed, with a single central bank in charge of monetary policy but no federal government to sit above it and run fiscal policy. In these models, mass involuntary long-term underemployment is impossible, as economies always tend towards full employment. And in these models, the distinction between a currency issuer and a currency user is either ignored or misrepresented. The neoclassical economists who continue to dominate our economic debates are still working, whether they realise it or not, with an essentially

[6] Leon Walras, author of Éléments d'économie politique pure (1874), replaced the famous 'invisible hand' of classical economics with a mythical neoclassical auctioneer. This auctioneer coordinates the demand for and the supply of every commodity in a market system, via a trial-and-error process called 'tatonnement', finding a vector of relative prices at which all markets within a barter system would clear. Walrasian general equilibrium theory, incorporating Say's Law, is fundamental to all orthodox macroeconomic theory. There is no role for money or realistic financial institutions in the Walrasian system.

barter model of a type of medieval artisan economy which never actually existed. On those occasions when they make sensible policy suggestions, it is despite that model and not because of it.

Economies have been based on money[7] for well over 5000 years (Graeber 2011), with the development of money and finance pre-dating and facilitating the emergence of markets and an exchange economy. This is not a new discovery. It has been widely understood outside economics for generations. Economies are monetary systems. Modern economies are sets of inter-locking balance sheets, evolving over time in ways that always have the potential to lead to growing financial fragility and eventual crisis. Modern capitalist economies are what Hyman Minsky called Wall Street economies. They are naturally unstable, with a tendency towards boom and bust, and to unemployment and inequality. They need to be managed and regulated. The dominant consensus over the past thirty years, that the wisest approach to economic management is to rely on changes in interest rates to stabilise the economy, with governments aiming to balance their budgets or run budget surpluses, has proved to be unsustainable. This should lead in the years to come to a paradigm shift away from neoclassical economics and towards MMT. Such a shift will enable policy debates to focus more on issues that really matter, like economic, racial, and ecological justice, free from the need for them to conform to a set of myths about government budgets.

14.3 The MMT Paradigm Shift

A new paradigm should predict and explain events which the alternative it is replacing has failed to predict or coherently explain. MMT economists and their colleagues predicted that the fiscal surpluses run by the Clinton administration during the late 1990s would not be sustained (Godley and Wray 2000); they predicted a major economic crisis in the mid-2000s (Godley and Zezza 2006); they identified the vulnerability of the eurozone to instability and crisis long before those vulnerabilities came to the surface in 2010 (Godley 1992); they explained that even large-scale quantitative easing by central banks would not be inflationary (Fullwiler and Wray 2010); and they have identified in 2020 that the responses of major governments to the economic consequences of the pandemic have been insufficient and poorly targeted (Kelton 2020).

Modern monetary theory is a lens for thinking about the economy, rather than a set of specific policy proposals, although as we shall see that lens is suggestive of policies that a government committed to equitable full employment and environmental sustainability might choose to pursue. It is primarily a description of the operations of monetary systems, and of the link between monetary system design

[7] For an explanation of what money is, see Mitchell-Innes (1913). For a modern discussion, see RBA 2018 "Money in the Australian economy", https://www.rba.gov.au/publications/bulletin/2018/sep/money-in-the-australian-economy.html

and the budgetary freedom of action or fiscal space of a federal or central government. A detailed understanding of monetary operations, and of the financial relationship between the government and the non-government sector within the monetary system is of central importance and must underlie any competent policy advice. This is true whether that advice is concerned with the risk of a financial crisis and the design of policies to stabilise the economy at full employment over time; or whether it is related to the risk of ecological breakdown and the management of real productive resources to build a sustainable economic system in time to avoid the worst-case scenarios for climate change.

Taking Australia as an example, the federal government is the monopoly issuer of the Australian dollar. There will always be a demand for Australian dollars in the private sector while tax liabilities to the government must be paid using that currency. It is this demand for Australian dollars in the private sector that allows the government to spend dollars into existence. Every dollar the government spends is a new dollar. *It is not tax-payers' money* – it is the government's money. The currency-issuing government does not need to get dollars from taxpayers before it can spend them; rather, taxpayers need to get dollars from the government before they can be used to pay taxes or make other payments to the federal government.

Taxation cannot come first and does not 'pay for' national government spending, in the sense that the government does not need to receive tax receipts before it spends new dollars into existence. In fact, it is precisely the other way around. The dollars in private bank reserve accounts at the central bank (such as the RBA in Australia) that are necessary for the payment of federal taxes, must either have been spent into existence by the government or lent into existence by its central bank before they can be used to pay federal taxes. You cannot pay taxes with non-existent dollars. Federal spending funds federal taxes. The taxes do not fund the spending.

This is hardly a new insight. The then chairman of the New York Federal Reserve, J. Beardsley Ruml, explained as long ago as 1946 that "taxes for revenue are obsolete", at least at the federal level (Ruml 1946). The government cannot run out of its own currency. The government does not need prior tax revenue to pay for its spending. Federal taxes have other functions: (a) to create a demand for the government's money; (b) to create room within the productive capacity of the economy for non-inflationary government spending (creating 'fiscal space'); (c) to discourage some activities, like smoking or the emission of greenhouse gases, and to encourage others, like renewable energy; and (d) to limit inequality.

Currency-issuing governments have the greatest discretion over how they use the power of the purse when they are full monetary sovereigns. A full monetary sovereign:

- issues its own currency;
- collects taxes in that currency;
- maintains a floating exchange rate; avoids borrowing in foreign currencies; and is not heavily dependent on imported necessities priced in foreign currencies.

A floating exchange-rate means that the government has no obligation to exchange its currency at a fixed rate for any foreign currency or commodity[8], or anything which is in limited supply. This is a necessary requirement for full monetary sovereignty. The maintenance of a fixed exchange-rate involves a commitment to use your reserves of foreign currency to purchase your own currency when necessary to support that exchange rate. Where foreign currency is concerned, the government is a currency user and not the currency issuer, and can quickly deplete its reserves in defence of the fixed exchange-rate, accumulate foreign-currency denominated debt, and/or be forced into a devaluation or a potentially severe currency depreciation[9] when it is no longer able to defend the fixed rate. This possibility limits the freedom the government has to use its budget to meet domestic policy objectives and implies that the government is not a full monetary sovereign.

A floating exchange-rate is not a sufficient condition for full monetary sovereignty when a government has a significant amount of foreign-currency-denominated debt. No government can be forced to default on debt which is to be repaid in its own currency. A government which has debt that is to be repaid in a foreign currency can be forced into default if it has insufficient foreign exchange reserves to make repayments; or into destructive austerity, exacerbating widespread poverty and impoverishing public services, if it requests support from the International Monetary Fund, as has occurred in many countries, and was the case for Argentina in 2018. Monetary sovereign governments do not have significant foreign-currency-denominated debts.

The monetary sovereignty of small, low-income economies is often compromised by a dependence on imported food, energy (especially oil), or other essential commodities priced in US dollars. This makes such economies particularly vulnerable to currency depreciation, since a depreciation raises the domestic-currency cost of these imports, raising the cost of living, and undermining social stability. This is another reason for a fear of floating exchange rates in such economies, with a resultant attachment to fixed exchange rates in terms of the US dollar or other reserve currencies.

Australia is obviously not a low-income economy and the Australia economy has over a long period shown itself to be highly resilient to exchange-rate volatility. The Australian dollar has been floating for nearly forty years. The federal government has no significant foreign-currency liabilities. This means that it is a fully

[8]A gold standard is where the government chooses to hold the price of gold constant, in terms of its currency, and promises to buy or sell gold whenever necessary to maintain a constant gold price. Where two countries are on a gold standard, the exchange rate between their currencies is fixed, depending on the price of gold in each country. The classical gold standard, including both the UK and the USA, existed only between the 1870s and 1914, and then briefly again between 1925 and 1931. After World War II, only the US dollar was on the gold standard, while other countries fixed their currencies against the US dollar, rather than directly against gold. The USA went off gold for the last time in 1971.

[9]A devaluation is when a government deliberately reduces a fixed exchange rate; a depreciation is when a floating exchange rate falls due to trading on the foreign exchange market.

monetary-sovereign currency issuer. The same thing is true of the governments of the USA, the UK, Japan, Canada and New Zealand, among others.

Such governments face no purely financial budget constraints. This does not mean there should be no limits to government spending, and it does not mean taxes are unnecessary, as we have already explained. It just means that when there are goods and services available for sale in return for domestic currency and when there are unemployed or underemployed workers available for hire who are prepared to work in return for that currency, the national government can always afford to buy those goods and services and to hire those workers, should it choose to do so. Unemployment and underemployment – collectively called labour underutilisation – are therefore a political choice of the government, and not an economic necessity imposed by a lack of dollars. Countries with monetary-sovereign governments could quickly eliminate involuntary unemployment if only their citizens would demand it, regardless of their existing unemployment rates.

A nation's productive capacity is limited by its supply of labour, skills, capital equipment, technology, infrastructure, natural resources, and institutional capacity. It is our productive capacity that limits what we can produce and consume, and not a lack of dollars. We admit this to students, at least implicitly, in the very first economics lecture they ever attend. When we discuss scarcity, choice and opportunity costs, we generally do so with respect to a diagram called a 'production possibility curve', which is supposed to represent the maximum quantities of goods and services it is possible to produce, in a fully employed economy. The location of the curve is determined by the availability of the real resources listed above and the state of technology. It has nothing to do with money. For a monetary-sovereign government, money is never a scarce resource. It is real resources and their current productive potential which are limited in supply.

14.4 The Role of Private Banks

The essential functions of commercial banks are to operate the payments system and to create and allocate credit. Financial deregulation in many countries during the 1980s and 1990s lowered barriers which had been put in place during the 1930s between commercial banking and high-risk investment banking, and encouraged banks to diversify into a wide variety of financial and related services in pursuit of earnings growth. While transformational banking reform seems at the moment unlikely, given the continuing political influence of the major banks, there is a strong case for a return to a more highly regulated, narrow banking system, where commercial banks concentrate once again on those two essential functions listed above (Wray 2016).

Some would go further. It is already the case that the payments system could operate without private banks. Central banks are planning for the introduction of e-currencies, and there is no reason transaction accounts could not be centralised, excluding the private banks from the payment system altogether. That leaves the creation of credit. The case for having private institutions play the major role in lending to the rest of the private sector rests on the supposition that they will do this more effectively than a governmental institution. If this is not the case, then there is a strong argument for the nationalisation of the main private banks. Given that they are too big to fail, that at least some of their liabilities are generally supported by a government guarantee, and that they are providing essential public services, it is not clear why they should reside in the private sector, in the absence of evidence that they make more efficient lending decisions.

However, if they are not to be nationalised, it is of great importance that private banks are regulated in such a way that they use their privileged position in the monetary system to contribute towards the development of a sustainable and equitable economy, by financing the capital investments necessary to bring that about, rather than adding to bubbles on property and equity markets, and creating unsustainable household debts (Ferguson 2019).

The fact that lending involves the creation of bank deposits, and that bank deposits make up almost the entire money supply according to most official monetary aggregates, is a point which is often raised by people when we talk about monetary sovereign governments being the monopoly issuers of their currencies. It causes a deal of confusion and sometimes leads people to advocate for well-meaning but unnecessary and perhaps ill-advised proposals for monetary reform, including the suggestion that private banks should no longer be permitted to create money when they lend.

It is virtually impossible to ban the private creation of money. Modern money is just a financial liability, or debt (or 'IOU'), which is widely accepted in exchange for goods and services. As Minsky said, anyone can create money. The problem lies in getting it accepted (1986, p. 228). I can create IOUs, as can Apple Inc, as can the Bank of America or the ANZ. My IOU for $50 is unlikely to be generally accepted in exchange for goods and services. This is mainly due to a lack of widespread confidence that I could be relied upon to convert them into government currency on demand. Apple Inc is in a much better position than me to issue tokens which might be generally accepted as having value – partly because they could be used in exchange for goods offered for sale by Apple, but also partly because a promise to pay $50 on demand issued by Apple would carry far more credibility.

IOUs issued by the Bank of America and ANZ have an even higher status, and their deposit liabilities are included in the official monetary aggregates of the USA and Australia, respectively. What gives a deposit at ANZ value in exchange is partly the fact that it can be used to repay a debt that you owe the bank. But more importantly, if you hold a deposit of Australian dollars at ANZ, you can rest assured that the bank will convert that deposit into physical currency (at par); or use its own deposit at the central bank (its exchange settlement account balance at the RBA) to make payments on your behalf to those who hold deposits at other banks, and to make payments on your behalf to the federal government (including the payment of tax liabilities). If the deposit is less than $250,000, the deposit comes with an explicit government guarantee. The currency-issuer stands behind the bank's

IOU. This is just one of the ways private banks and their IOUs enjoy a privileged position in our monetary system.

If private banks are to exist at all, they should play an important role in the sustainable development of the economy. Investment spending takes time and needs to be financed. Properly regulated and well-managed banks scrutinise business plans and provide funding to those ventures which are most likely to be successful and can be directed to prioritise activities consistent with a Green New Deal. Of course, they also provide finance to households in the form of mortgages and consumer credit. This is consistent with financial stability only if banks base lending decisions on reasonable expectations of future incomes and the ability to make repayments. There is every need to re-regulate and supervise the banking system to ensure that private banks operate in the public interest in the future and contribute towards neither financial nor ecological instability.[10]

14.5 The National Government's Budget

The notion that deficits and surpluses cancel out across the monetary system is not yet widely understood. It is an accounting fact of life which is of great significance, so it is worth explaining from first principles (Godley and Lavoie 2006). It is not only national governments that have budget deficits or surpluses. We all do. If I earn more than I spend this year, I will have a financial surplus. For me to run this surplus, someone else or some institution must be running a deficit, since my income is someone else's spending, and my spending is someone else's income.

It is common to aggregate across (1) the domestic private, (2) foreign and (3) government sectors, for the purposes of simplicity. Aggregating across the domestic private sector means adding together the financial positions of all households, business firms and other private-sector organisations. By doing this, you sacrifice some information, because if you are in substantial financial surplus, while I am rapidly going into debt, the private-sector sectoral balance will not show this, because our positions will cancel out. In this way, a sectoral-balance approach used naively could lead you to miss a build-up of financial fragility fuelled by inequality, because an apparently healthy consolidated private sector balance sheet may be hiding the growing indebtedness of many who are at risk of bankruptcy and default. However, the sectoral-balances approach is enormously useful, and the basis for the sectoral-balances chart. The chart is based on the sectoral-balances equation, which states that across the three sectors, financial balances always sum to zero.

$$\text{Domestic private} - \text{sector balance} + \text{government balance} + \text{foreign} - \text{sector balance} = 0$$

[10] Recommendations for the reform of banking regulation from a leading modern monetary theory economist can be found in Mosler (2011).

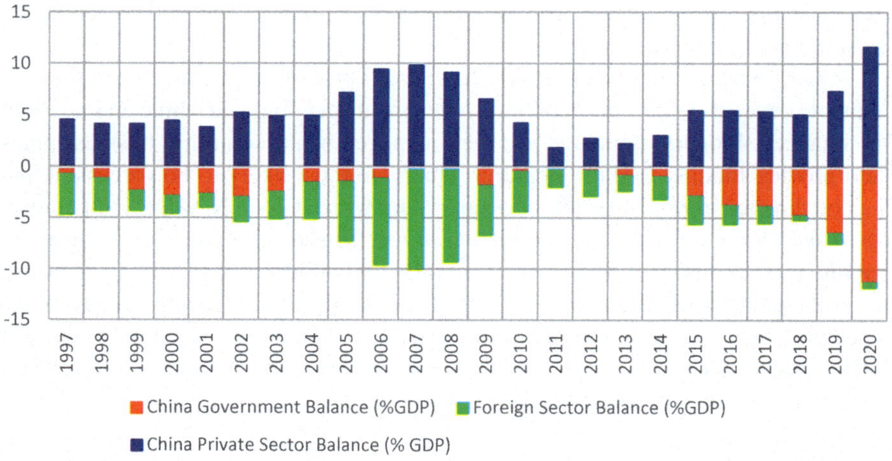

Fig. 14.1 Sectoral Balances for China 1997–2020, percent of GDP. (Data: IMF). (Source: all figures created by author)

Take, as an example, the following sectoral balances chart for China (Fig. 14.1).

Looking at the data for China between 1997 and 2020, the domestic private sector was in surplus across the entire period. These private surpluses were funded partly by foreign sector deficits, since China of course has run persistent trade surpluses in recent years, and partly by government deficits. If China had not run trade surpluses, it would have had to run larger government deficits to allow for the same level of private sector saving. For the domestic private sector to be in surplus, at least one of the other two sectors must be in deficit.

For a government to run a fiscal surplus (often called a budget surplus), then at least one out of the domestic private and foreign sectors must be in deficit. The foreign sector will be in deficit if the country is running a trade surplus (strictly speaking, a current account surplus on its balance of payments). A trade surplus means the rest of the world is running a deficit with this country. This country is exporting more, in value terms, than it is importing and therefore lending the difference to the rest of the world. If the country is not running a trade surplus, then the only way for the government to have a budget surplus is for the private sector to be in deficit. While the private sector can run deficits for a while, this is not the normal situation and is generally associated with rising private debt, often in the form of household debt. This was the case in the USA (Fig. 14.2) in the run up to the Global Financial Crisis, with the Clinton surpluses of the late 1990s and the inadequate fiscal deficits of the early 2000s having as their counterpart a private sector deficit driven by financial deregulation and growing financial fragility.

In the case of Singapore (Fig. 14.3), as with China the private sector has been in perpetual surplus in recent decades. However, the government has also been in surplus for most of the period since 1990. It has been Singapore's large trade surpluses over time which have funded the private and government surpluses. The foreign sector has been in deficit to Singapore, allowing the other two sectors to be in surplus.

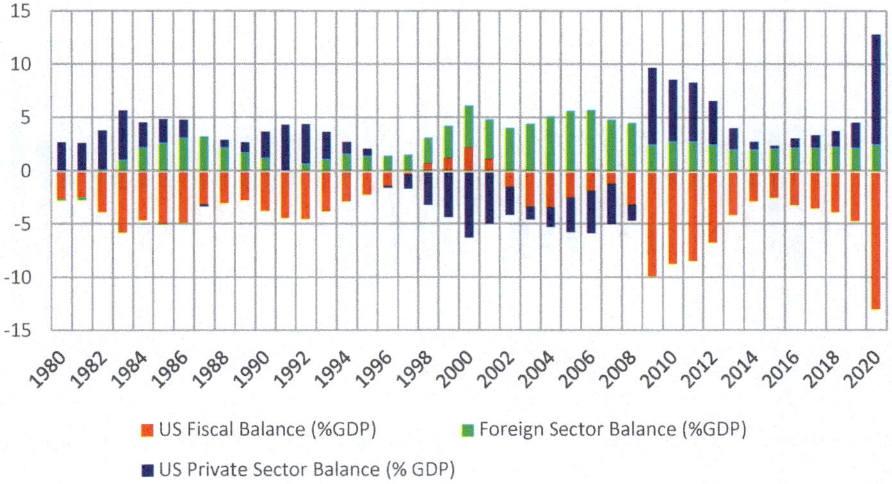

Fig. 14.2 Sectoral Balances for the USA 1980–2020, percent of GDP. (Data: IMF).

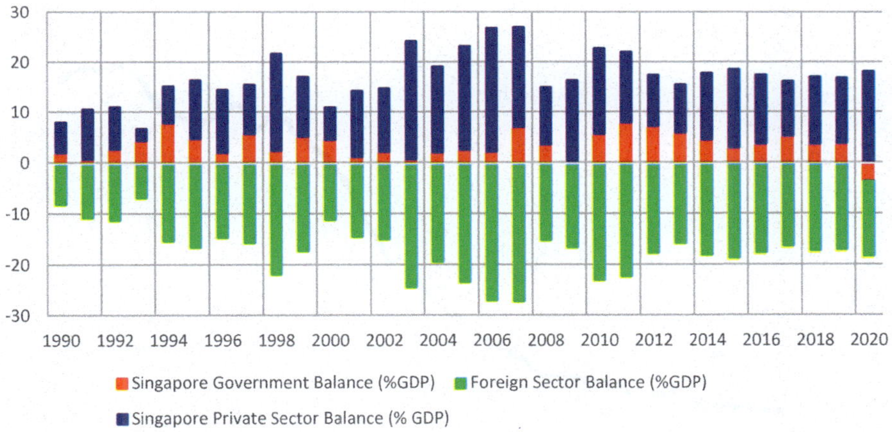

Fig. 14.3 Sectoral Balances for Singapore 1990–2020, percent of GDP. (Data: IMF).

In Australia (Fig. 14.4), the Howard government (1996–2007) ran budget surpluses for almost the whole time it was in power, while the country's trade deficit (again, strictly the current account deficit) reached record levels. So, the foreign sector was heavily in surplus at the same time the government was running a budget surplus. Only very large private deficits made this possible.

In those years, the borrowing was being done mainly by households, rather than businesses. Australia's household debt-to-GDP ratio went from a comparatively modest 40% to an almost world record level of more than 120% in just a few years (Fig. 14.5). Such a trend is unsustainable, and a consequence of an irresponsible fiscal policy. The government's much-trumpeted surpluses depended on the

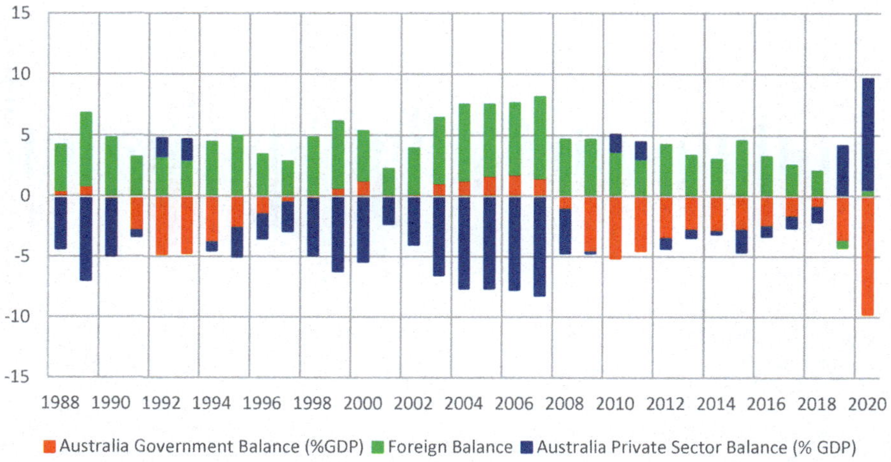

Fig. 14.4 Sectoral Balances for Australia 1990–2020, percent of GDP. (Data: IMF).

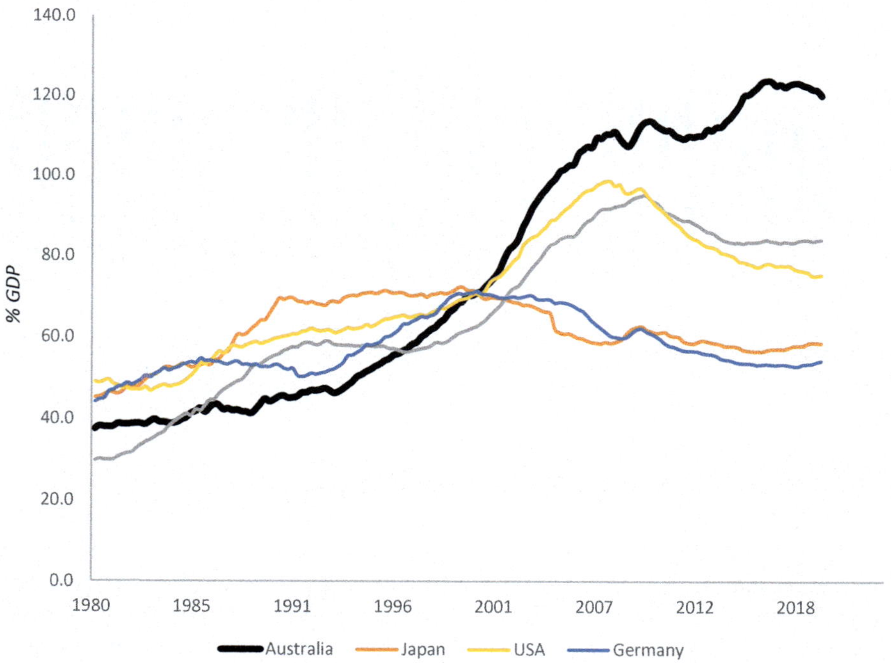

Fig. 14.5 Household Debt 1980–2019, percent of GDP. (Data: BIS).

encouragement of household borrowing in a deregulated and under-supervised financial system, which contributed towards increasingly unaffordable housing.

This is not to say that fiscal surpluses are never sustainable. Governments like those of Singapore, Norway or Germany can run budget surpluses indefinitely without driving their private sectors into deficit, but only by running trade surpluses with the rest of the world. As we have explained, a trade surplus means a foreign sector deficit on your monetary system, which can support a balanced budget or even a small budget surplus for the government, while meeting the net savings needs of the private sector. When you run a trade surplus, you are exporting the need for deficit spending to another country. But the essential point remains that for the government to run a budget surplus, somebody else somewhere must be deficit spending. In the case of Germany (Fig. 14.6), its growing current account surpluses during the early years of the eurozone had as their counterpart growing current account deficits elsewhere in the currency area, in countries like Spain (Fig. 14.7).

Given that the Spanish government was running budget surpluses just before the financial crisis of 2008, the deficit spending had to be happening in the private sector. It was associated with rising household debt, a property bubble, and growing financial fragility which contributed to the near collapse of the eurozone after 2010, and a prolonged economic slump in southern European economies.

Globally, the aggregate of all government deficits must equal the global private sector surplus. Private-sector financial surpluses are the norm in most countries because that sector prefers to net save. In aggregate, the 39 countries classified as advanced economies by the International Monetary Fund ran private sector surpluses continuously from 2001–20, almost going into a small deficit just before the financial crisis erupted (Fig. 14.8). If we want to avoid dangerous household debt and provide most people with economic security, the private sector will need to run surpluses almost all the time, based both on logic and on historical experience. In a

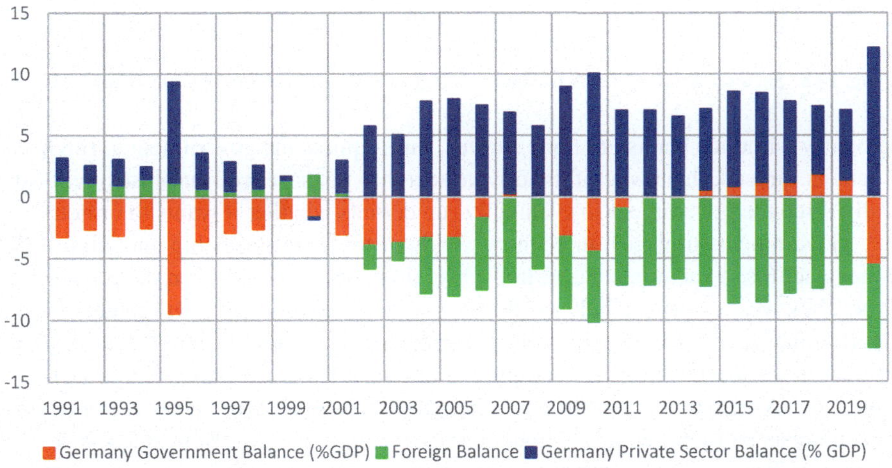

Fig. 14.6 Sectoral Balances for Germany 1991–2020, percent of GDP. (Data: IMF).

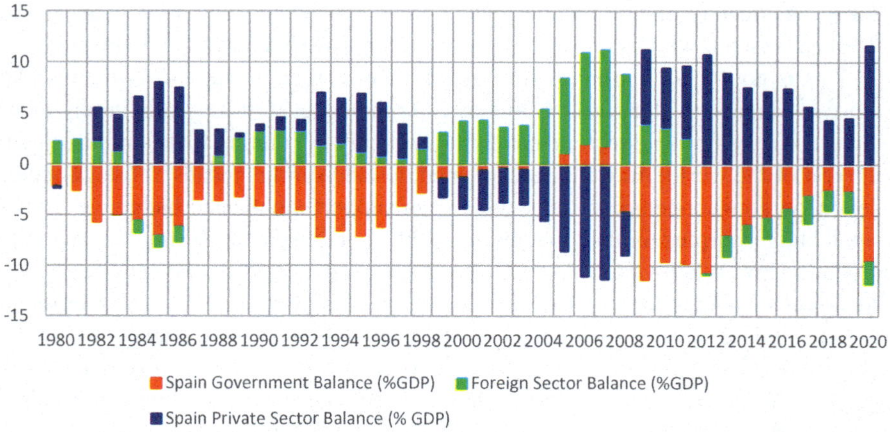

Fig. 14.7 Sectoral Balances for Spain 1980–2020, percent of GDP. (Data: IMF).

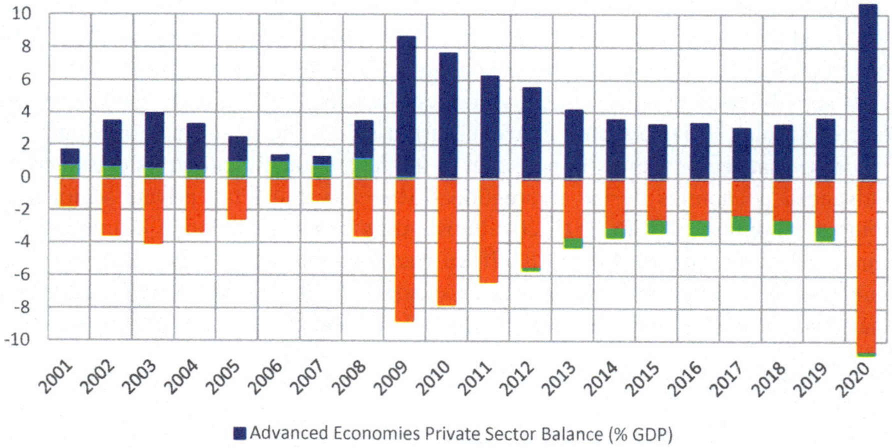

Fig. 14.8 Sectoral Balances for Advanced Economies 2001–20, percent of GDP. (Data: IMF)

country without a persistent trade surplus, this requires the government to run fiscal deficits. Indeed, the desire of the private sector to net save (run surpluses) will usually force the government to eventually run deficits whether it wants to or not.

Government surpluses, in countries without trade surpluses big enough to fund them, act like a vacuum cleaner, hoovering up financial assets from the private sector. They weaken private balance sheets, and in the absence of a willingness on the part of the private sector to take on more debt, drive economies into recession. This undermines the government's surplus as tax receipts crash and welfare spending increases. Given that this is the case, the question to ask is why so many politicians, economists and journalists continue to talk as if national government surpluses are a sign of fiscal rectitude, where 'balancing the budget over the economic cycle' is seen as good management? The answer lies partly in misconceptions about government debt.

14.6 Government Deficits and Debt

Under current arrangements, deficit spending adds to the government's net debt, which is conventionally (but misleadingly) referred to as the 'national debt'. This is because the government auctions treasury securities (bonds) to match the deficit. It is these bond sales that economists call 'debt'. But when bonds are issued to the private sector, they are purchased using currency the government or its central bank has previously spent or lent into the banking system. The government is 'borrowing' back its own currency. We have already explained that a currency-issuing government does not need to tax before it spends, and indeed that currency must be spent into the monetary system before it can be taxed back out of it. Similarly, the government does not need to raise money by issuing bonds before it can spend. The deficit spending must happen first, to supply reserves to the banking system which are then available to buy newly auctioned treasury bonds.

Bond issuance plays an important role for non-monetary sovereign governments. Governments which do not issue their own currencies, including members of the eurozone, are currency users, and are limited in their ability to spend by their tax receipts and their ability to borrow to finance deficits. For such governments, the conventional narrative applies, and they must tax or borrow before they can spend. Governments with currencies on a fixed exchange rate system or gold standard also need to issue bonds when they deficit spend. For example, on a gold standard, the government's reserves of gold limit the amount of currency it can issue, since the government has accepted a commitment to exchange its currency for gold at a fixed price.

This does not apply to a monetary sovereign government. Such governments do not need to issue bonds to fund spending. They do not need to issue bonds at all. They cannot run out of the currency they issue, and they have made no commitment to exchange that currency for anything they can run out of. They choose to issue bonds. That they generally continue to issue bonds even when they are running fiscal surpluses, as was the case in Australia in the early 2000s, suggests that they do not sell bonds purely to raise money. That governments continue to auction treasury bonds to private investors in the primary market at the same time as their central banks are buying them back on the secondary market, during periods of central bank quantitative easing, suggests the convention of bond issuance to match deficit spending is something of a muddle (Hail and Joy 2020).

One plausible motivation for bond issuance is in defence of the central bank's official interest rate target (Fullwiler 2006). Private banks use their reserves at the central bank to facilitate the millions of transactions which take place in the economy every day, and to allow the system to continue running smoothly, banks with excess reserves lend to banks which have run short of reserves. Central banks normally target the rate of interest on this short-term (often overnight) inter-bank lending. Government deficit spending puts additional reserves into the banking system, and if those reserves are left in the system, most banks have excess reserves, few if any banks need to borrow reserves, and the interest rate on inter-bank loans falls below the central bank's target. To defend the target, excess reserves due to deficit spending need to be removed from the system by swapping those reserves for newly issued treasury bonds. Viewed this way, the issuance of treasury bonds is best thought as contributing towards the monetary policy (interest rate management) of the central bank and not the fiscal policy (deficit spending) of the national government.

In Australia, when the policy of auctioning bonds to cancel out deficit spending was first introduced, in 1982, the central bank and the treasury announced that it was for precisely the reason given above, although few people are aware of this fact today (Australian Office of Financial Management 2011). Moreover, whenever quantitative easing has been introduced by a central bank – in Japan in 2001, the USA in 2008, the UK in 2009, and in Australia in 2020 – the central bank has either lost control of its policy rate or changed the interest rate it has targeted, because quantitative easing creates the same excess reserves in the banking system which would have resulted from deficit spending without bond issuance. This is because quantitative easing involves the central bank purchasing bonds (and sometimes other financial assets) from private banks and non-bank financial institutions, paying for those bonds by crediting private bank reserve accounts. The net effect is as though those bonds had never been sold to the private sector at all.

Another reason for issuing bonds is to offer private-sector investors a safe interest-bearing asset which they can add to their portfolios. Treasury bonds generally offer investors a better rate of return than bank deposits, while being completely free of default risk. Some have described this as corporate welfare for the finance sector (Mitchell 2010). It was pleas from fund managers which persuaded the Australian government to continue auctioning treasury bonds during the years of budget surpluses (Australian Treasury 2002).

A third possible reason is so that the interest rates on treasury bonds can act as benchmark interest-rates, with reference to which the rates of interest on private sector debts can be set.

None of these reasons are compelling. It would be more transparent to replace treasury bonds with term deposits at the central bank if there is a need to offer safe assets to fund managers and to provide a benchmark for financial markets. The advantage of this institutional change would be to clarify the fact that a monetary sovereign does not need to borrow and cannot be said to have a debt at all, in the normal sense of the term. The scale of the 'debt' is often used to scare people – the public and politicians alike – into tightening the purse strings (austerity), and to drive commitments to balance the budget, or move 'back into surplus'. But there is no reason to be scared. All it would require to 'repay the debt' is to type reserves into the banking system using keystrokes on a computer. And far from being a crazy idea, this could be combined with changes to the monetary system which would clarify the nature of government financial liabilities and the appropriate role for the government budget to play in the economy.

The government's debt is not the citizens' debt. It is simply dollars that the federal government has spent into existence and not yet taxed back out of existence again. It is not something which ever needs to be repaid. It is correctly defined as the net financial assets of the non-government sector. More informally, it is best thought of as the net money supply.

The orthodox response to this is to point out that higher levels of government debt lead to increasingly costly interest payments to investors who have purchased government debt securities, which it is said will either divert funding from desirable public goods in the future or impose a tax burden on future generations. Once again, this is misleading, and the result of a misunderstanding of the mechanics of the monetary system.

Government deficit spending today does nothing to limit the ability of the government to spend in the future. Like all dollars the government spends, dollars used to make interest payments are new dollars. They are mildly stimulative, because investors who own the bonds spend part of their interest income on goods and services, but only mildly so, because most of it is saved. Some economists fear that an increase in the government debt to GDP ratio will lead to a downgrading of the government's credit risk by Standard & Poor's, Moody's and other credit-rating agencies, and that this might force up the interest rate the government has to offer on its bonds. They further claim that if the interest rate on government bonds exceeds the growth rate of GDP, then even governments with no primary deficit, defined as the gap between government spending excluding net interest payments and tax receipts will see this ratio rise continuously, until eventually the bond market forces the government into austerity, by refusing to 'finance' its debt any longer (Fischer and Easterly 1990).

These economists have not yet understood the distinction between a monetary sovereign and a non-monetary sovereign government. They are entirely correct where non-monetary sovereigns are concerned, and until the European Central Bank stepped in to prevent the collapse of the euro in 2012, the above narrative applied to Greece and some other eurozone countries. Non-monetary sovereigns, remember, are currency users, not currency issuers. They must acquire currency before they spend it, and when they get into debt they risk financial distress and insolvency. There is always a risk of default for eurozone countries if they are not supported unconditionally by the currency-issuing European Central Bank. The economist Wynne Godley (1992) presciently warned about the likely consequences of this design flaw with the eurozone, seven years before the currency was even launched.

This is not the case for monetary sovereigns. Such governments can ignore credit-rating agencies. We have seen that they may have good reasons to issue interest-bearing government securities to match their deficit spending, but there is no risk they could ever be forced into default, on debt denominated in the currency they issue. As central banks like the Bank of Japan have shown in recent years, if this needed to be shown, the interest rates offered on treasury bonds can be a policy variable, if the central bank wants to use its balance sheet to make this so, and not determined by the bond market at all. The bond market has no real power over a monetary-sovereign government. During the 2020 pandemic, the Reserve Bank of Australia has demonstrated this anew, buying government securities in sufficient quantities to set and hold the yield on treasury bonds with a residual maturity of three years at 0.10% per year.

Now we have established that a monetary sovereign government does not have to raise revenue before it can spend, is never obliged to balance its budget, and cannot be held hostage by the bond market, it is time to consider what such a government should do with the fiscal freedom it enjoys.

14.7 What a Prudent Monetary-Sovereign Government Should Do

A responsible government should seek to balance its economy – not its budget. The fiscal stance of such a government should stabilise the economy at or near non-inflationary full employment.

While the government and its central bank can influence the net savings desires of the private sector at full employment, they cannot determine them. Similarly, while the government can influence the trade deficit (or surplus) at full employment, this also depends on the desire of the rest of the world to net save on the domestic monetary system, which the government cannot control. This implies that the appropriate level of the fiscal deficit is largely outside the government's control and determined endogenously by the behaviour of the private and foreign sectors. There are even circumstances when it may be appropriate for a government to run fiscal surpluses, particularly in countries with significant trade surpluses, in order to limit total spending to the productive capacity of the economy, but these circumstances will be rare. It is normal, natural and nothing to fear for the government to run a deficit. A government deficit is, after all, just a non-government surplus. And the non-government sector, including the domestic private and foreign sectors, will normally need to run surpluses.

The issue should be the purpose for which that deficit spending is being undertaken, and the location of the resulting surpluses in the private sector (Kelton 2020). It is clearly vitally important whether the government is investing in green infrastructure or in subsidies to fossil-fuel industries. It matters a great deal whether the dollars have been deposited into the bank accounts of billionaires or those who would otherwise be living in poverty. But whether deficit spending has been in the interests of the many or the few, and whether it has been in pursuit of a Green New Deal or the continuance of extractive and pollutive economic growth, it remains the case that the government's deficit is dollar-for-dollar equal to the increase in the net financial assets of the rest of us: what we have called the net money supply. A fiscal deficit this year adds to that net money supply. A fiscal surplus destroys dollars and reduces it.

A common source of concern is that irresponsible politicians would abuse the potential spending power that this knowledge provides, possibly leading to accelerating inflation. Some would prefer we pretend the government is not a currency issuer, as a means of imposing discipline on what might otherwise be profligate governments. This was the view of the pre-eminent US economist Paul Samuelson (1915–2009), who in an oft-referenced interview compared the myth that the US federal government could run out of its own currency to an old-fashioned religion[11], arguing that to dispel the myth would remove an essential source of discipline in a civilised society (Wray 2012b). On this view, these moral guardians of what the public can be trusted to know have to lie to protect society from itself. The hubris is breathtaking.

One reaction to this is that it is enough to say that MMT is technically correct, and this alone justifies its use by those of us who wish to understand macroeconomic issues and contribute towards policy development. Another is to assert that the consequences of this hubris have been insecure employment, high unemployment, ballooning private-sector debt, excessive and growing wealth inequality, deteriorating infrastructure, and a lack of action on climate change. In any case, in a democracy we elect politicians on the basis that they should act in the public interest, and should they fail to do so, it is up to us as voters to punish them for their failures. It is not up to the economics profession to mislead us.

Instead, economists should assist policymakers and the democratic process by advising on the potential consequences of fiscal proposals for a range of economic variables, including expected inflation. At the moment, bodies like the Congressional Budget Office (in the USA), the Parliamentary Budget Office (in Australia), and the Office of Budget Responsibility (in the UK), concentrate on the impact of spending proposals or changes in the tax system on the budget balance itself, rather than the consequences for output, employment and inflation. A change in focus, to the inflationary impact of a policy programme, and the provision of independent inflation forecasts relating to rival political programmes, could discipline governments and opposition parties without the use of an old-fashioned religion based on a fiscal myth, and without subcontracting responsibility for limiting inflation to unelected officials at the central bank.

But the following question remains: if we say that the government should use its power as a currency issuer to balance the economy (not the budget), and by balancing the economy we mean achieve and maintain non-inflationary full employment, how can we best go about achieving that outcome?

14.8 A Government Job Guarantee Scheme

A monetary-sovereign government can permanently eliminate involuntary unemployment via a job guarantee scheme, thus achieving true full employment (Minsky 2013): that is, 1% or 2% unemployment and zero underemployment. It is hard to

[11] Samuelson's discussion of the balanced budget myth can be found here: https://www.youtube.com/watch?v=4_pasHodJ-8

think of a more important policy innovation, given the destructive nature of unemployment and poverty (Tcherneva 2020).

A job guarantee involves the offer of a full- or part-time public sector job to anyone not otherwise in full-time employment who chooses to take up the offer, at an equitable base wage, with decent working conditions designed to set the floor for acceptable employment conditions elsewhere in the economy. Such a job guarantee would be funded by the national government, but administered locally, to meet local needs. Those needs would centre around caring for the environment and the community, while not competing with private firms. The job guarantee would also be separate from the regular public service. There is no shortage of ideas about suitable activities that could be incorporated into a job guarantee program. There are many useful tasks which are not currently being undertaken, as they are not profitable for the private sector and cannot under current arrangements be funded by local governments.[12]

Job guarantee spending would act as a superior cyclical stabiliser to our current system. In a downturn, people who had lost their jobs in the private and conventional public sectors would be able, if they chose, to join the job guarantee. The income they received as part of the job guarantee would support their spending, and therefore the local economy, and push the government budget further into deficit, just as this became necessary. Moreover, this would happen with no policy lag, and without any bureaucrats having to forecast and plan for a downturn in advance. In a recovery, people would transition from the job guarantee into better-paid jobs elsewhere in the economy, which would reduce the size of the deficit when it was appropriate to do so.

Meanwhile, those who would otherwise have become unemployed would have had the opportunity to remain in employment, carrying out socially productive and rewarding tasks, with the opportunity, but not the obligation, to engage in training in areas of local skill shortages and to be matched with potential employers, while in the program.

A job guarantee is nothing like compulsory schemes where people are forced to work for unemployment benefits. It is far from 'work for the dole'. It would be optional, providing a real work opportunity, rather than an obligation, to anyone who was unemployed. A job guarantee was discussed widely as part of the US 2020 presidential election campaign. Proposals have existed for such a scheme in Australia for many years (Cook et al. 2008). A large-scale job guarantee-like program has operated in India since 2005 (Kareemulla et al. 2013), and there have been a number of other successful public sector employment schemes with some of the features of the MMT job guarantee introduced around the world in the relatively recent past (Tcherneva 2012; Kaboub 2007).

While there is no 'one size fits all' job guarantee which could apply in every country, and while there would be variations in the jobs performed even in different

[12] The budgetary impact of a job guarantee would differ over time, depending on the state of the economy, but the net cost would normally lie between 1% and 3 or 4% of GDP. A costing in Australia during the Covid-19 pandemic was provided in Mitchell and Watts (2020).

parts of a single country, should such a scheme be introduced, the MMT job guarantee has certain essential features:

- It is a permanent scheme, and not a temporary crisis measure.
- It hires mostly those with less education and skills, at a wage that allows for full participation in society.
- It is available to all, compulsory for none, and allows the otherwise unemployed to remain and work in their local communities.
- It is scalable – growing and shrinking counter-cyclically.
- It stabilises the economy across the economic cycle, acting as an inflation anchor.
- The appropriate level of the fiscal deficit to bring about non-inflationary full employment is determined by the number of people who walk into a job guarantee office and take up a place in the program.
- It is an essential feature of any policy program aimed at combining social justice with ecological sustainability (Forstater 2003).

For all these reasons, a job guarantee is to be preferred to a stand-alone universal basic income (UBI). A UBI would have to be set at a level not sufficient to eliminate poverty; would require changes to the tax system which would be unlikely to meet with public support; or would be inflationary, due to the scale of transfer spending involved. Moreover, it would not guarantee everyone the opportunity of paid employment, and crucially has no counter-cyclical stabilising properties. A job guarantee is a core part of MMT: to ignore it is to support the only other alternative, which is the current 'buffer stock' of the involuntarily unemployed (Mitchell 1998). This is economically inefficient and morally bankrupt. There should be no surprise that all leading MMT economists advocate for a universal and permanent job guarantee. It is as much a part of MMT as the non-accelerating inflation rate of unemployment (NAIRU) is a part of modern neoclassical macroeconomics.

14.9 Paying for a Green New Deal

We have established that balancing the government's budget, or aiming for a surplus, never makes sense as a policy target. We have explained that what limits the sum of government and private sector spending which is consistent with a stable inflation rate is the productive capacity of the economy. The productive capacity of the economy depends on the availability of real resources and the state of technology. We have argued that non-inflationary full employment can be achieved with the assistance of a well-designed job guarantee. We have not yet engaged properly with what are clearly the most significant issues ever to confront economic policymakers – climate change, ecological sustainability more generally, and a just transition to sustainable prosperity for all.

Any suite of policies designed to address these issues is now generally labelled as a Green New Deal (GND). The 'Green' part of a GND is about transitioning energy systems away from fossil fuels and moving society away more generally from a dependence on highly extractive, pollutive and destructive industries. The 'New Deal' part is about addressing the causes and consequences of income and wealth inequality, broadening the opportunities available to people to live a full and rewarding life, and offering economic security to all.[13]

As with any set of proposals, the reflex reaction of an orthodox economist to a GND policy package is to ask: "how are you going to pay for it?" You would pay for a GND the same way you pay for all government spending. You would pass a budget through parliament, or congress, and that would be it. Each dollar the government then invested in the GND would be a new dollar. Where spare resources existed in the economy or could be diverted from parts of the economy which will naturally shrink or even disappear as part of a GND, and where there is no reason to expect the spending to be inflationary, that would be the end of the story. If government investments had to be on a scale that risked driving the economy beyond its productive capacity with subsequent rising inflation, then offsets would need to be found. One such offset might be higher taxes for those who could afford them, but that would not be the only available option, and would often not be the first preference (Nersisyan and Wray 2019). The question policymakers committed to a GND should address is not whether they can find the tax revenue to 'pay for' their proposals. It should be whether the real resources exist or can be created to build the green infrastructure, transition the power system, broaden access to health care and to higher education, eliminate homelessness and poverty, and whatever other priorities they might choose to set, across their chosen time horizon. It is this simple point – that a GND is about real resources and resource planning, and not about whether you can find the money to pay for it – that is the essential insight provided by an MMT perspective.

As Keynes (1936) once said, "the ideas which are here expressed so laboriously are extremely simple and should be obvious". The Parliamentary Budget Office in Australia, and the Congressional Budget Office in Washington D.C. on which it was modelled, should drop their focus on projected fiscal balances years into the future, and instead get down into the details of the economy's available real resources and discuss the inflation risks and environmental risks associated with using fiscal policy for proper public purposes. Politicians should stop talking about returning the budget to surplus and paying down debt and concentrate on things that actually matter – the delivery of ecological sustainability, equitable full employment, and excellent and resilient public services. We should stop basing our economic policies on an old-fashioned religion and start to see the economy through the superior lens of modern monetary theory.

Then, as Naomi Klein (2020) put it, in words which are reproduced on the front cover of Stephanie Kelton's *The Deficit Myth*, we will have "the tools we desperately need to build a safe future for all". It will be up to us to put them to use.

---

[13] A discussion of a Green New Deal from the perspective of a leading modern monetary theorist can be found here: https://www.youtube.com/watch?v=AaAlJXaWTEI

References

Australian Office of Financial Management (2011) A history of treasury bond tenders and performance. https://www.aofm.gov.au/publications/research/history-treasury-bond-tenders-and-performance. Accessed 13 Jun 2020

Australian Treasury (2002) Review of commonwealth government securities market. https://treasury.gov.au/sites/default/files/2019-03/CCSMReview.pdf. Accessed 13 Jun 2020

Bell (Kelton) S (2000) Do taxes and bonds finance government spending? J Econ Issues 34(3):603–620

Bell (Kelton) S (2001) The role of the State and the hierarchy of money. Camb J Econ 25(2):149–163

Cook B, Mitchell W, Quirk V, Watts M (2008) Creating effective local labour markets: a new framework for regional Economic Policy. Centre of Full Employment and Equity, Newcastle

Ferguson A (2019) Banking bad. HarperCollins, Sydney

Fischer S, Easterly W (1990) The economics of the government budget constraint. World Bank Res Obs 5(2):127–142

Forstater M (2003) Public employment and environmental sustainability. J Post Keynesian Econ 25(3):385–406

Fullwiler S (2006) Setting interest rates in the modern money era. J Post Keynesian Econ 28(3):496–525

Fullwiler S (2017) Modern Central Bank operations—the general principles. In: Rochon L-P, Rossi S (eds) Advances in endogenous money analysis. Edward Elgar, Cheltenham, pp 50–87

Fullwiler S, Wray LR (2010) Quantitative easing and proposals for reform of monetary policy operations. Levy Economics Institute of Bard College, Annandale-on-Hudson

Godley W (1992) Maastricht and all that. Lond Rev Books 14(19):3–4

Godley W, Cripps F (1983) Macroeconomics. Oxford University Press, Oxford

Godley W, Lavoie M (2006) Monetary economics—an integrated approach to credit, money, income, production and wealth. Palgrave Macmillan, New York

Godley W, Wray LR (2000) Is Goldilocks Doomed? J Econ Issues 34(1):201–206

Godley W, Zezza G (2006) Debt and lending: a Cri de Coeur. Levy Economics Institute of Bard College, Annandale-on-Hudson

Graeber D (2011) Debt: the first 5,000 years. Melville House, New York

Hail S (2018) Economics for sustainable prosperity. Palgrave Macmillan, New York

Hail S, Joy D (2020) Federal debt and modern money. Global Institute of Sustainable Prosperity. http://www.global-isp.org/policy-note-121/. Accessed 6 Jan 2021

Kaboub F (2007) Employment guarantee programs: a survey of theories and policy experiences. Levy Economics Institute of Bard College, Annandale-on-Hudson

Kareemulla K, Ramasundaram P, Kumar S, Rama Rao CA (2013) Impact of national rural employment guarantee scheme in India on Rural poverty and food security. Curr Agri Res J 1(1):13–28

Kelton S (2020) The deficit myth: modern monetary theory and the birth of the people's economy. Public Affairs, New York

Keynes JM (1936) The general theory of employment, interest and money. Macmillan, London

Klein N (2020) Review of the deficit myth. Modern Monetary Theory and How to Build a Better Economy by Stephanie Kelton. https://www.politics-prose.com/book/9781541736184. Accessed 6 Jan 2021

Knapp GF (1924) The State theory of money. Augustus Kelley, New York

Krugman P (2009) How did economists get it so wrong? New York Times Magazine, 2 September 2009

Lerner A (1943) Functional finance and the federal debt. Soc Res 10(1):38–51

Lerner A (1947) Money as a creature of the State. Am Econ Rev 37(2):312–317

Minsky HP (1986) Stabilising an unstable economy. Yale University Press, Boston

Minsky HP (2013) Ending poverty: jobs, not welfare. Levy Economics Institute of Bard College, Annandale-on Hudson

Mitchell W (1998) The buffer-stock employment model and the NAIRU: the path to full employment. J Econ Issues 32(2):547–555
Mitchell W (2010) Market participants need public debt. http://bilbo.economicoutlook.net/blog/?p=10404. Accessed 13 Jun 2020
Mitchell W, Watts M (2020) Investing in a job guarantee for Australia. Centre of Full Employment and Equity, Newcastle
Mitchell W, Wray LR, Watts M (2019) Macroeconomics. Macmillan, London
Mitchell-Innes A (1913) What is money? Bank Law J 30:377–408
Mitchell-Innes A (1914) The credit theory of money. Bank Law J 31:151–168
Mosler W (1996) Soft currency economics. Valance Economics, U.S. Virgin Islands
Mosler W (2011) Proposals for the banking system Huffington Post. https://www.huffpost.com/entry/proposals-for-the-banking_b_432105. Accessed 6 Jan 2021
Nersisyan Y, Wray LR (2019) How to pay for the green new deal. Levy Economics Institute of Bard College, Annandale-on-Hudson
Romer P (2016) The trouble with macroeconomics. https://paulromer.net/trouble-with-macroeconomics-update/WP-Trouble.pdf. Accessed 13 Jun 2020
Ruml JB (1946) Taxes for revenue are obsolete. Am Aff 8(1):35–39
Tcherneva P (2012) Beyond full employment: what Argentina's plan Jefes can teach us about the employer of last resort. In: Murray M, Forstater M (eds) Employment guarantee schemes job creation and policy in developing countries and emerging markets. Palgrave Macmillan, New York, pp 79–102
Tcherneva P (2020) The case for a job guarantee. Polity Press, Cambridge
Woodford M (1998) Doing without money: controlling inflation in a post-monetary world. Rev Econ Dyn 1(1):173–219
Wray LR (1998) Understanding modern money – the key to full employment and price stability. Edward Elgar, Cheltenham
Wray LR (2012a) Modern Money theory—a primer on macroeconomics for sovereign monetary systems. Palgrave Macmillan, New York
Wray LR (2012b) Paul Samuelson on deficit myths. http://neweconomicperspectives.org/2010/04/paul-samuelson-on-deficit-myths.html. Accessed 13 Jun 2020
Wray LR (2016) Why Minsky matters—an introduction to the work of a Maverick economist. Princeton University Press, Princeton

Chapter 15
Conclusion and Policy Options

Stephen J. Williams and Philip Lawn

Abstract This concluding chapter synthesises the information presented throughout the volume and restates the thesis from the introductory chapter: that humanity probably has a decade to fundamentally change its adversarial relationship with the planet – and each other – if disaster is to be avoided. Our current approach amounts to a 'war on nature' when 'harmony with nature' is required. To achieve this, the discipline of economics, especially, needs to change course towards the principles of ecological economics (which we may well call real-world economics, where biophysical limits are respected). We advance a range of policies to inspire discussion, and call on all governments to declare a global emergency in line with recent authoritative reports.

> A nation is achieving sustainable development if it undergoes a coevolutionary process that improves the total quality of life of every citizen, both now and into the future, while ensuring its rate of resource use does not exceed the regenerative and waste-assimilating capacities of the natural environment. It is also a nation that ensures the survival of the biosphere and all its evolving processes while recognising, to some extent, the intrinsic value of sentient non-human beings (Lawn 2016).

Humanity has largely ignored warnings of a looming disaster for at least half a century (see Chap. 1 of this volume). As a consequence, we now face a compounding existential emergency of the most serious kind (UNEP Report 2021). Instead of a planned and gradual shift to sustainable systems from the 1970s, we now need to urgently invoke emergency responses to avoid worst-case scenarios. Whether there is yet the collective wisdom and momentum to save humanity from mass suffering is moot. While there is increasing awareness of the interrelated problems of overshoot, both the speed and depth of an effective response is not yet evident to us. We may have ten short years for a fundamental paradigm shift, where the principles of

S. J. Williams
Silverwater, NSW, Australia

P. Lawn (✉)
Measuring Genuine Progress, Adelaide, Australia

Torrens University, Adelaide, Australia

S. J. Williams, R. Taylor (eds.), *Sustainability and the New Economics*,
https://doi.org/10.1007/978-3-030-78795-0_15

ecological economics are universally accepted and used to inform policy, remembering that policies can only be as good as the assumptions that underpin them.

Whether capitalism itself can adapt to a non-growth paradigm (and a degrowth paradigm in some high-income nations) is yet to be seen, but its existing form, which demands bigger profits and rising rates of resource throughput, is neither sustainable nor existentially desirable. Any successful response will need broad community support, either as the instigating mechanism or as an upshot of effective global leadership – but more likely a combination of both.

In the country where we live – Australia – the issue of sustainability and the risk of systemic collapse is essentially non-existent in current politics or the mainstream media. Arguing about the timing of net zero emissions, or permissible extraction quantities of water from the nation's main river system, or what the immigration quota should be, or the possible extinction of koalas – all conducted in a piecemeal manner – is not an holistic discussion about sustainability. As the Covid-19 pandemic continues with much physical, mental, and economic harm, the political discussion is mostly about returning to 'growth' as quickly as possible, with almost no understanding that growth is firmly implicated in the pandemic due to humanity's disruption of biological systems (Carrington 2020; Dasgupta 2021, p1; UNEP Report 2021).

The starting point for an effective response is for governments to follow the advice of relevant scientific experts, rather than mostly deferring to economic analysis that, since the rise of neoliberalism in the 1970s, has patently failed to deliver universal wellbeing. The Covid-19 experience has clearly shown that the nations which have navigated the pandemic best are those which have followed scientific advice. Climate change, the biodiversity crisis, and similar problems all require scientific – which is to say evidence-based – solutions. At the heart of any solution will be the need to accept and calculate the sustainable rate of resource throughput relative to biocapacity and to remain within scientifically estimated limits. This implies limiting the physical size of the economy to one that can be maintained by a sustainable rate of resource throughput.

The correct attitude must be this: instead of viewing a new paradigm centred around emergency responses as a sacrifice, we should see it as advancing national and global wellbeing and rush enthusiastically towards it. To not do so, is to invite disaster and inevitable collapse. As things currently stand, biophysical indicators show that the size of the global economy is well beyond what the Earth can sustain, while alternative indicators of wellbeing suggest that excessive growth is already suppressing the economic welfare of the world's seemingly most fortunate countries (see Chap. 11).

Resistance will undoubtedly come from the zealots of free enterprise, who largely operate beyond reason, but also from those who are milder in their preferences, but not astute enough to take action. The economics profession, seemingly incapable of self-correction, is undoubtedly a major barrier to reform, as is a commercial media wedded to capitalistic expansion. It has been said that science mostly advances one funeral at a time[1] – this seems doubly true for the economics profession that prefers patching the building rather than accepting that the foundations are unsound. This was the mainstream response to the impact on the global economy of the oil price spikes of 1973 and 1979 and the Global Financial Crisis of 2008. Well, we no longer have time for older orthodoxy to dominate the 'narrative'.[2]

This book argues for a new economic paradigm that harmonises with science, especially Earth science and the maturing concept of planetary boundaries (see Chap. 2 of this volume). The foundations of such an economics can be found in the now well-established literature of ecological economics. We have outlined what we consider to be the main principles of ecological economics in Chap. 11. But an overarching Earth-centred philosophy – which ecological economics largely is – still needs a macroeconomics that can explain things like the nature and history of money; how money is created in the modern world; the relationship between spending, inflation, and unemployment; the ecological as well as economic effects of monetary and fiscal policy; the role of taxes and their potential uses; and numerous other things. We find that modern monetary theory (MMT: see Chap. 14) provides the best explanation of how many of these elements of modern economies interact today. This still-emerging knowledge of today's operational monetary processes allows us to change things for the better: it is generally not a lack of money that prevents governments from doing good things, but a lack of knowledge about true constraints, especially the ceilings imposed by sustainably-managed real resources, which need to be efficiently allocated to their best use given the relative scarcity imposed by a respect for biocapacity and increasing human populations.

Finite real resources, sustainably managed, impose limits on what and how much can be produced. A new approach is therefore required centred around the redistribution of resources, as embodied in final goods and services, from those who enjoy an (unearned) and sometimes obscenely excessive claim on real wealth to those who do not have enough even to live, let alone thrive. The ubiquitous but nonsensical idea of continually increasing the size of the economic pie to ostensibly satisfy everyone – the 'rising tide lifts all boats' approach to raising material living standards – must be totally abandoned. This has been a groupthink delusion on a monumental scale, and possibly humanity's greatest blunder (see Box 15.1).

The most urgent task is to design and implement either an ascent (for low-income countries) or a descent (for high-income countries) into the safe operating space of the Earth's biocapacity (the doughnut in Raworth's metaphor: 2017). It follows that each country needs to determine and acknowledge its biocapacity, as well as its current resource throughput. Much of this work is already being done by organisations

---

[1] This is a common paraphrase of scientist Max Planck's 1950 principle.

[2] Besides the global 'schools strike for climate', another example of generational change is that 'Young Labor' in Australia (the state-based youth wings of the main progressive political party) has almost unanimously adopted a federal job guarantee as a policy platform, whereas the party itself seems stuck in dead-end neoliberal thinking, where a federal government that creates fiat money from nothing could not 'afford' to create jobs, given the assumed budget constraint, but where countless jobs urgently need to be done (Young Labor for Full Employment 2021).

Box 15.1: Why Are Countries Obsessed With Forever Increasing GDP?

1. **Governments fear recessions, which is an unplanned failure to achieve an increase in GDP, usually over two quarters.**
Governments should plan for a permanent quantitative steady state. Where this requires degrowth because the economy has grown beyond its maximum sustainable scale, appropriate planning and institutional setting can avert the negative outcomes normally associated with unplanned contractions, including rising unemployment, business failures, and poverty. Monetary-sovereign nations are in the best position to do this as their central governments can always employ the resources rendered idle by the transition process and reduce fluctuations in the economic cycle. Uppermost should be a job guarantee, which would guarantee abled people with meaningful work and a base-level income with which to thrive.
2. **Increasing GDP is associated with national security and an ability to maintain military capabilities.**
The global community must unite against our common foe, which is overshooting the planet's biocapacity. The endless pursuit of GDP growth incites countries to exploit others in a continual battle for scarce resources and trade advantages. Endless growth dramatically increases the chance of war and makes collapse a near certainty. High-income countries are the worst offenders.
3. **A high GDP is associated with high international prestige, especially at international forums.**
The new paradigm must value the pursuit of wellbeing above all else, which has little to do with GDP once material affluence reaches a sufficient level, but has everything to do with sustainability (avoiding an ecological deficit) plus good scores on health, education, relative income equality, low unemployment, and human rights.
4. **GDP is believed to be synonymous with national income, and therefore net wealth.**
National income is best defined as the monetary value of what can be consumed whilst keeping capital intact (see Glossary). Because human-made capital and natural capital are non-substitutable, it is necessary to maintain both forms of capital. Monetary payments generated from the destruction of natural capital, even when it is undertaken to increase the stock of human-made capital, does not represent 'true' income, since natural capital depletion limits future income by impairing the health of the planet.
5. **GDP is seen as a proxy for wellbeing.**
Simon Kuznets, who developed the modern GDP metric in the 1930s, warned against misusing it as a measure of wellbeing (see Chap. 8 of this volume). Both the costs and benefits of economic activity are added together to form GDP, whereas the costs should be subtracted, as is done

(continued)

Box 15.1 (continued)

with the Genuine Progress Indicator (GPI, see Chap. 11 of this volume). Even the GPI, which is confined to factors associated with the economic process, is not strictly a measure of total wellbeing, which is a more complex and broader concept that includes the services provided by ecological and social systems.

6. **Government debt is reduced as a percentage of GDP if the latter increases relative to the former.**
Monetary-sovereign governments that issue their own currency do not need to 'borrow' their own money from the private sector or other nations (see Chap. 14 of this volume) and should not borrow in foreign currencies. Bond issuance by monetary sovereigns does not involve borrowing – it is a means by which central banks manage interest rates on behalf of central governments. Nor is bond issuance the equivalent of debt issuance since a financial liability is only a debt if something *real* must be surrendered to extinguish the liability. A monetary sovereign gives up nothing real to extinguish its bond-generated liabilities, which are denominated in the currency it issues. Governments that are not monetary sovereigns should borrow prudently without thought to a GDP growth imperative (or Gross State Product in the case of state/provincial governments). Better still, if they are a central government, they should seriously consider reinstating their monetary sovereignty.
7. **Increasing GDP wards off demand-pull inflation by keeping aggregate supply ahead of rising aggregate demand.**
Although a high rate of inflation is undesirable, GDP should never be increased simply to ward off demand-pull inflation, particularly if it means boosting GDP beyond what can be ecologically sustained (or beyond what maximises economic welfare). In such circumstances, demand-pull inflation should be addressed in the appropriate way – by limiting aggregate demand, which may amount to reducing aggregate demand if a degrowth phase is required. Should this result in job losses, a Job Guarantee would fill the employment gap and ensure affected people receive a minimum living wage. A Job Guarantee would also insulate regions most likely to be impacted by reductions in aggregate demand.

like the Global Footprint Network and material flow monitors. International standards need to be agreed upon and improved over time. The harvesting of, and pollution into, common areas like oceans and the atmosphere need special attention, and the IPCC needs to improve its systems in respect to economic costs (see Chap. 10).

Once the maximum sustainable rate of resource *throughput* is ascertained, the next task is to avoid exceeding it, which is best accomplished by imposing quantitative limits (caps) on the rate of resource extraction and waste generation. This having been achieved, the extracted natural resources need to be efficiently *allocated* – that is, allocated to their best possible use. This requires market prices to fully account for all benefits and costs and for governments to provide adequate quantities of public goods; ensure people are sufficiently well informed to make beneficial decisions; and limit market power and the private capture of economic rents. These goals should be pursued while at all times maintaining a fair *distribution* of goods and services via a progressive tax-and-provisioning system that goes beyond providing a welfare 'safety net' to enable everyone to fully participate in society without being unduly influential in social affairs. Such a system would maintain full employment and impose maximum incomes to limit luxurious waste and political interference. Governments should also focus on calculating and tracking wellbeing, which could reach its maximum potential level with appropriate policy settings. We see no immediate limits to increases in things like human understanding via education and research, and general wellbeing.

As mentioned in the introduction to this book, the UN Sustainable Development Goals (SDGs) are one reasonable starting point for setting objectives to achieve ecological sustainability, distributional equity, and allocative efficiency, but need refinement in terms of acknowledging ecological limits to growth and the potential for overshoot relative to the supporting biosphere. The issue of population growth also needs to be addressed in any revision of the SDGs. The new economic paradigm that would emerge from this revision would reflect the need to minimise rather than maximise the rate of resource throughput and, therefore, the importance of developing new technologies that would reduce the resource-intensity and pollution-intensity of economic activity and increase the rate of materials recycling.

15.1 Herman Daly's Policies

In more precise policy terms, it is instructive to consider the ten policy recommendations for the United States by one of the founders of ecological economics, Herman Daly. Daly's top recommendation is for a cap-auction-trade system for basic natural resources and troublesome forms of waste:

> Caps limit biophysical scale by quotas on depletion or pollution, whichever is more limiting. Auctioning the quotas captures scarcity rents for equitable redistribution. Trade allows efficient allocation to the highest uses. This policy has the advantage of transparency. There is a limit to the amount and rate of depletion and pollution that the economy can be allowed to impose on the ecosystem. Caps are physical quotas, limits to the throughput of basic resources, especially fossil fuels. The quota usually should be applied at the input end because depletion is more spatially concentrated than pollution and hence easier to monitor. Also, the higher price of basic resources will induce their more economical use at each upstream stage of production, as well as at the final stages of consumption and recycling. Ownership of the quotas is initially public – the government periodically auctions them to individuals and firms. There should be no "grandfathering" of quota rights to previous users or "offshoring" of quotas for new fossil fuel power plants in one place by credits from plant-

ing trees somewhere else. Reforestation is a good policy on its own. It is too late for self-cancelling half measures – increased carbon sequestration and decreased emissions are both needed. The auction revenues go to the treasury and are used to replace regressive taxes, such as the payroll tax, and to reduce income tax on the lowest incomes. Once purchased at auction the quotas can be freely bought and sold by third parties, just as can the resources whose rate of depletion they limit. The cap serves the goal of a sustainable scale; the auction serves the goal of fair distribution; and trading allows efficient allocation – three goals, three policy instruments. Although mainly applied to non-renewable resources, the same logic works for limiting the off-take from renewable resources, such as fisheries and forests, with the quota level set to approximate a sustainable yield (Daly 2013).

Daly's other recommendations are as follows:

1. *Ecological tax reform.*
   Shift the tax base from value added (labour and capital) to 'that to which value is added', namely the entropic throughput of resources extracted from nature (depletion) and returned to nature (pollution).[3]
2. *Limit the range of inequality in income distribution with a minimum income and a maximum income.*
   Without aggregate growth, poverty reduction requires redistribution. Unlimited inequality is unfair; complete equality is also unfair. Seek fair limits to the range of inequality.
3. *Free up the length of the working day, week, and year.*
   Full-time work could be 4 days per week, or 30 h per week. Existing work could be shared between more people. Stop valorising the concept of hard work at the expense of caring and sharing.
4. *Re-regulate international commerce – move away from free trade, free capital mobility, and globalisation.*
   Countries should adopt compensating tariffs to protect, not inefficient firms, but efficient national policies of cost internalisation from standards-lowering competition with foreign firms that are not required to pay the social and environmental costs they inflict.
5. *Downgrade the World Trade Organization, World Bank, and International Monetary Fund.*
   Reform these organisations based on something like Keynes' original plan for a multilateral payments clearing union, charging penalty rates on surplus as well as deficit balances with the union – seek balance on the current account, and thereby avoid large foreign debts and capital account transfers.
6. *Move away from fractional reserve banking toward a system of 100% reserve requirements.*
   This would put control of the money supply and seigniorage (profit made by the issuer of fiat money) in the hands of the government rather than private banks.
7. *Stop treating the scarce as if it were free and the free as if it were scarce.*
   Enclose the remaining open-access commons of rival natural capital (e.g., the atmosphere, the electromagnetic spectrum, and public lands) in public trusts,

[3] This is supported by the UNEP Report (2021): "Shifting taxation from production and labour to resource use and waste is important to promote a circular economy that uncouples prosperity from pollution and favours job creation".

and price them by cap-auction-trade systems, or by taxes. At the same time, free from private enclosure and prices the non-rival commonwealth of knowledge and information.

8. *Stabilise population.*
   Work toward a balance in which births plus in-migrants equals deaths plus out-migrants.
9. *Reform national accounts – separate GDP into a cost account and a benefits account.*
   This is what the Genuine Progress Indicator does, so we achieve a net result that tells us something more meaningful in terms of welfare (Daly 2013).

It would be a brave ecological economist to take on the skill and experience of Daly, but in writing the above list, he did not have or accept the insights of MMT, where countries can and should be separated into those that have currency sovereignty and those that do not (see Chap. 14). Currency sovereignty – where a nation issues its own currency, floats it on international markets, and does not borrow in foreign currencies – gives the government maximum policy space because it maximises the government's command over real resources. It does not have to defend a currency peg (a promise to maintain a fixed exchange rate). Nor can it become insolvent in the currency it issues. It doesn't have to balance its budget, issue debt instruments (bonds, treasuries, or gilts), or take heed of credit-ratings agencies. It can set domestic interest rates at whatever level it wants through its central bank. It has full agency. It can, for instance, employ all idle labour should it choose to do so. It does not need to roll out the red carpet to foreign capital. It can create the domestic industries it needs within its resource limits. It can pursue resource independence to the maximum amount allowed by its natural endowments and biocapacity. It can build a 100% renewable energy system if the idle resources are available or, if they are not, can use taxation to free up and acquire the necessary resources.

We would therefore add at least one important policy to Daly's list – the implementation of a government-funded Job Guarantee program (Mitchell et al. 2019) to ensure full employment (1% or 2% of 'frictional' unemployment and zero underemployment): see also Chap. 14 of this volume. The Job Guarantee, while eliminating involuntary unemployment, would be an important inflation anchor since it would set the minimum wage for all employees and avoid triggering a wage-price inflation spiral. A Job Guarantee would also uphold the right to work as enshrined in Article 23 (1) of the Universal Declaration of Human Rights.[4]

Daly's recommendation of 100% reserve requirements for bank lending, while welcomed by many in the sustainability community, is debatable. If implemented,

[4] A motion was passed in the state parliament of Tasmania, Australia, on 23 September 2020, calling on that government to investigate how a government Job Guarantee (JG) could be adopted. Also, an Australian indigenous leader, Noel Pearson, believes a JG would be one of the best policies the Federal Government could adopt for his people (Pearson 2020). The Australian Greens political party supports a JG for those aged 18–29. In the US, in February 2021, Congresswoman Ayanna Pressley, and others, have introduced a House Resolution calling for a government-funded JG (Pressley 2021). See also footnote 2.

the 100% money recommendation would compel banks to lend what has previously been saved (bank deposits), which would always consist of base money created out of nothing and injected into the economy by the central government (net of taxation, which destroys base money). It is claimed that this would prevent banks from unjustly enjoying any seigniorage and place money back in 'public hands' by allowing the central bank to directly control the size of the money supply.

It is arguable whether this would restrict bank lending given that all bank deposits consist of past savings and each new loan – until the principal is repaid – manifest as new deposits/savings that, under the 100% money stipulation, would be available for re-lending.

We believe that it is not the size of the money supply *per se* that is an 'inflation' or 'growth imperative' concern, but the magnitude of aggregate spending – the conversion of money (virtual wealth) into final goods and services (real wealth) – that is the potential problem. Whilst it is true that more money unspent is more spending-in-waiting, taxation already exists to deal with any demand-pull inflation that might be sparked by excessive spending.

As for the growth imperative, which is the perceived need to increase GDP to match the increase in aggregate spending to avert demand-pull inflation, a cap-auction-trade system would deal with this by always preventing excessive spending from pushing the throughput of natural resources beyond the maximum sustainable rate. Sure, the prices of resource-access permits and pollution permits would rise, which would increase the cost of production and raise the price of final goods and services (i.e., be inflationary). However, this would serve as a form of taxation to help bring aggregate spending into line with the maximum sustainable GDP, which would stabilise under a cap-auction-trade system (the steady-state economy). As a penalty for excessive money creation, this would render excessive bank lending and the associated spending futile. We believe this would eventually stabilise lending levels and the money supply. Furthermore, it would direct lending towards the financing of low resource-intensive and low pollution-intensive activities and away from throughput-increasing activities, thus helping to increase the GPI as the nation's GDP remained relatively steady. To compensate the poor from any decline in spending power caused by lending-induced inflation, the government could always reduce the marginal tax rates on low incomes.

15.2 United Nations Environment Programme (UNEP)

The UNEP released a 'synthesis' report on 18 February 2021 titled 'Making Peace With Nature' (UNEP 2021). It is a strong call for root-and-branch reform – not unlike the thesis of this book – as a response to the acute emergency the world finds itself in. The executive summary includes a table (pp 33–41) that summarises many worthwhile reforms, with categories for different actors: government; intergovernmental organisations; financial organisations; the private sector; non-government organisations; individuals; scientific and educational organisations; and media and social networks.

There are too many recommendations to summarise here, but it is worth highlighting its call for a healthy environment to be recognised as a basic human right (p 33). If enshrined in a Bill of Rights, it is possible that citizens could challenge government inaction, bad policies, and bad laws that breach this right, in the relevant courts.

15.3 Alliance of World Scientists

A list of steps humanity can take to transition to sustainability can be found in Ripple et al. (2017). The website 'Scientists' Warning' (2020), in cooperation with the Alliance of World Scientists, lists six steps to tackle the climate emergency, under the headings: energy; short-lived pollutants; nature; food; economy; and population.

15.4 Other Policy Priorities Worth Considering[5]

15.4.1 General

- All countries declare an 'Earth System' emergency (climate, biodiversity, pollution) under the auspices of the United Nations and work co-operatively to limit and reduce global resource throughput to its maximum sustainable scale. Each country must ultimately limit resource throughput to its own biocapacity. Calling such a state of emergency should mean all government policies need to take account of the emergency.
- Embed the principles of ecological economics (see Chap. 11 of this volume) into all United Nations departments and agencies.
- Empower women everywhere, ensuring equal access to education and opportunity, and access to sexual health and family planning services.
- Quickly phase out all fossil fuel subsidies.
- All countries should co-operate in reserving large, interconnected portions (30% initially but probably moving to at least 50% of all habitats) of the Earth's land and oceans for nature, based on a scientific consensus and the need to protect biodiversity (Dinerstein et al. 2019; UNEP Report 2021).
- Make 'ecocide' a criminal offense under the jurisdiction of the International Criminal Court.
- Natural monopolies and essential services should be under government ownership and control rather than profit-making entities.

[5] Some of these are covered in the Daly and UNEP suggestions, but are worth repeating.

- Citizens should have the option of banking with a government bank that would set minimum standards for commercial banks to follow.[6]
- All citizens should have access to free basic healthcare and education as a basic human right, including vocational and higher education.
- The financial services sector needs to be heavily regulated, with much current speculative (non-productive) activity banned.
- All countries should unite to eliminate tax havens.
- Governments should confiscate economic rents (unearned income) via the tax system.
- Countries should try to maximise self-sufficiency, especially in essential commodities like food and energy, even though total self-sufficiency is neither possible nor desirable. As countries decrease self-sufficiency, they increase their vulnerability: nations that rely heavily on revenue from fossil fuel exports to finance the importation of final goods and services because of their weak secondary and tertiary trophic levels are obvious examples (Carbon Tracker 2021).
- Trade/exchange should be as local as possible. Unless transferring goods and technology to assist a distant impoverished nation, countries should prioritise regional trade before global trade.
- Failed or failing states hurt the global community, as does any injustice anywhere.[7] There must be a renewed emphasis on assisting struggling nations via debt forgiveness, appropriate technology transfer, family planning assistance, and related help.
- Nations should move in the direction of currency sovereignty within the limits of their unique economic circumstances, since currency sovereignty maximises policy options. Full currency sovereignty is usually hamstrung by a fear, sometimes real but usually perceived, of currency depreciation due to resource limitations and other vulnerabilities.
- Governments should promote the establishment of repair and recycling centres to help reduce the virgin resource-intensity of GDP. In cases where the centres have public goods and/or natural monopoly characteristics, they should be government owned and operated. A cap-auction-trade system, by increasing the cost of resource use and waste generation, would facilitate the proliferation of a repair-recycling industry within the private sector.

---

[6] Similarly, a government-financed Job Guarantee would set minimum employment standards for all employers to follow.

[7] In the words of US civil rights leader Martin Luther King: "Injustice anywhere is a threat to justice everywhere. We are caught in an inescapable network of mutuality, tied in a single garment of destiny. Whatever affects one directly, affects all indirectly" (King 1963).

15.4.2 Agriculture-Forestry

- Food production should move to regenerative systems that, firstly, preserve natural capital and the biogeochemical cycles upon which agriculture is based, and secondly, prioritise animal welfare (UNEP 2021).
- Small-scale producers in low-income countries need particular support and protection from the undesirable and often speculative forces of globalised agriculture.
- Unsustainable aquaculture systems urgently need to be reformed to operate in a manner consistent with sustainability principles. Again, a cap-auction-trade system would be of great assistance.
- The world's forests need to expand considerably via investments in natural capital, while forest harvesting needs to be based on best-practice sustainable methods (selective logging and extraction rates limited to regeneration rates). Global boycotts should be imposed on unsustainable forestry and countries with globally important forest regions (e.g., Brazil) should receive ecosystem service payments to preserve their forests and prevent deforestation. The largest monetary contributions should come from the world's high-income nations.

15.4.3 Corporations

- Laws governing the way in which corporations operate should change so that maximising shareholder returns does not conflict with society's sustainability and equity goals. This requires firms to successfully operate within the confines of strict sustainability parameters – best imposed via a cap-auction-trade system – and in an economy with an industrial relations system that guarantees decent and safe workplaces and equitably shares the nation's income between capital and labour. Large penalties should be imposed for any corporate transgressions. Above all, changes in the way corporations operate should result in the most profitable firms, and therefore the firms delivering the highest shareholder returns, being those which operate most efficiently within an appropriately regulated economy.
- Corporate power should be limited to prevent market domination, monopoly profits, and rent-capturing, and to avoid undemocratic outcomes.
- Where taxes, subsidies, and a cap-auction-trade system fail to induce desirable forms of economic restructuring, governments will need to regulate manufacturers more directly. We believe the imposition of a currently non-existent macro-level constraint would considerably reduce the need for micro-level regulation, which is still likely to be required with respect to product quality and safety.
- Privacy laws in respect to the internet need to be massively strengthened.

15.4.4 Education

- Citizens should learn basic sustainability and biophysical principles from an early age, rather than nonsense about the ever-increasing possibilities and benefits of eternal growth.
- Reforming academia is extremely difficult given the propensity of academics to defend the dominant paradigm within individual academic disciplines, no matter how much it deviates from real evidence. Economics is probably the worst offending discipline in this regard. Yet reform is vital, as acknowledged by Dasgupta (2021). Besides integrating academic disciplines within tertiary institutions, the academic review process would be improved by having academic journal articles assessed by at least one reviewer from an external discipline, preferably someone from the physical sciences in the case of a social sciences paper, and vice versa.

15.4.5 Energy

- All subsidies to fossil fuel interests must cease, with a rapid but planned phase-out of fossil fuel combustion.
- All nations must urgently move towards 100% renewable energy systems while reducing energy demand and rapidly improving energy efficiency. Clean-technology transfer should occur from high-income to low-income nations to benefit the global community by safeguarding the global commons.
- Petroleum-based plastics need to be phased out urgently and replaced with bioplastics or other organic materials.

15.4.6 Governance

- Nations should have independent corruption commissions operating at national and state/provincial levels.
- Donations to political parties need to be fully transparent in real time, with meaningful caps on donations.

15.4.7 Mining

- Mining companies should be rewarded for the service they perform (i.e., making available resources to be transformed into final products), but should not earn an economic rent from the ore they mine, which reflects the value created by Mother

Nature. When mining companies retain economic rents, the payment they receive upon the sale of their output exceeds the value of the real wealth *they* create. When this happens, people elsewhere receive payments below the value of what they produce. Hence, economic rents should be confiscated via taxation for equity reasons. If mining companies were required to cultivate renewable resource substitutes to replace depleted non-renewable resources, the cost would go a long way towards cancelling out the economic rents they often keep, which would reduce the need for taxation.

15.4.8 Trade

- Because of globalisation, much current trade is ecologically destructive and inefficient (not to be confused with low-cost). A lot of international trade also entails high-income nations appropriating the net wealth and natural resources of low-income countries (Dorninger et al. 2021).
- The world economy should be based on the principles of internationalisation, which would require international arrangements that balance trade at the national level; that allow national governments to impose compensating tariffs to safeguard efficient national policies of cost internalisation from standards-lowering competition; and impose appropriate restrictions on the mobility of international capital so that trade is governed by the desirable principle of *comparative advantage*.

In conclusion, humanity must embrace an entirely new definition of progress that employs different measurement tools and places international co-operation based on our shared vulnerability at the very forefront. At the same time, we must remain mindful of unique national circumstances and national aspirations for sustainable prosperity. To do this, we must urgently end our war on nature and our acceptance of social and individual injustices. If there is to be an international 'arms race' of some kind, it should be towards excellence and leadership in sustainability and equity. Provided we acknowledge our past mistakes and are prepared to dramatically alter the path we are on, there is every reason to believe in a prosperous future for all that is based on sufficiency rather than the brutal winner-take-all competition that characterises the present political economy. While the advancement of human knowledge, capability, and ethics may have no limit, the planet's finite resources and biological adaptability certainly do, as we are currently witnessing and are now paying the price for failing to respect.

References

Carbon Tracker (2021) Beyond Petrostates: the burning need to cut oil dependence in the energy transition, 11 February. https://carbontracker.org/reports/petrostates-energy-transition-report/. Accessed 16 Feb 2021

Carrington D (2020) Pandemics result from destruction of nature, say UN and WHO. The Guardian, June 17. https://www.theguardian.com/world/2020/jun/17/pandemics-destruction-nature-un-who-legislation-trade-green-recovery. Accessed 15 Feb 2021

Daly H (2013) Top 10 policies for a steady state economy. The Daly News, October 28. https://steadystate.org/top-10-policies-for-a-steady-state-economy/. Accessed 2 Feb 2021

Dasgupta Review (2021) The economics of biodiversity: the Dasgupta Review (headline messages). https://assets.publishing.service.gov.uk/government/uploads/system/uploads/attachment_data/file/957629/Dasgupta_Review_-_Headline_Messages.pdf. Accessed 13 Feb 2021

Dinerstein E, Vynne C, Sala E et al (2019) A global deal for nature: guiding principles, milestones, and targets. Sci Adv. https://doi.org/10.1126/sciadv.aaw2869

Dorninger C, Hornborg A, Abson DJ et al (2021) Global patterns of ecologically unequal exchange: implications for sustainability in the 21st century. Ecol Econ. https://doi.org/10.1016/j.ecolecon.2020.106824

King ML (1963) Letter from a Birmingham jail. Dr King's Legacy, National Civil Rights Museum, USA. https://mlk50.civilrightsmuseum.org/justice. Accessed 12 Feb 2021

Lawn P (2016) Resolving the climate change crisis. Springer, Heidelberg, p 68

Mitchell W, Wray LR, Watts M (2019) Macroeconomics. Red Globe Press, London

Pearson N (2020) Australians deserve a job guarantee. Cape York Partnerships media article. https://capeyorkpartnership.org.au/media-articles/australian-job-guarantee/. Accessed 19 Feb 2021

Pressley A (2021) Press release: Rep. Pressley, economists, advocates unveil historic federal job guarantee resolution, February 18. https://pressley.house.gov/media/press-releases/rep-pressley-economists-advocates-unveil-historic-federal-job-guarantee. Accessed 19 Feb 2021

Raworth K (2017) Doughnut economics: seven ways to think like a 21st century economist. Random House, London

Ripple WJ, Wolf C, Newsome TM et al (2017) World scientists' warning to humanity: a second notice. BioScience, December, pp 1026–1028. https://doi.org/10.1093/biosci/bix125

Scientists' Warning (2020) The six steps. https://www.scientistswarning.org/six-steps/. Accessed 21 Feb 2021

UNEP Report (2021) Making peace with nature: a scientific blueprint to tackle the climate, biodiversity and pollution emergencies (key messages and executive summary). https://wedocs.unep.org/xmlui/bitstream/handle/20.500.11822/34949/MPN_ESEN.pdf. Accessed 19 Feb 2021

Young Labor for Full Employment (2021) Tweet dated 14 February. https://twitter.com/YL4Jobs/status/1360784231035641857. Accessed 16 Feb 2021

Appendices

Appendix 1: Timeline of Key Events from the Big Bang

Age of the Universe 13.8 billion years.

Age of the Sun 4.6 billion years.

Age of Planet Earth 4.5 billion years.

Life Begins on Earth About 4 billion years ago.

Early Humans Appear About 2–3 million years ago.

Modern Humans Appear About 250,000 years ago.

Modern Humans Disperse from Africa About 70,000 years ago.

Modern Humans Reach Australia About 50–60,000 years ago (longest continual culture on Earth).

Modern Humans Spread across Europe About 40,000 years ago.

Holocene Epoch Arrives and Agriculture Begins About 12,000 years ago.

First Classical Civilisations About 5000 years ago.

Capitalism Begins About 1500 onwards.

S. J. Williams, R. Taylor (eds.), *Sustainability and the New Economics*, https://doi.org/10.1007/978-3-030-78795-0

Industrial Revolution 1760 onwards; widespread enclosure of common lands in UK-Europe.

First Labour Unions Early 1800s onwards.

Great Acceleration, Suggested Start of Anthropocene 1950s onwards.

Club of Rome Limits to Growth Warning 1972 onwards.

Global Financial Crisis 2007–09, Green New Deal gains momentum.

Covid-19 Pandemic and Recessions 2020 onwards.

Collapse of Global Civilisation ?

Death of Sun (and Earth) About 5 billion years from now.

Appendix 2: Glossary of Financial Terms

Philip Lawn, Stephen J. Williams

All Financial Intermediaries (AFIs) Authorised Deposit-taking Institutions (ADIs), plus registered financial corporations that are issuers of credit money (loans) and destroyers of credit money (repayment of loans).

Authorised Deposit-Taking Institutions (ADIs) Banks, credit unions and building societies, so a somewhat narrower grouping than AFIs.

Base Money Created only by the central bank, base money includes currency, bank holdings of notes and coins, plus bank reserves held with the central bank, plus other central bank liabilities to the private non-bank sector. Base money represents the money-like liabilities of the central bank.

Bank for International Settlements The central bank for all central banks, created in 1930 and based in Switzerland.

Bonds A general term for securities, issued by governments and corporations, but often referring to national-government securities (e.g., gilts in the UK). Essentially a kind of term deposit, except they can be on-sold at different prices in secondary bond markets. Bonds sold by a monetary-sovereign government have zero default risk and are therefore highly prized by the private sector. By convention, national-government bond sales, in the aggregate, are tallied and called national-government debt, as the government must eventually transfer back the principal of the bond, plus pay interest over the term of the bond. More accurately, national-government debt exists as interest-bearing financial assets held by the non-government sector. For a monetary-sovereign government, the bond sale is not a borrowing operation, but merely a transfer of reserves from the reserve account of a financial institution (held at the central bank) to a different account held at the central bank paying a slightly higher interest rate. The bond sale is a means of regulating reserves to control the official (overnight) interest rate. For a central government, bond sales are largely anachronistic and more direct, transparent, ways of achieving the central-bank interest rate should be pursued.

Broad Money M3, plus other deposit-like borrowings of All Financial Intermediaries (AFIs) from the non-AFI private sector (such as short-term debt securities).

Cash Currency (official notes and coins held by the private non-bank sector).

Central Bank The national government's bank that, amongst other things, issues the official base money of the nation on behalf of the government. Although the central bank will make some decisions autonomously, it is a part of the government as it is owned, regulated and monitored by government, with all key appointments made by government. Has intimate interactions with the government's Treasury (Exchequer). Typically sets a key interest rate that affects all other interest rates in the economy. A nation's commercial banks all have reserve accounts at the central bank. In the neoliberal period, governments have propagated the myth of central bank independence to shield themselves from harsh decisions sometimes made by the central bank.

Chartalism A precursor to MMT as it describes money as the state's invention to control the economy via spending and taxation, rather than something evolving from a barter economy. Credited to the work of German economist Georg Knapp, whose book *State Theory of Money* (1905) was pivotal.

Credit Money Money created from nothing by private banks (under licence) when a loan is issued, thus creating a deposit. The deposit must be exchanged for base money by the issuing bank if it is transferred by the borrower to someone with an account at a different bank; is withdrawn over the bank counter or from an ATM as cash (see base money); or is used to extinguish tax liabilities. Also called horizontal money to differentiate it from vertical money issued by the currency-issuing government. Credit money forms most of the money supply.

Currency 1. The official money of a nation (official unit of account). In some nations it is issued by the central government with a monopoly on such issuance, while a few nations use a foreign currency, such as the US dollar. Some nations in a grouping use a common currency (such as the euro, issued by the European Central

Bank). 2. In a narrow sense, the official money of a nation, held as notes and coins by the private non-bank sector.

Deposits Official money in electronic form, held in bank accounts: a liability for the financial institution where the account is held, while an asset for the account holder. Deposits are created from nothing when banks make loans (see credit money), but this new deposit is equal to the borrower's new liability (debt). The loan-deposit is extinguished when the loan is repaid. Deposits form most of a nation's currency.

Endogenous Money Endogenous money implies a money supply that cannot be controlled by the central bank. If a central bank has an interest rate target, it will insert and drain base money into and from the system, as required and in response to factors beyond the central bank's influence, to hit its target rate. However, the money supply will be endogenous because the central bank will have no control over whether it needs to insert or drain base money and how much it will need to insert/drain from the system. Almost all countries have an endogenous money supply.

Exchange Settlement Accounts The name given in Australia to reserve accounts held at the central bank by private banks through which they settle their obligations that accrue in the clearing of funds process.

Exogenous Money Exogenous money implies a money supply that can be controlled, via various means, by the central bank (i.e., the central bank can achieve a specific money supply target). Few countries have an exogenous money supply.

Fiat Money Money that is not backed by any valuable commodity, such as gold, and so is not convertible into anything but itself. Its value comes from government 'fiat' (authority, especially the authority to impose tax liabilities, which drives demand for the government's money). The central governments of monetary-sovereign nations all issue fiat money.

Fractional-Reserve Banking A less-than-common system whereby banks are required to retain on hand a specified fraction of their deposits (liabilities) as reserves with the central bank. It is often assumed that the banks lend out the rest, but banks create new money when they lend. Hence, banks do not require reserves in order to lend, nor do they need to ensure they have sufficient reserves prior to lending (i.e., lending is not reserve-constrained). Banks do need to ensure they have sufficient reserves to meet their settlement requirements, which they only do subsequent to lending and, should additional reserves be required, are obtained by either borrowing from a bank with excess reserve requirements or from the central bank. If a 100% fractional-reserve banking system existed, banks would only lend other people's savings (deposits) rather than create new credit money from nothing through the act of lending. It is falsely believed that a 100% system would limit spending (consumption) by limiting the money supply. However, the size of the money supply does not determine the level of spending, nor do the two necessarily move in the same direction. For example, more spending can take place with the same quantity of money if the existing money supply is spent more frequently – that is, if there is an increase in the velocity of money circulation. Adequate taxation, by destroying private-sector spending power, is the dominant means by which governments control private-sector spending.

Gold Standard The so-called Bretton-Woods system whereby the US dollar was pegged to the value of gold (and other currencies were pegged to the US dollar), fully abandoned by US president Nixon in 1971 as it became increasingly unworkable.

Government Debt The issuing of various government securities, collectively called bonds. For monetary-sovereign nations, this is not a borrowing operation. Nor is it debt in the conventional sense given that a monetary sovereign need not earn money through taxation, borrow money, or sell existing public assets to repay the principal of the bond, plus pay interest over the term of the bond. All outstanding bond liabilities of a monetary sovereign are paid in the same way as other forms of central-government spending – via the *ex nihilo* creation of base money.

Horizontal Money Another term for credit money, created by private banks. See vertical money.

Inflation The process whereby money loses some purchasing power over time as the price of general goods and services rises. Governments typically have an inflation target of about 2% because a high rate of inflation is seen as dangerous because it has undesirable distributional effects, distorts the resource allocation process, and increases uncertainty. Deflation – where prices fall over time – is also seen as dangerous for similar reasons. In the current neoliberal period, following the high inflation of the 1970s and 80's, inflation has been feared more than unemployment such that an 'inflation first' policy has commonly been adopted. Indeed, for neoliberals, a certain level of unemployment is deemed necessary to control inflation.

Cost-push inflation: A form of inflation triggered by price increases arising from a sudden increase in the cost of production. Cost-push inflation ensues when the initial increase in prices sparks a wage-price spiral, where increases in wages and prices continually push each other higher. Became a major problem in the 1970s when the price of oil rose suddenly and dramatically, which led to increased production costs for most goods and services that was quickly followed by rising wage demands of workers supported by the strong labour unions of the time.

Demand-pull inflation: A form of inflation where prices continue to rise due to the total spending of a nation (aggregate demand) continually exceeding the productive capacity of its economy (maximum aggregate supply). On rare occasions, demand-pull inflation is triggered by a collapse of a nation's production capacity (war damage, internal unrest), in which case maximum aggregate supply plunges below aggregate demand prior to any catch-up money creation by the central government (e.g., Germany's Weimar Republic (1921–23) and Zimbabwe (2007–09)).

Financial Leakages Savings, federal taxes and imports (in the latter case where money, via spending, goes abroad) are financial leakages from the domestic monetary system.

Hierarchy of Money A hierarchy representing the ease with which an IOU can be converted to base money, the only money that a monetary sovereign will accept as payment of taxes. To be considered modern money, and hence be part of the hierarchy of money, an IOU must have a nominal value fixed in the unit of account used by a monetary-sovereign to determine tax liabilities and legal debt obligations, such as the Australian dollar by the Australian Federal Government. Hence, the issuer of an IOU must be able to convert the IOU into an IOU that can ultimately be converted into base money, usually at some borrowing cost. For obvious reasons, base money sits at the top of the hierarchy of money. Credit money sits just below base money because banks are obligated to convert credit money to base money, on demand. The debt liabilities of non-banking institutions sit further down the hierarchy and can only loosely be regarded as modern money because there is either some risk in their conversion into base money, or some delay prior to conversion. While relatively new forms of non-state electronic money, such as cryptocurrencies, are accepted by some people as a medium of exchange, they are not forms of modern money and do not belong in the hierarchy of money. They are speculative assets, much like gold and silver, with no guaranteed value at the government's unit of account.

Legal Tender A synonym for cash.

M1 Money Currency (sense 2), plus current (cheque) deposits of the private non-bank sector at banks.

M3 Money M1, plus all other deposits of the private non-bank sector at banks (including certificates of deposit), plus all deposits of the private non-bank sector at credit unions and building societies.

Mint An entity given authority by a currency-issuing government to manufacture (mint) coins.

Money IOUs with value to someone, whether created by a national government or some other entity such as a state government, local government, community group, or private individual – but usually referring to something widely accepted, especially currency, as a medium of exchange. Currency-issuing governments, via their central banks, are constantly creating new base money from nothing via their (i) spending; (ii) their lending of base money to banks with insufficient reserves; and (iii) their purchases of government bonds from the non-government sector. They are also constantly destroying existing base money via taxation and temporarily removing base money via the sale of government bonds. Not all money can be considered modern money (see Hierarchy of money). Money is virtual wealth if it can be readily exchanged for real wealth, such as goods. The money possessed by an individual, whose spending will always be a negligible fraction of total spending within the economy, will exist as virtual wealth. For a nation, some of its money supply will not exist as virtual wealth since, in the unlikely collective attempt to convert the entire money supply into real wealth, the economy's maximum productive capacity would at some point be exceeded, whereupon conversion into real wealth would cease and demand-pull inflation would ensue.

Printing Money A mythical concept that says while a currency-issuing government can create new money from nothing, it should only take such action in exceptional circumstances because of the risk of inflation. The term 'printing' is used as a pejorative as it evokes images of a government printing copious quantities of cash every time it spends, thus resulting in wheelbarrows full of worthless banknotes. In reality, currency-issuing governments create new money from nothing whenever they spend, lend, or purchase government bonds, usually by electronically adding a deposit to someone's bank account and adding the equivalent value of base money to the reserve account of the recipient's bank. The exception is the issuing of a cheque, not the printing of cash. Currency-issuing governments only print money when people want to hold some of their money as cash, which they can obtain by converting a cheque or some or all of their account balance into cash over a bank counter or, in the latter case, from an ATM (subject to withdrawal limits).

Quantitative Easing (QE) The act of a central bank buying its own government bonds (so-called government debt), usually from the private sector in the secondary bond market, with the private sector having first bought the bonds in the primary bond market. Since the central bank is part of government, QE represents the government buying back its own bonds, and therefore adding reserves – previously removed by the original bond sales – back into the banking system. QE is thought by the mainstream to stimulate economic activity by increasing liquidity in the banking sector, but this is based on the false belief that the lending of credit money by banks is reserve-constrained. QE involves a financial asset swap (bonds for base money) and does not alter the net financial asset position of the non-government sector (financial assets less financial liabilities). It does not, therefore, increase the credit-worthiness of potential borrowers. Consequently, it does not lead to an increase in loan-financed spending. QE is sometimes erroneously called money 'printing'.

Reserves Base money deposited in the reserve accounts of banks that are held with the central bank.

Seigniorage Represents the difference in the value of the real wealth that the acquisition of an IOU can claim and the value of the real wealth foregone to obtain the IOU. Since a monetary-sovereign can create base money (IOUs) out of nothing, and since most base money exists as numbers in cyberspace, it forgoes virtually nothing to obtain real wealth. It thus has close to 100% seigniorage in the computer age. To obtain a financial claim on real wealth (money), most people and entities must forego real wealth equal in monetary value to the IOU obtained, usually by selling goods or their labour. Hence, most people and entities have zero seigniorage. Banks enjoy some seigniorage, not because they can create credit money from nothing, but because upon repayment of a loan, a bank receives interest above the repaid (destroyed) principal and therefore enjoys a claim on real wealth without foregoing the equivalent value of real wealth to obtain it (the opposite occurs for the borrower).

Taxation Does not provide a revenue stream to finance the spending of monetary-sovereigns. For monetary-sovereign governments, taxation is (i) a way of creating demand for the government's money; (ii) a way of depriving the private sector of some spending capacity, and a way of destroying currency to help control inflation; (iii) a way of reducing inequality via a progressive tax-and-transfer system; and (iv) a way of punishing undesirable activities (e.g., taxes on tobacco, alcoholic beverages, and carbon emissions). Of these four functions, taxation is best seen, from a macroeconomic perspective, as an inflation-controlling device. A monetary sovereign uses taxation to reduce sufficient private-sector spending power to avoid having to engage in a bidding (inflation-triggering) war with the private sector to acquire the real resources it needs to fulfil its policy agenda and other public goods-providing responsibilities.

Vertical Money Base money issued by a central government via its central bank. The highest (most authoritative) form of money in the hierarchy of money. See horizontal money.

Appendix 3: Glossary of General Economic Terms

Philip Lawn, Stephen J. Williams

Affluence For a nation, GDP per capita. Relevant when determining a nation's ecological footprint (F), where the footprint = population (P) x affluence (GDP/P) x resource-intensity of GDP (F/GDP). Hence, F = P x GDP/P x F/GDP. Affluence is not a synonym for wellbeing, which may decrease as affluence rises should median income fall due to growing income inequality and/or should social and environmental costs rise significantly due to increasing crime, unemployment, and the rapid deterioration of the natural and built environments.

Allocation The apportionment of the natural resources extracted from the ecosphere for economic purposes. In market economies, allocation within the private sector is largely undertaken within markets and guided by price signals. Prices are invariably determined by price-setting firms, albeit they are influenced by consumer preferences, resource scarcities, society's institutions, and government regulation, including price-altering taxes, subsidies, licence fees, and permit prices. Efficient allocation is where resources are put to their best (socially desirable) use.

Austerity An ideological position of recent decades, associated with neoliberalism, whereby national governments reduce spending to satisfy perceived budgetary imperatives, often with adverse impacts on the lives of the poor, and even the middle class (health, education, social housing). Sometimes undertaken while lowering taxes for the rich. Occasionally promoted by the International Monetary Fund and other institutions in the interests of 'budget repair'. The opposite of Keynesianism.

Biocapacity The amount of land available to provide the resources, absorb wastes, and still provide essential ecosystem services for a nation (or global civilisation) on an ongoing basis. It therefore represents ecologically sustainable resource supplies. If a nation's ecological footprint exceeds its biocapacity, it is operating unsustainably. According to some calculations (Global Footprint Network), humanity's footprint began to exceed the Earth's biocapacity in about 1970 and is now approaching twice its sustainable scale. Note that biocapacity is not identical to natural capital, although it is partly determined by the latter.

Cap-Auction-Trade System A proposed method for quantitatively controlling a nation's natural resource extraction and pollution generation, especially the generation of greenhouse gases. The government sets absolute caps on the rate of resource extraction and pollution to keep them within the ecosphere's regenerative and waste-assimilative capacities. In the case of greenhouse gas emissions, the cap decreases each year to meet international obligations. The caps are administered by issuing a limited number of short-lived permits, which are periodically auctioned by a government agency, thus letting market forces set permit prices (in effect, a tax). The permits are subsequently tradeable – which allows permit prices to fluctuate – so that permitted emissions are generated by the best-use sectors. Properly executed, cap-auction-trade systems provide certainty on absolute levels of natural resource use and pollution (ecological sustainability); assist in allocating resources to their best possible use (allocative efficiency); and, by destroying some spending power of permit buyers, allow governments to reduce tax rates on low incomes to compensate the poor for consequential increases in goods prices (distributional equity).

Capitalism An economic system, beginning in Europe during the sixteenth century, based on the accumulation of financial and human-made capital (e.g. factories, ships), predominantly privately-owned, where the allocation of extracted natural resources and human labour is largely determined by market forces. Because of the way virtually every capitalist system has evolved (i.e., with inadequate sustainability- and equity-guaranteeing institutions), capitalism has relied upon compound growth for its stability, has allowed pecuniary reward to become increasingly at odds with the generation of use value (often inequitably distributed), and has long produced sub-par outcomes despite the ongoing emergence of dazzling new technologies. In its present form, capitalism is ultimately a life-destroying (destabilising) exercise based on absurd assumptions of infinite growth and continued social tolerance of rising income-wealth inequality.

Circular Economy The idea that materials can and should be continually reused or recycled to dramatically reduce the use of virgin materials and limit or eliminate waste. Therefore in direct contrast with the dominant linear economy of material depletion and waste generation. Taken to an extreme, it aspires to zero waste. While reducing waste to sustainable levels is necessary, and the recycling of waste materials can reduce the required extraction of virgin materials for economic purposes, the concept of a circular economy fails to account for the second law of thermodynamics (the entropy law). The entropy law forbids the 100% recycling of non-organic materials, due to within-process degradation, and any recycling of energy. Recognising that the economic process ultimately involves the linear throughput of matter-energy highlights inevitable limits to the growth of an economy forever dependent on a non-growing ecosphere.

Collapse A relatively quick (probably over decades) unplanned and chaotic change in a society from complexity to a simpler structure due to the unsustainable nature of the society, caused by a mixture of ecological and/or economic problems. The opposite of sustainability. Usually associated with a significant increase in morbidity and mortality. Extremely common over the long historical record.

Consumption The effective destruction of durable goods (typically over years) and non-durable goods (more-or-less instantly) resulting in waste of an equivalent amount (first law of thermodynamics), although repair, repurposing, reuse, and recycling can postpone the conversion of goods to waste. Organic matter can be recycled by biological processes (composting, but with the release of CO_2 or methane), while the recycling of inorganic matter requires significant energy inputs and degrades quality during the recycling process. In short, there are limits to recycling goods on a large and continuous scale, not least because of the prohibitive energy requirements. Given its potential ecological consequences, consumption is best thought of as a 'necessary evil', without which the use value of physical goods could not be enjoyed. Contrary to the dominant view, anything that maintains the enjoyment from physical goods at reduced rates of consumption should be considered an advancement, even though it would likely reduce GDP.

Cost The true *social* cost of spending is best measured in terms of the foregone benefits from the use of any purchased real resources (opportunity cost). The *financial* cost of spending depends on the relevant entity. For a monetary-sovereign government, the financial cost is virtually zero because its ability to create unlimited currency for its own spending purposes means it foregoes virtually nothing of real value to obtain useful resources (100% seigniorage). Most individuals and entities enjoy zero seigniorage because, to obtain the necessary currency to claim something of real value, they must forego something of equal real value to obtain the claim-worthy currency, usually by selling goods or their labour.

Debt (Borrowing) A way of bringing spending forward in time, but usually with a financial penalty (interest). The opposite of savings, which allows one to delay spending from prior income.

Decoupling (Green Growth) In conventional parlance, decoupling is the idea that the resource-intensity of GDP can be sufficiently decreased through technological improvements, substitution, and productivity gains to either reduce the rate of resource throughput as GDP rises (referred to as *absolute* decoupling) or slow the rate of growth in resource throughout as real GDP rises (referred to as *relative* decoupling). A heavy reliance on absolute decoupling is mostly advanced by those intent on continued economic expansion (frequently labelled 'green growth'), who often cite the flawed Environmental Kuznets Curve as empirical support. In the short run, throughput rates can be reduced, even as real GDP increases (absolute decoupling), provided the resource-intensity of GDP is decreased at a faster rate than real GDP grows. Because of thermodynamic limits, throughput rates can only be reduced in the long run via reductions in real GDP (degrowth). In addition, operation of a steady-state economy that is no larger than its maximum sustainable scale is necessary for the rate of resource throughput to remain ecologically sustainable.

Degrowth A policy to deliberately but equitably reduce the size of an economy (usually of a nation, but also globally) because it is larger than its ecologically sustainable size. Degrowth may also be desirable if the economy is larger than its optimum size, noting that the optimum size will be smaller than the maximum sustainable size. Degrowth would stop at some desirable scale, and the economy would enter a steady-state size of dynamic equilibrium with negligible changes in quantity, but improving quality. Degrowth is different to a recession, which is an unplanned and chaotic economic contraction in a growth-dependant economy.

Jason Hickel: "A planned reduction of excess energy and resource use in rich nations to bring the economy back into balance with the living world, while reducing inequality and improving people's access to the resources they need to live long, healthy, flourishing lives."

Distribution The relative share of total income amongst the citizens of a particular jurisdiction. In ecological economics, distribution is equitable if each citizen can claim sufficient goods and services to thrive; plus the order-of-magnitude disparity in income of the richest and poorest individuals is deemed socially acceptable. Resources, which are required to produced goods and services, are always limited and need to be sustainably managed. Overpopulation creates undesirable distributional challenges by accelerating the rate of environmental destruction, intensifying competition for resources, and increasing the likelihood of wars. Because overpopulation affects the yet-to-be-born by altering the distribution of resources across time, it also has an intergenerational impact – invariably an undesirable one.

Doughnut Economics Based on the concept of planetary boundaries, and built around seven principles, UK economist Kate Raworth symbolises the safe operating space for humanity as the area of a doughnut. If an economy grows indefinitely, it will eventually move outside the outer circumference of the doughnut and into ecologically unsustainable territory. If the economy renders people impoverished, then it is in the void inside the inner circumference of the doughnut, where people do not receive what they need to thrive. It follows that all nations should design their economies to stay within the doughnut, with a focus on fair distribution and regeneration, where humans can flourish without destroying the planet's life-support systems. However, some impoverished countries have ecologically unsustainable economies due to overpopulation. Does their economy sit inside or outside the doughnut? Raworth's book *Doughnut Economics* was published in 2017.

Ecocide What humans are currently doing to the planet by triggering what is often called the sixth mass extinction of life.

Ecological Economics A change in the fundamental concept of economics, ecological economics emerged in the late-1980s and is based on the foundations created by economists such as Kenneth Boulding (*The Economics of the Coming Spaceship Earth,* 1966), Nicholas Georgescu-Roegen (*The Entropy Law and the Economic Process,* 1971), and Herman Daly (*Steady-State Economics*, 1977). Ecological economics conceives the economy for what it is – a physical provisioning system that exists inside society, which is in turn inside and dependent on a fragile and non-growing planet Earth and its critical life-support systems. It emphasises ecological limits on resource throughput and waste production and the need to operate within such limits to avoid degrading the entire system. Mainstream economics at the time, and still, places society within the economic system, where the latter takes precedence with little regard for life-sustaining ecological processes. Should not be confused with environmental economics, which is mostly based on mainstream assumptions.

Ecological Footprint Represents the area of land (global hectares) required to generate the resources, absorb the wastes, and provide the critical ecosystem services needed to sustain economic activity at its current level – for an individual or grouping. Must be compared to the available biocapacity to determine if overshoot (ecological unsustainability) is occurring.

Economic Growth In conventional terms, seen as an increase in economic activity, typically of a country, usually measured by GDP, in a given period. In mainstream economics, it is assumed that a country should strive for, and can achieve, economic growth by at least a few percent every year. However, this exponential increase leads to a plotted growth curve that becomes almost vertical. For instance, a 2% annual growth rate, if maintained, means a doubling in size every 35 years. Increasing GDP has everywhere been accompanied by rising rates of material throughput, regardless of technological progress and the transition by many countries to a services-based economy. Some ecological economists confine their use of the term 'economic growth' to cases when GDP growth increases benefits faster than costs (i.e., where the marginal benefits of growth exceed the marginal costs). If GDP growth increases costs faster than benefits, we have *un*economic growth.

Economic Rents In economics parlance, economic rents are unearned income, as in the term 'rent-seeking behaviour'. More precisely, it is an amount paid above the minimum necessary to bring a factor of production into operation. Economic rents often arise when there is a strictly limited supply of something, such as the number of player contracts for a professional football team, vacant land within a convenient distance of a city CBD, and remaining stocks of a non-renewable resource. Elite sportspeople, celebrities, land-owners, mining companies, and many corporate executives are paid economic rents. If a government sets a cap on pollution levels and sells the rights to pollute in the form of pollution permits, it creates an economic rent – desirable in that the permit price (i) encourages less pollution-intensive activities; and (ii) captures the economic rent, which the government can subsequently destroy. Private-sector capture of economic rents does not advance general wellbeing – indeed, it is inequitable – and economic rents should be confiscated via government taxation. Such action would not distort the resource allocation process because all factors of production would continue to be paid the minimum required to be supplied (their true worth). There are some who believe that a simple but blunt way to confiscate economic rents is to impose a maximum income limit (i.e., impose a 100% marginal tax rate on any income above the maximum).

Economic Welfare A monetary measure involving the subtraction of the costs of economic activity from the benefits (whereas GDP adds them all together). Now measurable by a metric called the Genuine Progress Indicator (GPI), which is often presented in both aggregate and per capita terms. The GPI is not identical to wellbeing (quality of life), which is a broader, qualitative concept rather than a monetary one. Nonetheless, the GPI can be used to ascertain whether GDP growth is 'economic' or 'uneconomic' (see economic growth). 'Standard of living' is another term with a more monetary (income-consumption) connotation.

Efficient Allocation The allocation of the incoming resource flow is efficient if the natural resources entering the economy are put to their best possible use (i.e., are directed towards the production of physical goods with the highest use values).

Entropy The second law of thermodynamics, whereby matter-energy tends to disorder and homogeneity without the continual input of new energy. Nature provides low-entropy (ordered) matter – such as trees, fish, and mineral ores – which humans, via the economic process, eventually convert into high-entropy (disordered) waste, even though they can be briefly turned into useful products such as furniture, houses, etc. Entropy plus the first law of thermodynamics limit available resources and preclude 100% materials recycling and any prospects for a truly circular economy.

Environmental Kuznets Curve A plotted curve falsely indicating that environmental harm will at first increase, then taper off, and eventually decline as GDP rises, much like the similarly discredited Kuznets Curve of the 1950s and '60s with respect to income inequality. While some discrete harms can decline as GDP rises due to substitution, regulation, etc. (e.g., the decline in acid rain in the US northeast), more important metrics, such as total resource throughput, have provided no evidence of throughput rates decreasing as GDP has risen.

Exports When a country foregoes *real* wealth by sending it abroad (the export) in exchange for *virtual* wealth denominated in the currency of the importing country, which is sometimes money but often the government bonds of the importing country, as in the case of China exporting to the USA. (Note: China is giving up something real and useful; the USA is giving up something that its federal government can create from nothing with computer keystrokes. So, ask yourself: who is the winner in that trading relationship?) Although exports contribute to a country's GDP, they are costs to the exporting country – real wealth foregone. This doesn't mean that a country should not export, but it should do so only to finance its imports. That way, export costs equal import benefits. Relying heavily on export revenue to generate additional domestic income can be high risk as the importing country may change its preferences, or seek to punish the exporter for political reasons, therefore exposing the exporting country to sudden declines in GDP. Such declines in GDP can be offset if a nation's central government is a monetary-sovereign always prepared to increase its spending to redirect the redundant resources and unemployed workers, which can in part be facilitated by a government job guarantee program. Fluctuating exchange rates also make a heavy reliance on trade very risky.

Externalities In mainstream economics, things that are considered outside (external) to the economy, and so are unpriced, such as when a firm pollutes the atmosphere or ocean at a cost to society but at no cost to itself (the pollution thus being the externality). Although mainstream economics recommends taxing the offending firm, it fails to acknowledge that, at best, 'internalisation' of the externality does little more than reduce the pollution-intensity of economic activity, as beneficial as this is. It does nothing to prevent the volume of economic activity rising sufficiently to overwhelm the reduction in pollution-intensity, which causes overall pollution levels to rise (the so-called 'rebound' effect). The reason for this is that price-altering taxation merely alters the allocation of the incoming resource flow that ultimately becomes pollution (first and second laws of thermodynamics). Pollution levels are a throughput concern, not an allocation problem. Hence, limiting pollution levels requires the imposition of caps (see cap-auction-trade system).

Extinction Rate There is always a natural rate of species extinction, periodically punctuated by mass extinctions (with five mass extinctions generally acknowledged, the most-recent including the loss of the dinosaurs about 66 million years ago, together with about 70% of all plants and animals). Currently, humans are almost certainly causing a sixth mass extinction event, with the rate of extinctions

100–1000 times higher than the natural rate. This has profound implications for human wellbeing and sustainability. High levels of biodiversity equate to a healthy planet and, therefore, healthy agricultural and natural resource-generating systems.

Fiscal Balance For a monetary-sovereign government, 'fiscal balance' is a preferred term to 'budget', since such a government has no need to monitor the gap between its spending and income (although tax receipts are not really income). That said, a monetary-sovereign government still needs to ensure the real resources it acquires via its spending are put to their best possible use. The usual term 'budget deficit' is best referred to as a 'fiscal injection' (more base money injected via government spending than destroyed via taxation); a 'budget surplus' is best referred to as a 'fiscal drain' (more base money destroyed than injected).

Framing The use of language, subtly or not-so-subtly, to influence what the reader/listener thinks, but especially in a distorting, political way rather than using more neutral language. For instance, a writer may say the economy is set to experience healthy growth of X% this year, implying a benefit, when it may be more correct to say the economy is presently bloated and any further increase in size would result in uneconomic growth. 'Budget reform' and 'budget repair', when applied to a monetary sovereign, imply a budget constraint when, in fact, it confronts no such constraint at all. 'Tax relief' primes the reader to assume the relevant taxpayer cohort are overburdened with tax. 'Structural adjustment' is a sugar-coated way of saying that a poor country that seeks to borrow money from the IMF must apply policies to restructure its economy in a seemingly beneficial way, when it only benefits rich nations to the detriment of the borrowing country. Framing is largely unavoidable – it can be good or bad, depending mainly on the extent to which the language used reflects reality.

Full Employment 1–2% unemployment rate (workers moving between jobs) and zero underemployment. The number of hours per week that should constitute full-time work is moot. Employment is a human right under the United Nations charter. Some central banks are charged with achieving full employment, where it is usually honoured in the breach rather than the observance.

Genuine Progress Indicator (GPI) A monetary metric designed to measure the net benefits of economic activity – i.e. economic welfare – although not an indicator of total wellbeing given that the latter extends well beyond the former. Unlike GDP, calculation of the GPI entails subtracting the costs of economic activity from the benefits and including informal activities, such as the shadow economy and unpaid household labour. Like GDP, the GPI can be measured on a per capita basis. The GPI should be used in conjunction with other new metrics (some still to be designed) to chart human progress and sustainability.

Government Job Guarantee (JG) A proposal guaranteeing full employment (zero involuntary unemployment). Usually associated with modern monetary theory, whereby a national government provides a real job, with training if required, for any working-age citizen who cannot secure a job in the private sector or conventional public sector. Remuneration is set at a minimum living wage. Ideally a transitory job until the person finds work in the regular economy. The types of jobs available are determined at the local government level, but most are not jobs performed by the regular public sector or private sector, and none are performed unless JG participants are suitably qualified to perform them. The JG scheme sets the minimum (socially acceptable) pay and conditions that all employers must meet. By doing so, the JG never provokes a wage-bidding war with the private sector, meaning it is never hyper-inflationary.

Green New Deal (GND) A set of policies from the progressive side of politics that seeks strong government intervention to address climate change and other ecological problems in a socially just way. To achieve its goal, most GND proponents advocate increasing state ownership of essential services and greater government spending. In terms of policy approach, the GND stands in direct opposition to neoliberalism and its market-based (small government is best) thinking. The idea gained momentum with the global financial crisis of 2007 onwards, with specific policy proposals varying from nation to nation. Key issues of concern include the extent to which GND policies conform to what is strictly required to achieve a safe climate and sustainable resource throughput, including the need for degrowth in high-income nations rather than the mythical 'green growth' often advocated by GND proponents. In 2021, most versions of a GND do not address or conform with sustainable resource throughput. The term 'just transition' is sometimes used to imply similar, or identical, policies to a GND.

Gross Domestic Product (GDP) A monetary measure of *final* goods and services *produced* in a country during a certain period (therefore a flow, rather than a stock). Not a measure, or even a proxy measure, of human wellbeing, despite many pretending it is, although it is a good proxy for resource throughput. Adds both the costs and benefits of economic activity. Aggregated globally to measure Gross World Product. Excludes the informal (unpaid and shadow) economy. In mainstream economics, GDP is regarded as national income, but this is questioned by some ecological economists because it fails to conform to John Hicks's (1946) definition of income – namely, the maximum amount that can be consumed over a period while maintaining the capacity to consume at least as much in the next period. A nation cannot simultaneously consume its entire GDP and maintain its capacity to consume (i.e., its productive capacity) because it must set aside some of its GDP for defensive and rehabilitative purposes, to replace worn out machinery, and overcome any losses of natural capital. In the Hicksian sense, GDP always overstates national income.

Gross National Product (GNP) Similar to GDP, but does not include domestic production by foreign-owned firms, although it does include production from domestically-owned firms operating abroad. Now usually called Gross National Income.

Gross World Product The sum of every country's GDP. Hence, the monetary value of all final goods and services produced within the global economy.

Imports When a country acquires *real* wealth from a foreign country (the import) in exchange for *virtual* wealth denominated in its own currency, which is sometimes money but can be the government bonds of its central government. Whereas exports are costs, imports are benefits (although the importing country is also importing future waste). Relying on the importation of a natural resource or product can be high risk as the exporting country may choose to consume the resource/product itself or punish the importer for political reasons. Such reliance can be difficult to avoid in the case of natural resources but can be avoided in the case of final products by maintaining the domestic capacity to produce them. Notwithstanding the potential benefits of international trade, a country should always be as self-sufficient as possible, which requires strength at all three economic trophic levels.

Jevons Paradox An extreme rebound effect that occurs when, to begin with, an increase in technical efficiency (reduced resource-intensity of economic activity) increases the relative supply of the natural resources whose efficiency of use has increased. This reduces the relative cost of natural resource use and stimulates the search for more uses of the now cheaper natural resources. Ultimately, without a cap placed on the rate of natural resource use (cap-auction-trade system), more natural resources are consumed, albeit more efficiently. Hence, the paradox – instead of an increase in technical efficiency reducing the quantity of natural resources consumed, as expected, it results in its increase.

Keynesianism The revolutionary economic principles of UK economist John Maynard Keynes (1883–1946), that were largely a response to the problems caused by the Great Depression. Keynes' principles became dominant in wealthy countries immediately before WWII and in the decades after, but were replaced by neoliberal principles from about the mid-1970s onwards, partly due to, or facilitated by, the stagnation (high inflation and high unemployment) triggered by skyrocketing oil prices. Keynesianism is associated with expansionary fiscal policy – usually increased government spending rather than reduced taxation – to keep unemployment low. Conversely, neoliberalism favours monetary policy (interest-rate adjustments) and a 'necessary' buffer stock of unemployed labour to control inflation.

Kuznets Curve A hypothesis formulated in the 1950s and 60's, arguing that, as per capita GDP grows, income inequality initially increases, but eventually decreases, so that a plotted relationship between the two variables takes the look of an inverted U-shaped curve. Not seen in reality without government redistribution measures to decrease inequality. Increasingly being undermined by globalisation, where the high mobility of financial capital has allowed transnational corporations to bypass national redistribution mechanisms, thus placing downward pressure on wages and resulting in stagnant real wage growth in high-wage countries.

Macro-Allocation Refers to the apportionment of the incoming resource flow between the private and public sectors. The macro-allocation process is efficient when the ratio of the private sector to the public sector provision of goods and services is socially desirable. Unlike the micro-allocation process, which is largely governed by price signals generated in economic markets, the macro-allocation of resources is determined within political markets (democratic elections).

Mainstream Economics A view of the economy (real and financial) as the economics profession would like it to be rather than what it really is. Based on a flawed body of largely ideologically-motivated false assumptions and therefore incapable of generating valid conclusions and useful policy-prescriptions. In part responsible for producing the current existential crisis. Characterised by groupthink, inbreeding, wilful blindness, and an inability to self-correct. Useful as an example of how humans can sometimes engage in dangerous mass delusions and wishful thinking.

Marginal Benefit The benefit of producing one extra unit of output or, at the macroeconomic level, an extra dollar's worth of GDP. As more goods are produced, the additional benefits increase at a diminishing rate (i.e., the marginal benefit declines).

Marginal Cost The cost of producing one extra unit of output or, at the macroeconomic level, an extra dollar's worth of GDP. As more goods are produced, the additional costs increase at an increasing rate (i.e., the marginal cost increases). Beyond a certain point, the marginal costs of additional output begin to exceed the marginal benefits and the growth in output becomes 'uneconomic' (see Uneconomic growth).

Material Footprint Analysis An estimate of total resource use, by weight, usually of a country, that includes the resources used to produce and transport imported goods to avoid false decoupling claims derived from the 'offshoring' of resource-intensive forms of production. Tracks closely with GDP. It is also instructive to calculate and compare total global resource use with Gross World Product, which similarly track closely. A maximum global sustainable resource use has been estimated at about 50 billion tonnes per year. In 2020, the world used 100 billion tonnes.

Maximum Sustainable Scale of the Economy Where the physical scale of the global economy is such that, in order to be sustained, humankind's ecological footprint exactly equals the planet's biocapacity. A global economy requiring a footprint less than biocapacity is smaller than the maximum sustainable scale; a global economy requiring a footprint greater than biocapacity is larger than the maximum sustainable scale. Applies equally at the national and sub-national levels. Because something must be sustainable to be optimal, the optimum scale of the economy is necessarily less than the maximum sustainable scale.

Micro-Allocation Micro-allocation refers to how the incoming resource flow is apportioned to produce individual goods and services. The micro-allocation process is efficient when there is a desirable relative share and quantity of different goods and services produced. Within the private sector, the micro-allocation process is largely governed by price signals, many of which are altered (for good or for bad) by government taxes, subsidies, licence fees, and permit prices.

Modern Monetary Theory (MMT) A revolutionary approach to macroeconomics that emerged in the late 1990s to challenge the dominant (neoclassical) model (see Appendix 4 for a comparison table). MMT is *descriptive* in that it provides a more realistic understanding of how modern monetary systems work. In doing so, it focuses on real resources as setting limits to economic activity rather than national-government budget outcomes. Although not prescriptive, it offers greater policy options to national governments that issue their own currency, since they can always use their unlimited fiscal capacity to acquire and mobilise idle resources on sale in their own currency. Some MMT theorists like to call MMT a lens, since we never see reality directly or unmediated – but in that case all versions of macroeconomics are a lens. More recently, some proponents claim MMT also has a *prescriptive* arm – for instance, GND-type policies – but this expansion would likely distract from MMT's claim to be the most technically correct version of macroeconomic reality that should be used by all policymakers. (For more, see Chap. 14 of this volume.)

Monetary-Sovereign Government A national government that (i) issues its own currency; (ii) presides over a floating exchange rate; (iii) does not substantially borrow foreign currencies; and (iv) has its own central bank. Such a government has no *budgetary* constraint on its spending, although what it can do with its spending is subject to real resource constraints. Such constraints can be circumvented in the short run if a nation opts to liquidate some of its natural capital to make available additional natural resources, as is the case with all countries whose Ecological Footprint exceeds its Biocapacity. However, this is unsustainable. In the long run, real resource constraints are defined by the ecologically sustainable use of available resources. Either way, spending by a monetary-sovereign that pushes total spending beyond the prevailing real resource constraint is typically reflected in demand-pull inflation. A monetary-sovereign can never become insolvent in its own currency and credit ratings of its fiscal position are meaningless.

Natural Capital The ecosphere, which is the source of all resources, the sink for all wastes, and the generator of all life-support services. Some criticise the use of the word 'capital' when referring to the ecosphere as it can be misinterpreted as anthropocentric and tending to obscure the aesthetics of the natural world and its right to exist for its own sake. Such criticism can be dealt with by adopting the biocentric position that some natural capital should exist 'untouched' because of its intrinsic and aesthetic values and because some natural capital provides critical, non-substitutable services, while the remainder can be utilised for human purposes, albeit always in a sustainable manner.

Neoclassical Macroeconomics The pro-market version of macroeconomics that replaced Keynesianism from the mid-1970s onwards. Based on a neoliberal ideology that centres around the belief that all governments face a budget constraint and should balance their budgets, much like a household; that small government is best; and that fighting inflation should take precedence over eliminating unemployment. Patently false in many respects, with an appalling record of predicting eventualities and of failing to provide policy prescriptions to achieve ecologically sustainable, equitable, and genuine welfare-enhancing outcomes.

Neoliberalism An economic ideology promoted by Hayek and Freedman (Chicago school) against Keynesianism. Advocated small government, the primacy of monetary policy over fiscal policy, a 'fight inflation first' strategy, low taxation, deregulation, privatisation, low tariffs, free trade, and unfettered price signals to equalise supply and demand. Associated with the fantasy of 'trickle-down' wealth from rich to poor. In practice, has produced extreme forms of wealth- and income-inequality, where markets exist, not to facilitate allocatively efficient outcomes, but to socialise costs and privatise benefits.

Offsetting An indirect pollution-reduction strategy, mostly related to carbon pollution. The strategy allows an entity that has reduced its emissions as much as practicable to offset the remainder by funding an emissions-reduction activity elsewhere (done by itself or another entity), so that the former's net emissions are reduced to zero or a permitted level. It relies on 'additionality' – that is, on the understanding that the emissions offset would not have occurred if not for the offsetting activity. Fraught with potential error and the need for strong regulation and transparency, and therefore best seen as a last resort. Capping pollution to a sustainable level in the first place via a cap-auction-trade system is likely to be more effective.

Opportunity Cost If there are only enough resources to produce X or Y, not X and Y, the opportunity cost of producing X is the net benefit foregone by not producing Y. Recognition of opportunity cost is important in a world of limited resources where 'best use' corresponds to 'most desirable use'.

Overpopulation Too many individuals consuming too much too quickly. Can be rectified by reducing population numbers and/or reducing per capita consumption (throughput) so that the relevant biocapacity is not exceeded. In the case of humans, it is essential to estimate biocapacity and throughput rates to remain within ecological limits because, once limits have been exceeded, altering fertility rates and changing cultural attitudes towards reproduction to once again operate within limits can take time and their impact can take decades longer. Changing net immigration numbers can also be problematic, especially in growth-dependant economies not yet restructured to overcome their reliance on population growth, and where rising refugee numbers can demand humanitarian responses from already overpopulated countries.

Overshoot The degree to which resource throughput by humans exceeds the planet's biocapacity. According to the Global Footprint Network, a symbolic date can be identified every year after which all human economic activity enters overshoot, depleting 'natural capital' and therefore further reducing the planet's biocapacity. In 2020, the date was 22 August. Overshoot can also be described as the number of Earth-planets needed to support the global economy at prevailing rates of throughput, which is 1.6 planets in 2020.

Planetary Boundaries Thresholds in nine key geobiological functions affected by human activity that we exceed at our peril: climate change; ocean acidification; chemical pollution; nitrogen and phosphorous loading; freshwater withdrawals; land conversion; biodiversity loss; air pollution; and ozone layer depletion (see Chap. 2 of this volume).

Population Growth The result of births exceeding deaths (natural increase), or when net immigration offsets a natural decrease. Although affected by the Total Fertility Rate (TFR), where 2.1 births per woman eventually creates a stable replacement rate, populations can still increase for decades with TFR <2.1 because of demographic momentum – the combined effect of a movement of large numbers of the previously-young into reproductive cohorts and increases in life expectancy. Those who prioritise GDP growth usually espouse population growth because, all else equal, the latter drives the former. In 2020, the world's population was increasing by about 82 million per year. When population numbers are high, a decrease in the *rate* of increase may not achieve much in terms of reducing the *absolute* increase because of the large base population. Increased life expectancy is universally welcomed but, unless accompanied by a reduction in fertility rates, populations increase markedly. Many societies are seeking to reduce their fertility but are not receiving the sexual and reproductive help they desperately require.

Precautionary Principle Various versions of this exist, such as: "When an activity raises threats of harm to human health or the environment, precautionary measures should be taken even if some relationships are not fully established scientifically. In this context the proponent of an activity, rather than the public, should bear the burden of proof." (Wingspread Statement, 1998). Similar versions have been used in litigation.

Rebound Effect Occurs when an expected reduction in resource use due to efficiency gains is not realised, usually because the resource supply increases and/or the price decreases. In some cases there may be a reduction in the use of the resource, but not as much as expected given the efficiency gain. In other cases the efficiency gain may equal the new higher use level, with no net change in throughput. In other cases there may even by a net increase in the use of the resource, known as Jevons Paradox.

Recession Narrowly defined, two consecutive quarters of *unplanned* economic contraction (i.e., where GDP fails to increase). Should inadequate redistribution mechanisms exist, a recession is almost always associated with rising unemployment, bankruptcies, etc. Should be distinguished from degrowth, which is a planned contraction of an economy that has overshot its maximum sustainable scale where care is taken to protect the vulnerable and the focus of economic activity switches to qualitative improvements. Unlike a recession, the economic welfare emanating from well-managed degrowth can rise by increasing the difference between economic benefits and social and environmental costs.

Resilience The ability of a system to recover relatively quickly after damage, such as an ecosystem after a natural disaster: the greater the health of an ecosystem, including high biodiversity, the quicker it will recover. Similarly, a resilient economy is one with strong primary, secondary, and tertiary sectors (sometimes called trophic levels), abundant natural capital, adequate stocks of highly-productive human-made capital, a well-educated and trained workforce, low private-sector debt, low inequality, little if any unemployment, easy access to high-quality public goods (especially in health and education), and high levels of wellbeing.

Saving A way of delaying spending from prior income. Usually comes with a financial reward (interest), although income taxes will reduce the magnitude of the reward and inflation will erode the purchasing power of both the savings and the after-tax reward.

Steady-State Economy An economy that does not significantly increase or decrease in size, especially in terms of resource throughput, which is ideally kept to an ecologically sustainable rate. Can continue to improve the human condition through better outcomes in terms of health, education, environment, quality of goods, reduced income inequality, full employment, and so on. Sometimes called a state of dynamic equilibrium. Certainly not stultifying.

Stocks and Flows A stock is a quantity at a point in time (such as the number of unemployed people today); a flow is an addition to, or a reduction in, a stock over a given period. The GDP of a year (production), as an addition to the stock of all goods, is a flow over that year (inflow). Consumption/depreciation in a year, as a subtraction to the stock of all goods, is also a flow over that year (outflow). If GDP exceeds consumption/depreciation, the economy grows. If GDP is exceeded by consumption/depreciation, the economy shrinks. With a steady-state economy, inflows and outflows are the same. Although GDP is well measured and forecasted in macroeconomics, outflows and stocks of goods are not – the result of a GDP-fetish.

Sustainability Human endeavour pursued in a way that can continue indefinitely. The opposite of collapse. A situation where the resource throughput required to fuel dynamic socio-economic processes remains within the biocapacity of the planet. This requires adherence to three sustainability precepts: (i) renewable resources: extract them at a rate no greater than they can regenerate; (ii) non-renewable resources: extract them at a rate no greater than we can cultivate (substitute) renewable resources to replace them (thus keeping the total quantity of resource-providing natural capital intact); (iii) generate waste at a rate no greater than the ecosphere can safely assimilate it (thus keeping the total quantity of waste-assimilating natural capital intact). The pursuit of sustainability requires a shift, in a timely manner, to using renewable resources, but there always remains a limit to the scale of resource use and the technical capacity to transform them into goods and services. A wise society would operate within ecological limits at all times (i.e., ensure its Ecological Footprint does not exceed Biocapacity), rather than venture into unsustainable territory in the hope that technological breakthroughs might remedy the situation. Although degrowth offers a solution to nations with economies larger than their maximum sustainable scale, degrowth is a second-best outcome best avoided by remaining within long-run sustainable limits.

Sustainable Development In a nutshell, sustainable qualitative improvement. More broadly: "A nation is achieving sustainable development if it undergoes a coevolutionary process that improves the total quality of life of every citizen, both now and into the future, while ensuring its rate of resource use does not exceed the regenerative and waste-assimilating capacities of the natural environment. It is also a nation that ensures the survival of the biosphere and all its evolving processes while recognising, to some extent, the intrinsic value of sentient non-human beings." (Philip Lawn 2016).

Technical Efficiency Technical efficiency (E) is equal to the ratio of matter-energy embodied in final goods (Q) to the matter-energy embodied in the natural resources (R) used to produce them (i.e., $E = Q/R$). Because of the first and second laws of thermodynamics, 100% technical efficiency of production is impossible. Hence, technical efficiency is always less than one ($E < 1$). Technological progress, by increasing the technical efficiency of production, allows more output to be produced from the same rate of resource throughput, or alternatively, allows the same quantity of output to be produced from a lessened rate of throughput. Since E is always <1, technological progress does not permit more to be produced from less indefinitely. Technical efficiency is equivalent, in concept, to the resource intensity of economic activity, albeit it is measured differently. The resource intensity of economic activity is typically measured as the ratio of Ecological Footprint (F) to GDP (i.e., resource intensity = F/GDP). Ceteris paribus, an increase in the technical efficiency of production also increases allocative efficiency (see Efficient allocation), although allocative efficiency need not require maximum technical efficiency. For example, given the option of allocating a given rate of throughput to produce 1000 *useless* widgets or 500 *useful* gadgets, the more desirable and therefore more efficient option would be 500 gadgets even though it is technically less efficient. However, an increase in technical efficiency allowing 600 useful gadgets to be produced from the same rate of throughput would increase allocative efficiency.

Tipping Points When changes in a system, especially natural systems, reach a largely unpredictable critical point such that positive feedbacks cause an acceleration or cascade of changes that are irreversible and unknowable in their consequences. In climate change, warming temperatures due to CO_2 emissions could reach a tipping point where thawing of frozen ground unleashes yet more greenhouse gases, causing accelerated warming. Since we cannot precisely calculate tipping points, the precautionary principle should always apply.

Thermodynamics Physical laws, especially the first and second laws, that govern limits to the transformation of matter and energy, and the production of waste. Applicable to the economic process given that it involves the transformation of matter-energy for economic purposes. The laws of thermodynamics ultimately determine the maximum sustainable size of an economy – that is, the largest physical size that an economy can be and still be sustained by a rate of throughput limited to the ecosphere's regenerative and waste assimilative capacities.

Trophic Levels Just as a massive amount of vegetation sustains a large number of herbivores, which in turn sustain a smaller number of omnivores and then carnivores, in the case of a national economy, primary (agriculture-resources), secondary (manufacturing) and tertiary (services) levels should form a balanced whole where a deficiency in one level is likely to create a weakness in the whole. A reliance on trade to fill a void creates risk and potential volatility and uncertainty. The idea that a mature economy is one dominated by the services sector and is therefore low-polluting is erroneous, as evidenced by the fact that economies with the highest ratio of services to GDP (e.g., Australia, USA, Canada, Japan, and European nations) have the highest per capita rates of energy use and greenhouse gas emissions.

Throughput The input of low-entropy matter-energy (R = natural resources), their conversion into goods and services (Q) using labour (L) and capital equipment (Kh), and the subsequent output over time of high-entropy matter-energy (W = waste). For sustainability, the total volume of R, as well as certain stocks, need to remain within the planet's biocapacity. Technical efficiency increases when more productive L and Kh reduces the R required to produce the same Q, which also reduces the L and Kh requirement.

Uneconomic Growth A point reached where the rise in GDP increases the costs of economic activity faster than it increases benefits (i.e., reduces economic welfare). In economic parlance, it is where the marginal cost of GDP growth begins to exceed the marginal benefits, hence growth becomes 'uneconomic'.

Unemployment Rate The percentage of the labour force not engaged in any form of paid employment but actively seeking it. The official unemployment rate is only part of a suite of important labour-market statistics, that include underemployment, hidden unemployment, the participation rate, long-term unemployment, the quality of new jobs, over-qualified workers, and youth unemployment. Taken together, the broad labour underutilisation rate (and the consequent suffering) is always much higher than the official unemployment rate. Category definitions are usually set by the International Labour Organization: for instance, you are only officially unemployed if you are actively seeking work and you have not worked more than 1 h per week in the reference period. In the neoliberal era, a certain level of unemployment (or labour underutilisation) is deemed necessary to control inflation, although the sacrificial unemployed are told it is their own fault and are subject to punitive measures. Conveniently, labour underutilisation is used as a tool to moderate wage increases.

Appendix 4

Comparison of neoclassical theory vs. modern monetary theory as applied to monetary-sovereign national governments.

Stephen J. Williams

Neoclassical theory	Modern monetary theory
Does not prioritise theory of money.	Prioritises theory of money.
Money evolved for convenience in barter economies.	Money is a creature of the state, not private barter economies.
Little or no focus on money-sovereign nations v. non-sovereign nations.	Essential focus on money sovereignty leading to expanded policy options.
Concern with G'ment's ability to prudently fund its spending programs.	Concern with available real resources in context of sustainable use.
G'ment must fund spending with tax receipts, borrowing, asset sales.	G'ment does not 'fund' its spending, never needs to borrow its own money.
Cost of a public program is the financial cost; public money is always limited.	Cost of a public program is the real resources used, where real resources are limited.
Central bank is independent of G'ment.	Central bank is part of G'ment.
G'ment is like a household, has budget constraint: 'Sound finance' should be followed.	G'ment nothing like a household, no budget constraint (except inflation, sustainability): 'Functional finance' should be followed. Budget should be renamed as the 'fiscal statement' as budget implies a financial limit.

(continued)

Neoclassical theory	Modern monetary theory
G'ment should target a balanced budget or even surplus.	Budget outcome (usually beyond the control of G'ment) should not be a target. Balance the economy, not the budget.
Budget deficits accumulate year on year and lead to higher taxes, interest rates, inflation, possibly insolvency.	Budget deficits (better named as a fiscal injections) do not accumulate; G'ment should spend up to full employment and no further.
G'ment can 'print' new money in exceptional circumstances, but this is addictive and inflationary.	G'ment spends new money into existence (creates money) whenever it spends ('printing' is misleading metaphor used to scare us).
G'ment borrowing (bond sales) 'crowds out' private-sector borrowing, investment.	No 'crowding out' as banks will lend to any credit-worthy customer by lending new money (credit/horizontal money) into existence (banks do not lend deposits). This new credit money is destroyed when principal is repaid by borrower.
G'ment must sell bonds (borrow) to match deficit, therefore at mercy of bond vigilantes who largely set interest rate.	Bond sales are optional, used to soak up excess reserves (created by deficit spending) so central bank can achieve overnight interest rate target, which in turn regulates all other interest rates. G'ment, via central bank, sets interest rate.
G'ment collects taxes, then spends them.	G'ment spends, then collects taxes (as an offset). Federal taxes are destroyed, not spent.
Monetary policy is best tool to control economy.	Fiscal policy is best tool to control economy.
There is a natural rate of unemployment where inflation is stabilised (NAIRU).	No natural rate of unemployment, but there is usually some labour underutilisation without a G'ment full-employment program.
Market determines unemployment rate.	G'ment determines unemployment rate.
Use buffer stock of unemployed to control inflation.	Use fiscal policy, buffer stock of employed workers in job guarantee pool (on minimum wage) to control inflation.
Market, independent central bank sets interest rates.	G'ment (via its central bank) can set interest rates at any level, including zero.
Usually a need for foreign financial capital.	G'ment and its agents (private banks) can provide all financial capital.
G'ment should respect credit-rating agencies.	G'ment should ignore credit-rating agencies.
G'ment should fear sudden and deep currency depreciation and power of financial markets.	G'ment can implement capital controls in emergency if speculators attack currency.
Little fear of resource depletion due to human ingenuity, substitution.	Management of real resource constraint determines wealth, inflation, sustainability.
G'ment is at mercy of international forces, large corporations, financial sector.	Citizens, via elected G'ment, are masters of country's destiny.

The manufacturer's authorised representative in the EU is Springer Nature Customer Service Centre GmbH, Europaplatz 3, 69115 Heidelberg, Germany. If you have any concerns regarding our products, please contact ProductSafety@springernature.com

Printed and bound by CPI Group (UK) Ltd, Croydon, CR0 4YY
30/09/2026
02225593-0010